高职高专立体化教材　计算机系列

JSP 编程技术
(第 2 版)

杨学全　主　编

清华大学出版社
北　京

内 容 简 介

JSP是一种动态网页技术标准，利用这一技术可以快速构建跨平台的、先进安全的动态网站。本书全面、翔实地介绍了使用JSP进行Web应用开发所需的编程知识与技术，既有理论，又有编程实践，主要内容包括：JSP技术概述、Web开发基础、JSP语法基础、JSP内建对象、使用JavaBean、文件访问、JSP中使用数据库、Servlet技术、基于Servlet的MVC模式、JSP中使用XML、网上报名系统开发案例等。

本书将JSP编程的基本知识与过程性知识、基本理论和开发实践有机整合，适合“项目驱动，案例教学，启发式学习”的教学方法；融“教、学、做”于一体，提供了所有例题及项目的源代码、电子课件和习题等资源。

本书不仅可以作为大学计算机及相关专业的教材，也可供各类培训、计算机从业人员和程序设计爱好者参考使用。

图书在版编目(CIP)数据

JSP编程技术/杨学全主编. --2版. --北京：清华大学出版社，2015（2017.2重印）
(高职高专立体化教材　计算机系列)
ISBN 978-7-302-38936-1

Ⅰ. ①J…　Ⅱ. ①杨…　Ⅲ. ①JAVA语言—网页制作工具—程序设计—高等职业教育—教材
Ⅳ. ①TP312

中国版本图书馆CIP数据核字(2015)第005671号

责任编辑：桑任松
封面设计：刘孝琼
责任校对：周剑云
责任印制：何　芊

出版发行：清华大学出版社
　　网　　址：http://www.tup.com.cn，http://www.wqbook.com
　　地　　址：北京清华大学学研大厦A座　　**邮　　编**：100084
　　社 总 机：010-62770175　　**邮　　购**：010-62786544
　　投稿与读者服务：010-62776969，c-service@tup.tsinghua.edu.cn
　　质 量 反 馈：010-62772015，zhiliang@tup.tsinghua.edu.cn
　　课 件 下 载：http://www.tup.com.cn,010-62791865
印 装 者：三河市吉祥印务有限公司
经　　销：全国新华书店
开　　本：185mm×260mm　　**印　张**：26.5　　**字　　数**：641千字
版　　次：2009年4月第1版　2015年3月第2版　　**印　　次**：2017年2月第3次印刷
印　　数：4501～6000
定　　价：45.00元

产品编号：057595-01

《高职高专立体化教材　计算机系列》

丛　书　序

一、编写目的

关于立体化教材，国内外有多种说法，有的叫“立体化教材”，有的叫“一体化教材”，有的叫“多元化教材”，其目的是一样的，就是要为学校提供一种教学资源的整体解决方案，最大限度地满足教学需要，满足教育市场需求，促进教学改革。我们这里所讲的立体化教材，其内容、形式、服务都是建立在当前技术水平和条件基础上的。

立体化教材是“一揽子”式的(包括主教材、教师参考书、学习指导书、试题库)完整体系。主教材讲究的是“精品”意识，既要具备指导性和示范性，也要具有一定的适用性，喜新不厌旧。那种内容越编越多，本子越编越厚的低水平重复建设在“立体化”的世界中将被扫地出门。与以往不同，“立体化教材”中的教师参考书可不是千人一面的，教师参考书不只是提供答案和注释，而是含有与主教材配套的大量参考资料，使得老师在教学中能做到“个性化教学”。学习指导书更像一本明晰的地图册，难点、重点、学习方法一目了然。试题库或习题集则要完成对教学效果进行测试与评价的任务。这些组成部分采用不同的编写方式，把教材的精华从各个角度呈现给师生，既有重复、强调，又有交叉和补充，相互配合，形成一个教学资源有机的整体。

除了内容上的扩充外，立体化教材的最大突破还在于在表现形式上走出了“书本”这一平面媒介的局限，如果说音像制品让平面书本实现了第一次“突围”，那么电子和网络技术的大量运用，就让躺在书桌上的教材真正“活”了起来。用 PowerPoint 开发的电子教案不仅大大减少了教师案头备课的时间，而且也让学生的课后复习更加有的放矢。电子图书通过数字化使得教材的内容得以无限扩张，使平面教材更能发挥其提纲挈领的作用。

CAI(计算机辅助教学)课件把动画、仿真等技术引入了课堂，让课程的难点和重点一目了然，通过生动的表达方式达到深入浅出的目的。在科学指标体系控制之下的试题库，既可以轻而易举地制作标准化试卷，也能让学生进行模拟实践的在线测试，提高了教学质量评价的客观性和及时性。网络课程更厉害，它使教学突破了空间和时间的限制，彻底发挥了立体化教材本身的潜力，轻轻敲击几下键盘，你就能在任何时候得到有关课程的全部信息。

最后还有资料库，它把教学资料以知识点为单位，通过文字、图形、图像、音频、视频、动画等各种形式，按科学的存储策略组织起来，大大方便了教师在备课、开发电子教案和网络课程时的教学工作。如此一来，教材就“活”了。学生和书本之间的关系，不再像领导与被领导那样呆板，而是真正有了互动。教材不再只为老师们规定，什么重要什么不重要，而是成为教师实现其教学理念的最佳拍档。在建设观念上，从提供和出版单一纸质教材转向提供和出版较完整的教学解决方案；在建设目标上，以最大限度满足教学要求为根本出发点；在建设方式上，不单纯以现有教材为核心，简单地配套电子音像出版物，而是以课程为核心，整合已有资源并聚拢新资源。

网络化、立体化教材的出版是我社下一阶段教材建设的重中之重，以计算机教材出版为龙头的清华大学出版社确立了“改变思想观念，调整工作模式，构建立体化教材体系，大幅度提高教材服务”的发展目标，并提出了首先以建设“高职高专计算机立体化教材”为重点的教材出版规划，希望通过邀请全国范围内的高职高专院校的优秀教师，共同策划、编写这一套高职高专立体化教材，利用网络等现代技术手段，实现课程立体化教材的资源共享，解决国内教材建设工作中存在的教材内容更新滞后于学科发展的状况。把各种相互作用、相互联系的媒体和资源有机地整合起来，形成立体化教材，把教学资料以知识点为单位，通过文字、图形、图像、音频、视频、动画等各种形式，按科学的存储策略组织起来，为高职高专教学提供一整套解决方案。

二、教材特点

在编写思想上，以适应高职高专教学改革的需要为目标，以企业需求为导向，充分吸收国外经典教材及国内优秀教材的优点，结合中国高校计算机教育的教学现状，打造立体化精品教材。

在内容安排上，充分体现先进性、科学性和实用性，尽可能选取最新、最实用的技术，并依照学生接受知识的一般规律，通过设计详细的可实施的项目化案例(而不仅仅是功能性的小例子)，帮助学生掌握要求的知识点。

在教材形式上，利用网络等现代技术手段实现立体化的资源共享，为教材创建专门的网站，并提供题库、素材、录像、CAI 课件、案例分析，实现教师和学生在更大范围内的教与学互动，及时解决教学过程中遇到的问题。

本系列教材采用案例式的教学方法，以实际应用为主，理论够用为度。教程中每一个知识点的结构模式为“案例(任务)提出→案例关键点分析→具体操作步骤→相关知识(技术)介绍(理论总结、功能介绍、方法和技巧等)”。

该系列教材将提供全方位、立体化的服务。网上提供电子教案、文字或图片素材、源代码、在线题库、模拟试卷、习题答案、案例动画演示、专题拓展和教学指导方案等。

在为教学服务方面，主要是通过教学服务专用网站在网络上为教师和学生提供交流的场所，每个学科、每门课程，甚至每本教材都建立网络上的交流环境。可以为广大教师信息交流、学术讨论、专家咨询提供服务，也可以让教师发表对教材建设的意见，甚至通过网络授课。对学生来说，则可以在教学支撑平台所提供的自主学习空间中进行学习、答疑、操作、讨论和测试，当然也可以对教材建设提出意见。这样，在编辑、作者、专家、教师、学生之间建立起一个以课本为依据、以网络为纽带、以数据库为基础、以网站为门户的立体化教材建设与实践的体系，用快捷的信息反馈机制和优质的教学服务促进教学改革。

前　言

近日，国务院印发了《关于加快发展现代职业教育的决定》，《决定》提出要牢固确立职业教育在国家人才培养体系中的重要位置，以服务发展为宗旨，以促进就业为导向，适应技术进步和生产方式变革以及社会公共服务的需要，培养数以亿计的高素质劳动者和技术技能人才。要深化产教融合、校企合作、工学结合，推动专业设置与产业需求对接、课程内容与职业标准对接、教学过程与生产过程对接、毕业证书与职业资格证书对接、职业教育与终身学习对接，提高人才培养质量，强化职业教育的技术技能积累作用。

在加快职业教育发展，提高人才培养质量的新形势下，必须加强课程建设与改革，推动课程内容与职业标准的对接，教学过程与生产过程对接；经过职教工作者们多年的探索和实践，基于工作过程的课程开发理论得到了发展和应用，课程开发取得了突出的成果。高职高专院校逐步构建了以技术应用能力培养为主线，以就业为导向，基于工作过程的计算机应用类专业课程体系。

高职高专教材是为教师、学生和课程服务的，是知识的载体。它必须体现高职高专课程开发建设的新思想；必须根据职业岗位(群)的任职要求，参照国家职业资格标准开发和建设，使其具有职业性；必须将知识的学科性和工作的过程性有机地整合，体现其综合性；必须适用“教、学、做”一体化的课程教学模式，使其具有实用性。一句话：教材要教师用着好，学生学得好，学了用得上。

本书在第1版的基础上，采纳了读者和同行的建议，使用了JDK、Tomcat及开发工具的主流版本，同时延续了前版的章节体系。本书是一本以职业技术能力培养为主线，采用项目驱动模式的案例教材。教材融“教、学、做”于一体，注重基本知识与基本技术讲解(教)，给出具有实用价值的案例供学生模仿(学)，通过课程设计强化学生能力的培养(做)。本书适用于计算机应用类专业或非计算机专业的JSP编程技术课程教学。全书共分为11章，从基本概念和实际应用出发，由浅入深、循序渐进地讲述JSP编程的基础知识、JavaBean技术、Servlet技术、MVC模式和Web应用开发案例等内容；通过对本书内容的学习，读者可以快速、全面地掌握基于MVC模式的JSP编程技术；建议教学时数为72学时，也可根据教学的具体情况删减内容。

作为“项目驱动、案例教学”模式的教材，本书具有以下特点。

(1) 内容选择合理、时序安排科学。本书以Web应用程序开发能力培养为主线，根据岗位技术能力需要选择教材内容——JSP基础、HTML及页面布局、JavaScript与正则表达式、JavaBean、Servlet技术、MVC模式、XML以及基于MVC模式的Web应用开发等；根据工作过程和认知规律安排内容时序为“JSP基本知识→JSP+JavaBean模式应用→JSP+JavaBean+Servlet模式应用”，将文件操作、数据库访问、XML等编程技术合理地分配到模式1和模式2中，强调知识的层次性和技能培养的渐进性，最终为基于MVC模式框架开发打好JSP编程基础。

(2) 案例典型，代码规范，能力良构。本书以培养基于MVC模式的Web应用开发能力为目的，设置了具有代表性的例题、习题和案例，比如设置了购物车、留言板、文件上传下载、分页显示、页面布局、文件操作和数据库应用等案例；示例代码采用了 Sun 的模式 1 和模式 2，代码规范、实用；强调学生在例题、案例设置的工作情景中学习，潜移默化地培养学科性知识与工作过程性知识有机整合、理论与实践相结合、具有良好结构的 JSP 编程能力。

本书由杨学全老师主编，河北农业大学张悦、张春艳、苑萌萌等参加了部分章节的编写工作。刘海军教授审稿。

衷心感谢河北大学博士生导师徐建民教授、保定职业技术学院刘海军教授，他们的辛勤工作使我们受益匪浅。

衷心感谢所有关心本书编写的师长和朋友。

编写一本优秀的教材是一件非常不容易的事情，很多因素都会影响到教材的质量。尽管此书多次修改，每次修改都考虑如何突出职业能力培养这条主线，如何突出教材的高职特色等问题；尽管本书的定稿经过了多人的努力，但是我们还是感觉不太尽如人意，唯恐对不起关心和支持我们编写这本教材的朋友们，对不起孜孜求学的学子们。由于作者水平有限，加之时间仓促，书中难免有错漏之处，敬请同行们批评指正，我们将不胜感激。

编　者

目　录

第1章　JSP技术概述

学习目的与要求：

JSP技术是目前Web应用程序开发的主流技术之一。本章主要学习Web程序设计模式与原理，搭建JSP运行环境，JSP页面与JSP运行原理和JSP集成开发环境等内容。通过本章的学习，学生要了解B/S模式Web应用程序架构，JSP运行原理，JSP与JavaBean、Servlet的关系，理解Web服务目录的概念，掌握搭建JSP开发环境的方法以及集成开发环境MyEclipse的安装和配置，能够编写一个简单的JSP页面，并能通过客户端浏览器访问页面。学生必须认真学习本章，体验学习JSP的成功。万事开头难，良好的开端就是成功的一半。

1.1　Web程序设计模式与运行原理

用户在学习JSP编程技术之前，需要对Web程序设计模式有所了解。Web程序或网站的运行方式不同于单机或C/S模式的Windows应用程序，本节主要从Web服务、浏览器/服务器模式与动态网页技术三个方面对其简述。

1.1.1　Web服务器与动态网页

互联网中有数以亿计的网站，用户可以通过浏览这些网站获得所需要的信息。例如，用户在浏览器的地址栏中输入http://www.sina.com.cn，浏览器就会显示新浪网的首页，查看新闻等信息。那么新浪网首页的内容是存放在哪里的呢？新浪网首页的内容是存放在新浪网服务器上的。所谓服务器就是网络中的一台主机，由于它提供Web、FTP等网络服务，因此称其为服务器。用户的计算机又是如何将存在网络服务器上的网页显示在浏览器中的呢？当用户在地址栏中输入新浪网地址(URL，即“统一资源定位符”)的时候，浏览器会向新浪网的服务器发送HTTP请求，这个请求使用HTTP，其中包括请求的主机名、HTTP版本号等信息。服务器在收到请求信息后，将回复的信息(一般是文字、图片等网页信息，也就是HTML页面)准备好，再通过网络发送到客户端浏览器。客户端的浏览器在接收到服务器传回的信息后，将其解释并显示在浏览器的窗口中，这样用户就可以进行浏览了。整个过程如图1.1所示。

在这个“请求—响应”过程中，如果在服务器上存放的网页为静态HTML网页文件，服务器就会原封不动地返回网页的内容。如果存放的是动态网页，如JSP、ASP、ASP.NET等文件，则服务器会执行动态网页，执行的结果是生成一个HTML文件，然后再将这个HTML文件发送给客户端浏览器，客户端浏览器将其解释为用户见到的页面。

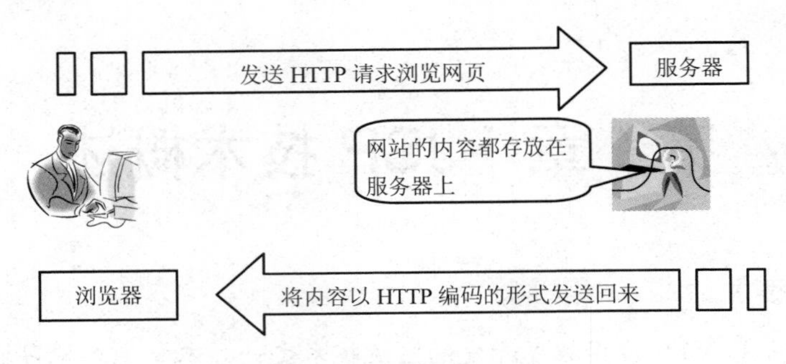

图 1.1　Web 服务过程

因此，静态网页和动态网页的根本区别在于服务器端返回的 HTML 文件是事先存储好的还是由动态网页程序生成的。静态网页文件里只有 HTML 标记，没有程序代码，网页的内容都是事先写好，存放在服务器上的；动态网页文件不仅含有 HTML 标记，并且还含有程序代码，当用户发出请求时，服务器由动态网页程序即时生成 HTML 文件。动态网页能够根据不同的时间、不同的用户生成不同的 HTML 文件，显示不同的内容。

1.1.2　浏览器/服务器结构及其优点

随着网络技术的不断发展，单机的软件程序已经难以满足网络计算的需求，因此，基于网络的软件架构应运而生。早期常用的网络架构为“客户/服务器”(即 Client/Server，简写为 C/S)模式。使用这种架构编写的软件分为客户端和服务器端两部分，需要分别在客户机和服务器上进行安装。这种模式在用户数据录入等方面很有优势，也降低了系统的通信开销，但是也有一定的缺点，如开发和维护成本较高，可移植性较差等。

互联网的普及使得用于上网浏览的浏览器已经成为操作系统中不可缺少的一项，浏览器的功能也越来越强大，甚至可以取代“客户/服务器”架构的客户端软件，成为统一的客户端。这样，程序员就可以只编写运行在服务器上的软件，浏览器代替 C/S 模式中的客户端软件，客户通过浏览器与服务器端软件进行交互并得到运算结果，这种软件架构就是“浏览器/服务器”(即 Browser/Server，简写为 B/S)模式。B/S 模式主要是利用了不断成熟的 WWW 浏览器技术，结合动态网站制作技术，通过通用浏览器实现了原来需要复杂的专用软件才能实现的强大功能，节约了开发成本，是一种全新的软件系统构造技术。随着互联网络的不断发展，B/S 架构已经成为当今应用软件的首选体系结构。

B/S 模式的应用程序相对于传统的 C/S 模式的应用程序来讲无疑是一个巨大的进步。它的主要优点如下。

1. 开发、维护成本较低

C/S 模式软件，当客户端的软件需要升级的时候，所有客户端都必须进行升级安装或者重新安装，而 B/S 模式的软件只需要在服务器端发布，客户端浏览器无须维护，因而极大地降低了开发和维护成本。

2. 可移植性高

C/S 模式软件，不同开发工具开发的程序，一般情况下互不兼容，主要运行在局域网中，移植困难。而 B/S 模式的软件运行在互联网上，提供了异种网、异种机、异种应用服务的联机、联网服务基础，客户端安装的是通用浏览器，不存在移植的问题。

3. 用户界面统一

C/S 模式软件的客户端界面由所安装的客户端软件所决定，因此不同的软件客户端界面不同，而 B/S 模式的软件都是通过浏览器来使用的，操作界面基本统一。

1.1.3　JSP 与其他 Web 开发技术

在简单介绍了 Web 服务器、动态网页和 B/S 模式 Web 应用程序结构的优点之后，读者要问哪些语言和技术可用于 B/S 模式的 Web 应用程序开发？目前使用较多的技术有 JSP、ASP、ASP.NET、PHP 等。本节对它们进行简单的介绍和比较。

1. JSP 技术及其优点

JSP 全称为 Java Server Pages，是 Sun 公司倡导、多家公司参与，于 1999 年提出的一种 Web 服务技术标准。它的主要的编程脚本为 Java 语言，同时还支持 JavaBeans/Servlet 等技术，利用这些技术可以建立安全、跨平台的 Web 应用程序。JSP 技术具有以下优点。

1) 跨平台性

由于 JSP 的脚本语言是 Java 语言，因此它具有 Java 语言的一切特性。同时，JSP 也支持现在大部分平台，拥有“一次编写，到处运行”的特点。

2) 执行效率高

当 JSP 第一次被请求时，JSP 页面转换成 Servlet，然后被编译成*.calss 文件，以后(除非页面有改动或 Web 服务器被重新启动)再有客户请求该 JSP 页面时，JSP 页面不被重新编译，而是直接执行已编译好的*.class 文件，因此执行效率高。

3) 可重用性

可重用的、跨平台的 JavaBeans 和 EJB(Enterprise JavaBeans)组件，为 JSP 程序的开发提供方便。例如，用户可以将复杂的处理程序(如对数据库的操作)封装到组件中，在开发中可以多次使用这些组件，提高了组件的可重用性。

4) 将内容的生成和显示进行分离

使用 JSP 技术，Web 页面开发人员可以使用 HTML 或者 XML 标记来设计和格式化最终页面。生成动态内容的程序代码封装在 JavaBean 组件、EJB 组件或 JSP 脚本段中。在最终页面中使用 JSP 标记或脚本将 JavaBean 组件中的动态内容引入。这样，可以有效地将内容生成和页面显示进行分离，使页面的设计人员和编程人员可以同步进行，也可以保护程序的关键代码。

2. 其他 Web 开发技术

1) ASP 技术

ASP 是 Active Server Pages 的缩写，是微软在早期推出的动态网页制作技术，包含在 IIS(Internet 信息服务)中，是一种服务器端的脚本编写环境，使用它可以创建和运行动态、交互的 Web 服务器应用程序。在动态网页技术发展的早期，ASP 是绝对的主流技术，但是它也存在着许多缺陷：由于 ASP 的核心是脚本语言，决定了它的先天不足，它无法进行像传统编程语言那样的底层操作；由于 ASP 通过解释执行代码，因此运行效率较低；同时由于脚本代码与 HTML 代码混在一起，不便于开发人员进行管理与维护。随着技术的发展，ASP 的辉煌已经成为过去，微软也已经不再对 ASP 提供技术支持和更新，ASP 技术目前处于被淘汰的边缘。

2) PHP 技术

PHP 从语法和编写方式上来看与 ASP 类似，是完全免费的，最早是一个小开放源码的软件，随着越来越多的人意识到它的实用性而逐渐发展起来。Rasmus Lerdorf 在 1994 年发布了 PHP 的第一个版本。从那时起它就飞速发展，在原始发行版上经过无数的改进和完善，现在已经发展到 5.5 版。PHP+MySQL+Linux 的组合是最常见的，因为它们都可以免费获得。但是 PHP 的弱点也是很明显的，例如 PHP 不支持真正意义上的面向对象编程，接口支持不统一，缺乏正规支持，不支持多层结构和分布式计算等。

3) ASP.NET

ASP.NET 是微软继 ASP 后推出的全新动态网页制作技术，目前最新版本为.NET 4.0。在性能上，ASP.NET 比 ASP 强很多，与 PHP 相比，也存在明显的优势。ASP.NET 可以使用 C#(读音为 C Sharp)，VB.NET，Visual J#等语言来开发，程序开发人员可以选择自己习惯或熟悉的语言进行开发。ASP.NET 依托.NET 平台先进而强大的功能，极大地简化了编程人员的工作量，使得 Web 应用程序的开发更加方便、快捷，同时也使得程序的功能更加强大，是 JSP 技术的有力竞争对手。

1.2 搭建 JSP 的运行环境

用户学习 JSP 或者使用 JSP 开发 Web 应用程序，必须搭建一个 JSP 运行环境。JSP 运行环境至少要具备三个基本条件：一是要在用户的计算机上安装 Java 的 JDK，并进行环境变量的设置；二要在计算机上安装 JSP 引擎，例如 Tomcat、J2EE、WebLogic、WebSphere 服务器；三要在计算机上有浏览器。搭建 JSP 运行环境是 JSP 的初学者必须要学习的基础知识，本节将引导读者一步步完成。

1.2.1 安装和配置 JDK

Sun 公司提供了一个免费的 Java 软件开发工具包 JDK(Java Development Kit)，该工具包包含了编译、运行及调试 Java 程序所需要的工具，此外还提供了大量的基础类库，供编写程序使用，它是开发 Java 程序的基础。2009 年 4 月 20 日，Oracle 宣布正式以 74 亿美元的价格收购 Sun 公司，Java 商标从此正式归 Oracle 所有。

1. JDK 的安装

Sun 公司为不同的操作系统平台，如 Windows、Unix/Linux 等提供了相应的 Java 开发包。用户可到 Oracle 公司站点 http://www.oracle.com 下载最新的适用于相应操作系统的开发包。本书中使用 Windows 32 位操作系统环境下的 Java 开发包 jdk-8u5-windows-i586.exe。书中的实例程序均在此版本下运行通过，所使用的操作系统为 Windows 7 旗舰版。

下载完成后，运行 jdk-8u5-windows-i586.exe 安装文件，本书安装目录为：C:\Program Files\Java。安装完成后在 C:\Program Files\Java 目录中会有 jdk1.8.0_05 和 jre8 两个子目录，jdk1.8.0_05 为 Java 开发工具目录，jre8 为 Java 运行环境目录。

2. JDK 的配置

安装完 JDK 后，需要在 Windows 操作系统中为 JDK 设置几个环境变量，以便系统能够自动查找 JDK 的命令和类库。对于 Windows 7 系统，从“控制面板”→“系统安全”→“系统”→“高级系统设置”可调出“系统属性”对话框，单击“系统属性”对话框的“高级”选项卡，然后单击“环境变量”按钮，弹出“环境变量”设置对话框，如图 1.2、图 1.3 和图 1.4 所示。

分别添加如下的环境变量。

变量名：JAVA_HOME，变量值：C:\Program Files\Java\jdk1.8.0_05。

变量名：CLASSPATH，变量值：

.;%JAVA_HOME%\lib\dt.jar;%JAVA_HOME%\lib\tools.jar;

变量名：PATH，变量值：%JAVA_HOME%/bin;

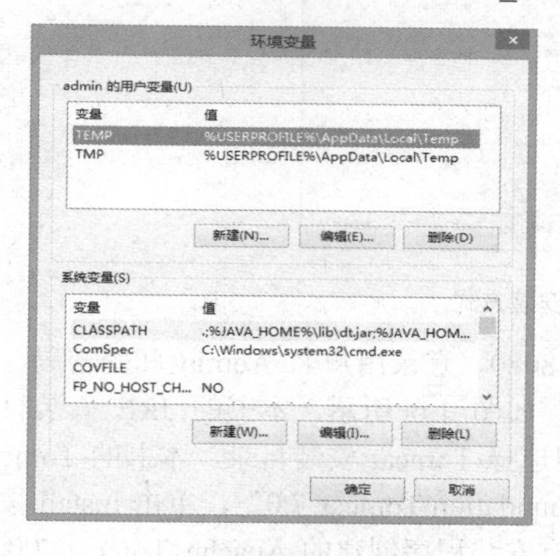

图 1.2　“环境变量”对话框

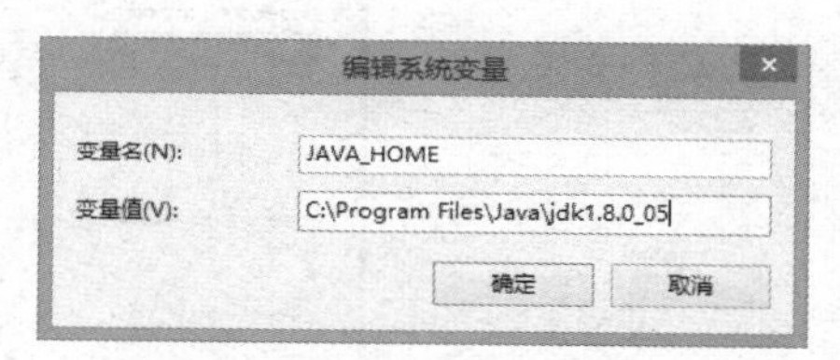

图 1.3　设置 JAVA_HOME

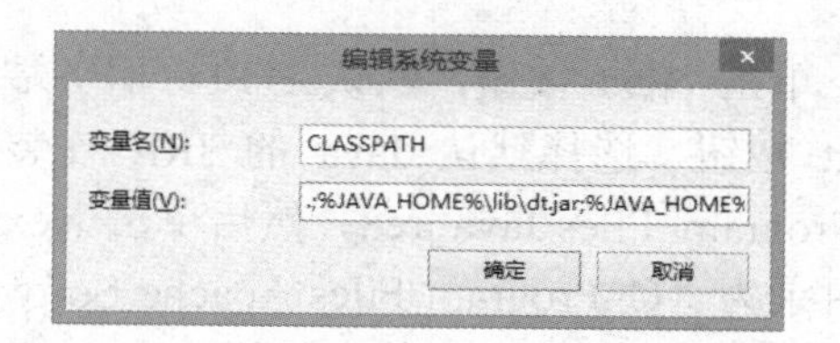

图 1.4　设置 CLASSPATH

注意： 如果用户已经有需要设置的环境变量，如“PATH”变量，可选中该变量进行编辑操作，将需要的变量值追加在后面即可，值与值之间用“;”分隔，切记不要把原来的值覆盖。CLASSPATH 变量中的“.;”不能少，其含义是在当前目录寻找类库。为了验证用户环境变量设置是否正确，可用记事本编写一个简单的 Java 程序，对其进行编译、执行，用来验证 JDK 的安装。

1.2.2　安装和配置 Tomcat

自 1999 年 JSP 发布以来，到目前为止出现了各种各样的 JSP 引擎。如 Tomcat、J2EE、WebLogic、WebSphere 等引擎。一般将安装了 JSP 引擎的计算机称为一个支持 JSP 的 Web 服务器，它负责运行 JSP 程序，并将执行结果返回给浏览器。Tomcat 是一个免费的开源 JSP 引擎，也称为 Jakarta Tomcat Web 服务器。目前 Tomcat 能和大多数主流 Web

服务器一起高效的工作。

1. 下载和安装 Tomcat

用户可以到 http://tomcat.apache.org/站点免费下载 Tomcat 7.0。在主页面中的 Download 里选择 Tomcat 7.0，然后在 Binary Distributions 里的 Core 中选择 zip(pgp, md5)、tar.gz(pgp, md5)或 32-bit/64-bit Windows Service Installer (pgp, md5)。本书下载的是 32-bit/64-bit Windows Service Installer(pgp，md5)，文件名为：apache-tomcat-7.0.53.exe。apache-tomcat-7.0.53.exe 是专门为 Windows 开发的 Tomcat 服务器。

双击 apache-tomcat-7.0.53.exe 文件，出现安装向导，单击 Next 按钮，出现“授权”界面，接受授权协议后，用户可以选择 Normal、Minimun、Custom 和 Full 安装形式，本书选择“Full”安装模式，如图 1.5 所示。

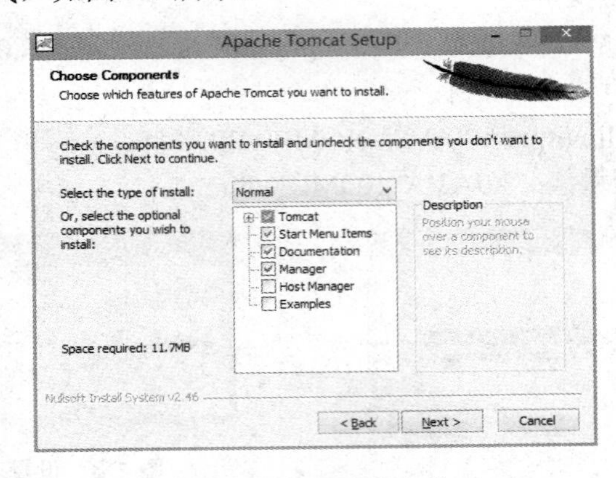

图 1.5　选择安装方式

单击 Next 按钮，默认 HTTP 服务端口号 8080、登录用户名 Admin 和空密码。单击 Next 按钮，选择默认 Java 的 JRE 安装目录，如图 1.6 所示，本书的 JRE 安装目录为 C:\Program Files\Java\jre8。然后单击 Next 按钮选择 Tomcat 安装目录，本书的 Tomcat 安装目录为“C:\Program Files\Apache Software Foundation\Tomcat 7.0”，单击 Install 按钮开始 Tomcat 的安装。安装完成后，在 C 盘中会有安装程序创建的 Apache Tomcat 7.0 菜单组，产生的目录结构如图 1.7 所示。

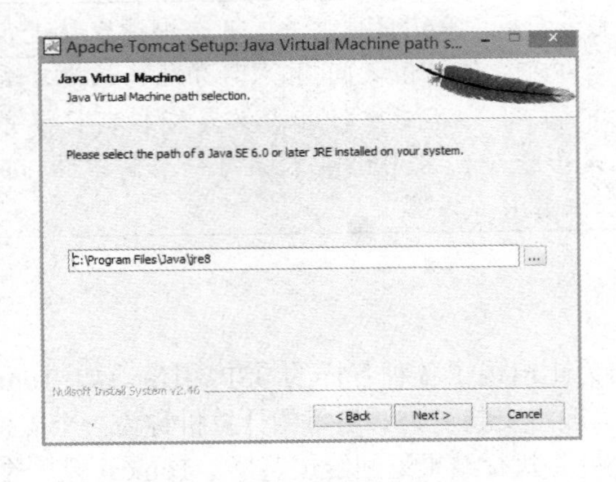

图 1.6　选择 JRE 安装目录

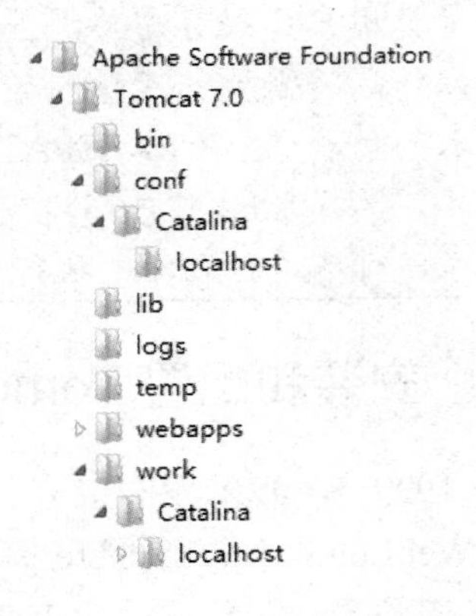

图 1.7　Tomcat 7.0 目录结构

2. 测试安装是否成功(运行测试页)

在“计算机管理”→“服务和应用程序”→“服务”中右击 Apache Tomcat 7.0 Tomcat 7 启动，即可启动 Tomcat 7.0 服务器。

Tomcat 服务器占用的默认端口是 8080，如果 Tomcat 使用的端口已经被占用，则 Tomcat 将无法启动。打开 IE 浏览器或 Mozilla Firefox 浏览器，在浏览器地址栏中输入“http://localhost:8080”并按 Enter 键，如果浏览器中出现如图 1.8 所示的页面，则说明用户的 Tomcat 已经正确安装。

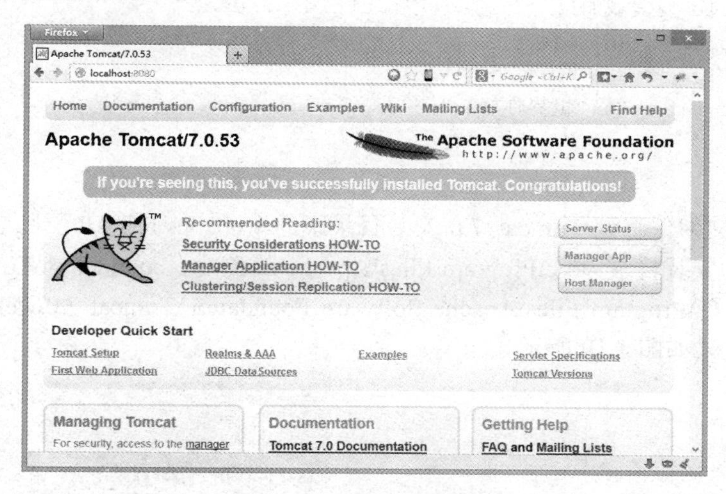

图 1.8　Tomcat 7.0 测试页面

8080 是 Tomcat 服务器默认端口。用户可以通过修改 Tomcat 7.0 安装目录下 conf 子目录中的 server.xml 配置文件来更改端口号。用记事本打开 server.xml 文件，找到下列内容部分：

```
<Connector port="8080" protocol="HTTP/1.1"
connectionTimeout="20000" redirectPort="8443" />
```

将 port="8080"更改为 port="80"，保存文件后并重新启动 Tomcat 服务器即可。此时，用户在浏览器中输入 URL 地址时可省略端口号，例如输入“http://127.0.0.1”，即可看到如图 1.8 所示的测试页面。本教材使用默认的 8080 端口。

1.3　JSP 页面与 JSP 运行原理

JSP 页面的组成、Web 服务目录和执行原理是读者学习 JSP 的基础，本节对其做简单的介绍。读者在这里了解 JSP 页面和执行原理即可，详细内容将在后续章节介绍。

1.3.1　第一个 JSP 页面

一个 JSP 页面是由普通的 HTML 标记和 JSP 标记，以及通过“<%”、“%>”标记加入的 Java 程序片段组成的页面。JSP 页面按文本文件保存，文件名要符合 JSP 标识符的规定，即文件名可以是字母、数字、下划线或美元符号组成，并且第一个字符不能是数字，文件扩展名为 jsp。用户可以用记事本来编辑 JSP 文件，文件保存的编码选择 ANSI，如图 1.9 所示。

【例 1.1】 编写一个简单的 JSP 页面。

```
myfirst.jsp
<%@ page contentType="text/html;charset=GB2312" %>
<html>
<head><title>这是我的第一个 JSP 程序</title></head>
<body bgcolor=cyan>
<h2>这是我的第一个 JSP 程序</h2>
<h3><% out.println("世界你好！");%></h3>
</body>
</html>
```

myfirst.jsp 文件保存到 Tomcat 7.0 安装目录下的 webapps\ROOT 子目录中，本教材假设 Tomcat 7.0 的安装目录为 C:\Program Files\Apache Software Foundation\Tomcat 7.0，则页面保存目录为：C:\Program Files\Apache Software Foundation\Tomcat 7.0\webapps\ROOT 目录。页面运行效果如图 1.10 所示。

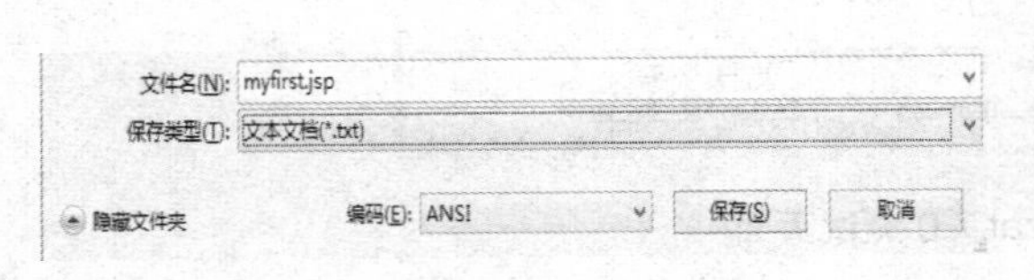

图 1.9　文件保存类型

图 1.10　页面运行效果

> **注意：** JSP 技术是基于 Java 语言的，所以 JSP 文件名区分大小写。Myfirst.jsp 和 myfirst.jsp 是两个不同的文件。JSP 页面文件保存编码还可以是 utf-8 等编码。

1.3.2　设置 Web 服务目录

从用户的角度看，Web 服务目录就是用户浏览器能够访问的页面所在目录。如果要发布网页，必须将编写好的 JSP 页面放到 Web 服务器中某个 Web 服务目录中。

1. 根目录

理解 Web 服务的根目录，需要从 Web 服务器和客户浏览器两个角度分析。

从服务器的角度分析，Tomcat 7.0 安装目录中的 webapps\ROOT 子目录称为 Tomcat 7.0 Web 服务的根目录。在本教材的 Tomcat 7.0 安装方式下，根目录在服务器上的物理路径为 C:\Program Files\Apache Software Foundation\Tomcat 7.0\webapps\ROOT。除非特别说明，本教材所指的 Tomcat 安装目录均指 C:\Program Files\Apache Software Foundation\Tomcat 7.0。

从客户浏览器的角度分析，在地址栏中输入的 URL——“协议://ip 地址或域名:端口号/目录/页面.jsp”中，“端口号”后面的“/”就是客户端看到的根目录。

用户访问 Web 服务器根目录中的 JSP 页面，在客户端浏览器中输入 URL 地址只需输入“http://ip 地址或域名:端口号/页面名称”即可。例如前面的例 1.1 中，访问根目录下的 myfirst.jsp 页面只需地址栏中输入“http://127.0.0.1:8080/myfirst.jsp”。例 1.1 myfirst.jsp 页面保存位置就是 Web 服务器的根目录。访问效果如图 1.10 所示。

2. Web 服务子目录

在 Tomcat 服务器安装目录中的 webapps 子目录下，除了 ROOT 以外，还有 docs、examples、manager 等子目录，这些子目录称为 Web 服务子目录。用户可以在客户浏览器中访问这些 Web 服务子目录中的页面，只需输入“http://ip 地址或域名:端口号/子目录/页面名称”即可，例如访问 webapps\examples 目录中的 index.html 页面，在浏览器地址栏中输入“http://127.0.0.1:8080/examples/index.html”，URL 中的 examples 为 Web 服务子目录。

除了 Tomcat 安装时创建的 Web 服务子目录外，用户可以在 webapps 目录创建新的 Web 服务子目录。例如在 webapps 目录中创建 myapp 子目录，将例 1.1 的 myfirst.jsp 文件保存 myapp 子目录中，在客户浏览器地址栏中输入“http://127.0.0.1：8080/myapp/myfirst.jsp”即可访问 myapp 服务目录下的 myfirst.jsp 页面。

3. 建立虚拟 Web 服务目录

除了在安装目录中的 webapps 目录下创建 Web 服务子目录外，用户还可以将 Tomcat 服务器的某个目录指定为 Web 服务子目录，并为其设定虚拟 Web 服务子目录名称，将实际的目录物理路径隐藏，用户只能通过虚拟目录访问该 Web 服务子目录。

建立虚拟 Web 服务目录，可通过修改 Tomcat 安装目录下的 conf 子目录中的 server.xml 配置文件来实现。例如，将 E:\programJsp\ch1 指定为新的 Web 服务子目录，虚拟目录名称为 ch1。用户首先在 Tomcat 服务器的 E 盘创建 programJsp\ch1 目录，然后用文本编辑器打开 server.xml 文件，在<Host>、</Host>节之间加入如下内容：

```
<Context path="/ch1" docBase="E:\programJsp\ch1" debug="0"
reloadable="true" />
```

上面代码中的 path="/ch1"为虚拟目录名称，docBase="E:\programJsp\ch1" 为 Web 服务子目录的物理路径。修改 server.xml 文件前，应该先停止 Tomcat 服务器，否则将发生拒绝访问错误。保存后重新启动 Tomcat 服务器，并将例 1.1 中的 myfirst.jsp 文件复制到 E:\programJsp\ch1 目录中，用户就可以通过虚拟目录 ch1 访问 myfirst.jsp 页面，浏览器中输入的 URL 地址为：http://127.0.0.1:8080/ch1/myfirst.jsp。

注意： server.xml 文件内容是区分大小写的，不能将 Context 写成 context，或者将 docBase 写成 docbase。同时路径中不能存在中文，否则 Tomcat 服务器将无法启动。在客户端输入 url 时，Web 服务子目录也是区分大小写的，如：http://127.0.0.1:8080/ch1/myfirst.jsp，不能写成 http://127.0.0.1:8080/ CH1/myfirst.jsp。每次修改 server.xml 文件后，必须重新启动 Tomcat 服务器，配置才能生效。

4. 相对服务目录

Web 服务目录下的子目录称为 Web 服务目录下的相对服务目录。例如在根目录 ROOT 中建立了一个 image 子目录，image 就是根目录下的相对服务目录，访问 image 的 URL 地址为 http://127.0.0.1:8080/image。在虚拟目录 ch1 中建立的子目录 image，image 就是虚拟目录 ch1 下的相对服务目录，访问它的 URL 地址为 http://127.0.0.1:8080/ch1/image。

1.3.3 JSP 的运行原理

如例 1.1 所示，用户在客户端浏览器中输入"http://127.0.0.1:8080/ch1/myfirst.jsp"，浏览器就会显示页面内容。那么包含 HTML 标记、JSP 标记和<% %>的 JSP 页面是如何显示到客户浏览器中的呢？回答这个问题就要了解到 JSP 的运行原理。

用户在客户端浏览器中输入"http://127.0.0.1:8080/ch1/myfirst.jsp"，就会对 Web 服务器上的 myfirst.jsp 页面产生请求。当服务器上的 myfirst.jsp 页面第一次被请求时，JSP 引擎首先转译 JSP 页面文件，形成一个 Java 文件(本质上是一个 Java Servlet 的 Java 文件，关于 Servlet 后续章节要介绍，在这里读者可以简单将其理解为执行在服务器端的 Java 小程序)，这个 Servlet Java 文件的文件名是 myfirst_jsp.java，存储在 Tomcat 安装目录的 work\Catalina\localhost\ch1\org\apache\jsp 子目录中，然后 JSP 引擎调用 Java 编译器编译这个文件形成 Java 的字节码文件 myfirst_jsp.class，存放在相同的目录中，编译完成之后，JSP 引擎就会执行 myfirst._jsp.class 字节码文件响应客户的请求，执行 myfirst_jsp.calss 的结果是发送给客户端一个 HTML 页面。当这个页面再次被请求时，JSP 引擎将直接执行这个编译了的字节码文件来响应客户的请求。当多个用户请求同一个 JSP 页面时，Tomcat 服务器为每个客户启动一个线程，该线程负责执行常驻内存的字节码文件来响应客户请求。JSP 的运行原理如图 1.11 所示。

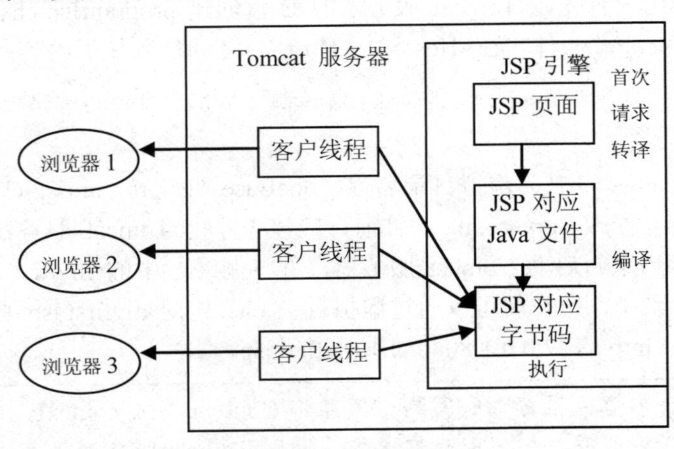

图 1.11 JSP 运行原理

说明： 如果对 JSP 页面进行了修改，Tomcat 会生成新的字节码。JSP 引擎编译 Java 文件需要 Java 的 JDK 环境，运行 Java 字节码文件需要 Java 虚拟机解释执行。

下面是 JSP 引擎生成的 myfirst_jsp.java 文件内容，读者可以从 Tomcat 7.0 安装目录中的 work\Catalina\localhost\ch1\org\apache\jsp 目录中找到该文件及其对应的字节码文件。

```
myfirst_jsp.java
package org.apache.jsp;

import javax.servlet.*;
import javax.servlet.http.*;
import javax.servlet.jsp.*;

public final class myfirst_jsp extends
    org.apache.jasper.runtime.HttpJspBase
    implements org.apache.jasper.runtime.JspSourceDependent {

  private static final JspFactory _jspxFactory =
    JspFactory.getDefaultFactory();

  private static java.util.List _jspx_dependants;

  private javax.el.ExpressionFactory _el_expressionfactory;
  private org.apache.AnnotationProcessor _jsp_annotationprocessor;

  public Object getDependants() {
    return _jspx_dependants;
  }

  public void _jspInit() {
    _el_expressionfactory = _jspxFactory.getJspApplicationContext
(getServletConfig().getServletContext()).getExpressionFactory();
    _jsp_annotationprocessor = (org.apache.AnnotationProcessor)
getServletConfig().getServletContext().getAttribute(org.apache.
AnnotationProcessor.class.getName());
  }

  public void _jspDestroy() {
  }

  public void _jspService(HttpServletRequest request,
    HttpServletResponse response)
        throws java.io.IOException, ServletException {

    PageContext pageContext = null;
    HttpSession session = null;
    ServletContext application = null;
    ServletConfig config = null;
    JspWriter out = null;
    Object page = this;
    JspWriter _jspx_out = null;
    PageContext _jspx_page_context = null;
    try {
      response.setContentType("text/html;charset=GB2312");
      pageContext = _jspxFactory.getPageContext(this, request, response,
                null, true, 8192, true);
      _jspx_page_context = pageContext;
```

```
      application = pageContext.getServletContext();
      config = pageContext.getServletConfig();
      session = pageContext.getSession();
      out = pageContext.getOut();
      _jspx_out = out;

      out.write("\r\n");
      out.write("<html>\r\n");
      out.write("<head><title>这是我的第一个 JSP 程序</title></head>\r\n");
      out.write("<body bgcolor=cyan>\r\n");
      out.write("<h2>这是我的第一个 JSP 程序</h2>\r\n");
      out.write("<h3>");
      out.println("世界你好！");
      out.write("</h3>\r\n");
      out.write("</body>\r\n");
      out.write("</html>\r\n");
    } catch (Throwable t) {
      if (!(t instanceof SkipPageException)){
        out = _jspx_out;
        if (out != null && out.getBufferSize() != 0)
          try { out.clearBuffer(); } catch (java.io.IOException e) {}
        if (_jspx_page_context != null)
            _jspx_page_context.handlePageException(t);
      }
    } finally {
      _jspxFactory.releasePageContext(_jspx_page_context);
    }
  }
}
```

下面是客户端浏览器看到的源代码:

```
<html>
<head><title>这是我的第一个 JSP 程序</title></head>
<body bgcolor=cyan>
<h2>这是我的第一个 JSP 程序</h2>
<h3>世界你好!
</h3>
</body>
</html>
```

分析客户端 HTML 代码和服务器端 Java 文件代码，可以看到 out.write("...")用于向客户端输出 HTML 代码，JSP 页面对应 Java 字节码文件的主要工作如下。

(1) 把 JSP 页面中的 HTML 标记发给客户端浏览器。

(2) 负责处理 JSP 页面中的 JSP 标记，并将处理结果发给客户端浏览器。

(3) 负责执行“<%”和“%>”之间的 Java 程序片，并将执行结果发给客户端浏览器。

1.3.4 JSP、JavaBean 和 Java Servlet 的关系

Java Servlet 是 Java 语言的一部分，它提供了一组用于服务器端编程的 API。习惯上称使用 Java Servlet API 的相关类和方法所编写的 Java 类为 Servlet 类，Servlet 类生成的对象为 Servlet 对象。Servlet 对象可以运行在配置有 JSP 运行环境的服务器上，访问服务器的各种资源，这极大地扩展了服务器的功能。

JSP(Java Server Page)是晚于 Java Servlet 产生的，它克服了 Java Servlet 的缺点，并是以 Java Servlet 技术为基础的 Web 应用开发技术标准。JSP 提供了 Java Servlet 几乎所有的好处，是 Java Servlet 技术的成功应用，不过 JSP 只是 Java Servlet 技术的一部分，而不是 Java Servlet 的全部。JSP 可以让 JSP 标记、Java 语言代码嵌入 HTML 语句中，这样能显著地简化和方便网页的设计和修改，但 JSP 页面最终会编译成 Servlet 执行来响应客户端的请求。

JavaBean 被 Sun 公司定义为一个可重用的软件组件。实际上 JavaBean 就是一种 Java 类，通过封装属性和方法成为具有某种业务逻辑处理能力的类，它一般负责 Web 应用系统的业务逻辑处理部分。JavaBean 类实例化的对象简称为 Bean。JSP 提供访问 JavaBean 组件的 JSP 动作标记。JSP 动作标记简单、方便，有效地分离了 JSP 页面的表示部分和业务逻辑、数据处理部分，因此使程序设计人员和页面设计人员可以同时工作。

较小规模的 Web 应用可以采用 JSP+JavaBean 模式。JSP+JavaBean 模式中，JSP 负责页面的显示，页面预处理和跳转控制，JavaBean 负责业务逻辑和数据处理。对于规模较大的 Web 应用，就需要采用 JSP+JavaBean+Servlet 模式。JSP+JavaBean+Servlet 模式中，JSP 负责页面显示(View)，JavaBean 负责业务逻辑和数据处理(Model)，Servlet 负责预处理和分发页面的请求(Controler)。关于这些模式的具体应用将在后续章节中讲述。

1.4 集成开发环境简介

集成开发环境可以有效地提高 Web 应用的开发效率，减轻程序员的劳动强度。优秀的集成开发环境能够使程序员如虎添翼，事半功倍，所以读者有必要了解目前流行的 Web 应用集成开发环境。比较常见的开发环境有 Eclipse/MyEclipse、JBuilder、NetBeans、WebSphere 等开发环境。本节将对 Eclipse/MyEclipse 做简单介绍。

1.4.1 MyEclipse

MyEclipse 企业级工作平台(MyEclipse Enterprise Workbench，简称 MyEclipse)是对 Eclipse IDE 的扩展，是 J2EE 开发插件的综合体，是功能丰富的 J2EE 集成开发环境。它包括了完备的编码、调试、测试和发布功能，完整支持 HTML、JSP、Struts、JSF、CSS、JavaScript、SQL、Hibernate 等。用户使用它可以在数据库和 J2EE 的开发、发布，以及应用程序服务器的整合方面提高工作效率。

MyEclipse 是收费的开发工具，一般下载的 MyEclipse 内部已经有一个 Eclipse 存在了。安装 MyEclipse，需要从 MyEclipse 官方网站或其他搜索引擎下载安装文件，并购买

注册码。

1.4.2 开源的 Eclipse

Eclipse 最初是 IBM 公司的一个软件产品，2001 年 11 月其 1.0 版发布。2003 年发布了 2.1 版，立刻引起了业界的轰动。之后 IBM 已将出巨资开发的 Eclipse 作为一个开源项目捐给了开源组织 Eclipse.org，其出色的独创性平台，吸引了众多大公司加入到 Eclipse 平台的发展中来。到书稿完成时为止，Eclipse 的最新版本为 V4.3。

Eclipse 是一个通用的工具平台。它提供了功能丰富的开发环境，允许开发者高效地创建一些工具并集成到 Eclipse 平台上来。Eclipse 的设计思想是：一切皆为插件。Eclipse 的核心非常小，其他所有的功能都以插件的形式附加到这个核心之上。Eclipse 的插件是动态调用的，也就是说插件被使用时调入，不再被使用则在适当时候自动清除。

用户可以去 Eclipse 的官方网站免费下载 Eclipse 工具包，也可以使用 Google 搜索工具搜索下载 Eclipse 工具包。常用的 Eclipse 有两种版本，Releases 版是稳定版本，StableBuilds 版本比较稳定。Eclipse 的安装非常简单，它属于纯绿色软件，只需将安装文件解压就可以运行 Eclipse。现在最新的 JBuilder、WebSphere、MyEclipse 等开发工具都是将 Eclipse 作为集成框架基础开发而成的。Eclipse 开发工具的详细使用方法，请读者参见有关资料。

1. 安装 Eclipse

运行 Eclipse 的安装文件，本教材使用的安装文件是 eclipse-jee-kepler-SR2-win32.zip，将其解压到指定磁盘即可完成安装，本教材使用的安装目录为 C:\。

2. 运行 Eclipse

安装完成后，启动 Eclipse，首先要选择 Eclipse 的工作空间，工作区里存放项目文件和一些配置信息，这里选择 C:\Users\admin\workspace 作为工作空间，并选中 Use this as the default and do not ask again 复选框，可保证下一次启动 Eclipse 不再提示选择工作空间。单击 OK 按钮就可启动。工作区的选择界面如图 1.12 所示。

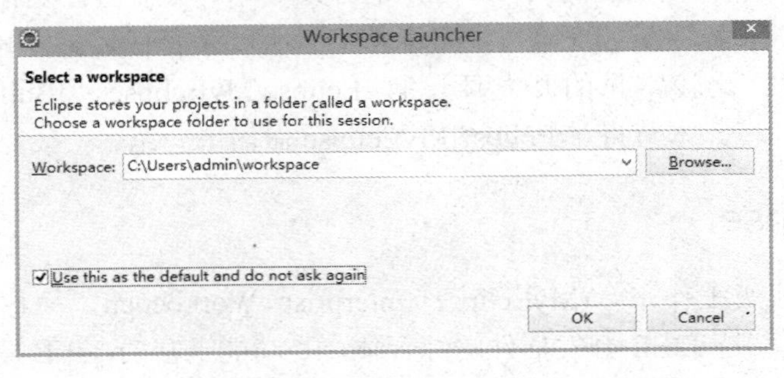

图 1.12 选择 Eclipse 的工作区

3. 配置 Eclipse 的 JDK 和 Tomcat 环境

选择工作区，启动 Eclipse 后，用户可以为该工作区配置 JDK 和 Tomcat 环境。选择主菜单中的 Windows→Preferences(首选项)命令，弹出如图 1.13 所示的首选项对话框。在左侧树形控件中选择 Server→Runtime Environments，在对话框的右侧编辑区单击 Add 按钮。

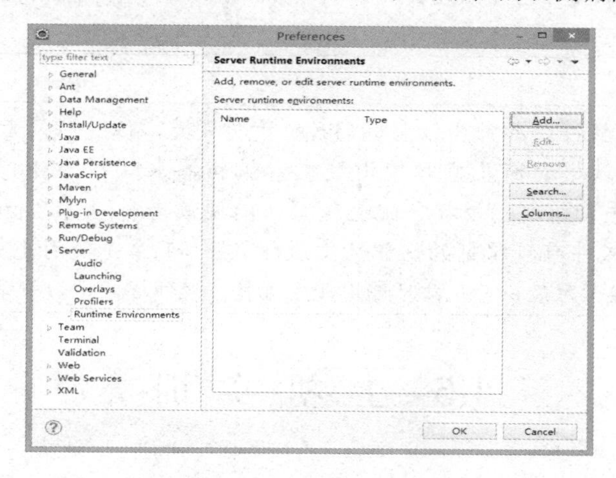

图 1.13　首选项对话框

在弹出的对话框中选择 Apache Tomcat v7.0，并单击 Next 按钮，进入 Tomcat 服务器配置界面，如图 1.14 所示。

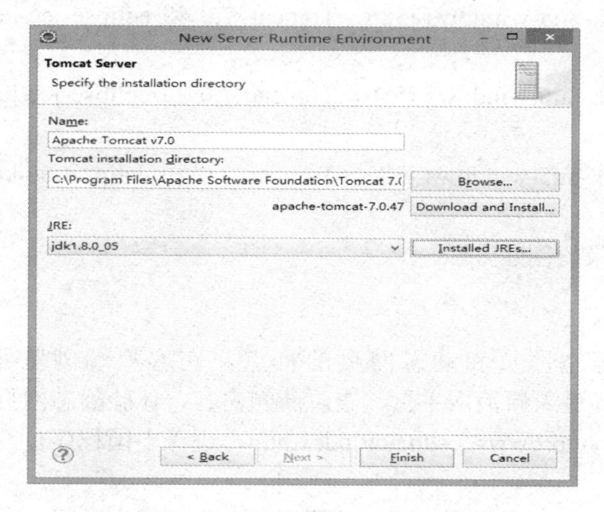

图 1.14　配置 Tomcat 和 JDK 环境

在图 1.14 所示界面中选择 Tomcat 服务器的安装目录，本教材使用的安装目录为 C:\Program Files\Apache Software Foundation\Tomcat 7.0，并选择本机安装的 JRE 版本 jdk1.8.0_05。然后单击 Finish 按钮即可完成 Tomcat 服务器和 JDK 的配置。

在使用 Eclipse 4.3 开发 Web 应用程序，需要在 Eclipse 4.3 中启动 Tomcat 服务器。如图 1.15 所示，在 Server 窗口中单击按钮或按钮运行或调试运行 Tomcat 服务器。

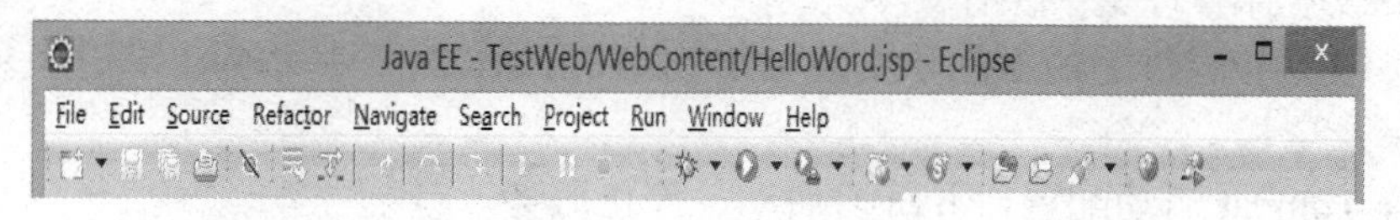

图 1.15　启动 Tomcat 服务器

注意：如果用户的计算机上已经启动了一个 Tomcat 服务器，在 Eclipse 中启动 Tomcat 就会失败，用户可以关闭已经启动的 Tomcat，然后再启动即可。

说明：Eclipse 是目前软件企业使用较多的 Java 开发环境，但是从读者学习的角度讲，建议读者在学习教材的前几章时使用纯文本编辑器来编辑 JSP 页面、JavaBean 和 Servlet，通过手动编写、编译、配置项目，以便读者对 JSP、JavaBean、Web 服务目录以及配置文件有较深刻的认识。本教材的第 11 章是使用 Eclipse 开发环境完成的，建议读者在学习教材后几章时使用开发工具，以便提高学习效率和熟悉开发工具。

1.5 上机实训

实训目的

- 体验静态网页和动态网页的特点。
- 了解 JSP 的运行原理。
- 掌握下载 jdk-8u5-windows-i586、Tomcat 7.0 和 eclipse-jee-kepler-SR2-win32 的方法。
- 掌握安装 jdk-8u5-windows-i586、Tomcat 7.0 和 eclipse-jee-kepler-SR2-win32 的方法。
- 掌握配置 jdk-8u5-windows-i586、Tomcat 7.0 和 eclipse-jee-kepler-SR2-win32 的方法。
- 掌握虚拟目录的创建方法。

实训内容

实训 1 体验静态网页和动态网页的特点，在客户端浏览器地址栏中输入：http://www.sina.com.cn/进入新浪网主页，查看新闻内容，体验静态网页的特点。在客户端浏览器地址栏中输入 http://www.csdn.net/index.htm，进入中国程序员网，注册会员，体会动态网页的特点。

要求：在“程序员网”注册成功后，使用用户名登录程序员网，查询相关 JSP 的学习资料。

实训 2 根据教材提供的地址或使用搜索引擎，下载 jdk-8u5-windows-i586、Tomcat 7.0 和 eclipse-jee-kepler-SR2-win32。

要求：

(1) 注意版本号和适用平台。

(2) 将下载软件整理集中保存。

实训 3 根据教材讲述的方法，安装 JDK 和 Tomcat，配置基本的 JSP 开发环境。

要求：

(1) 设置 Tomcat 启动配置，并练习启动、停止服务。

(2) 通过 Java 命令测试 JDK 安装，Tomcat 的测试页测试 Tomcat 的安装。

(3) 编写一个简单的 JSP 页面，测试 Tomcat 安装，体验 JSP 页面的编写过程。

实训 4 创建 Web 虚拟服务目录。

要求：

(1) 创建第 1 章的虚拟目录 ch1，物理目录为 E:\programJsp\ch1。

(2) 将 myfirst.jsp 文件复制到 E:programJsp\ch1 目录中，用浏览器测试 myfirst.jsp 页面。

实训 5 安装 Eclipse，建立工作区，配置 Eclipse 的 JDK 和 Tomcat。

(1) 安装 Eclipse。

(2) 创建工作区 E:\programJsp\ch1，配置工作区的 JDK 和 Tomcat。

实训总结

通过本章的上机实训，学生应了解静态网页和动态网页的特点；了解 JSP 运行原理；掌握下载 jdk-8u5-windows-i586、Tomcat 7.0 和 eclipse-jee-kepler-SR2-win32 的方法；掌握安装 jdk-8u5-windows-i586、Tomcat 7.0 和 eclipse-jee-kepler-SR2-win32 的方法；掌握配置 jdk-8u5-windows-i586、Tomcat 7.0 和 eclipse-jee-kepler-SR2-win32，搭建 JSP 开发环境的方法；掌握虚拟目录的创建方法。

1.6 本章习题

思考题

(1) 为什么要为 JDK 设置环境变量？

(2) Tomcat 和 JDK 是什么关系？

(3) 什么是 Web 服务根目录、子目录、相对目录？如何配置虚拟目录？

(4) 什么是 B/S 模式？

(5) JSP、JavaBeans 和 JavaServlet 之间的关系？

(6) 集成开发环境能为程序员做什么？

(7) 使用 Eclipse 开发 JSP 程序，需要做哪些配置？

(8) MyEclipse 和 Eclipse 之间有什么样的关系？

拓展实践题

(1) 通过网上书店购书，体验 Web 应用程序的特点。

(2) 下载安装 Apache Web 服务器，尝试集成 Tomcat 与 Apache Web 服务。

第 2 章　Web 开发基础

学习目的与要求：

HTML/XHTML 是网页设计语言，CSS 是描述页面外观的层叠样式表，DIV+CSS 模式是当前页面布局的主流技术。本章主要学习 HTML/XHTML 文档结构、常用标记，页面常见表格制作技术，DIV+CSS 页面布局原理与实现技术。通过本章的学习，学员要掌握 HTML/XHTML 常用标记的使用，掌握页面表格的制作技术，掌握定义 CSS 选择器的方法。理解层布局原理及其具体实现技术，对于熟悉这部分内容的读者，可以跳过本章直接学习后面的内容。

2.1　HTML 简介

编写 JSP 程序，需要在 HTML 代码中插入 JSP 标记、Java 程序片、Java 表达式等，因此需要读者掌握基本的 HTML 知识。

2.1.1　什么是 HTML

HTML(Hyper Text Marked Language，超文本标记语言)是描述网页的标记语言。目前的 HTML 大约有一百多个标记，这些标记描述 HTML 文档中数据的显示格式，它们可以定义文本、图形、表格的格式，指向其他页面的链接，以及提交数据的表单等。HTML 网页是 HTML 描述的文本文件。HTML 文件由 Web 服务器发送给客户端浏览器，客户端浏览器按 HTML 描述的格式将其显示在浏览器窗口内，呈现给读者多姿多彩的页面。HTML 文件通过 HTTP，使得 HTML 文件可以在互联网上顺畅地进行文件交换和访问。

HTML 文件是纯文本文件格式，可以用文本编辑器进行编辑制作，如记事本、Editplus 等，也可以使用专业的网页编辑工具编辑制作，如 Dreamweaver 等网页制作工具来完成。

2.1.2　什么是 URL

URL 是 Uniform Resource Locator 的缩写，中文称之为统一资源定位器。URL 是 Internet 中资源的简单命名机制。URL 由三部分构成：协议、主机 DNS 名或 IP 地址和文件名。例如 URL：

```
http://www.sina.com/index.html
```

其含义为：协议为 HTTP，主机为 www.sina.com.cn，文件名为 index.html。

2.1.3　HTML 文件结构

1. HTML 的标记与属性

HTML 标记是用“<”和“>”括起来的标识符，括号中间的标识符为标记名称，如标记<body>，标记名为 body。HTML 标记通过指定某块信息为段落或标题等来标识文档某个部分。HTML 的标记分为单标记和成对标记两种。成对标记有开始标记<标记名>和结束标记</标记名>，并配套使用，例如开始标记<html>和结束标记</html>配套使用。成对标记只作用于开始标记和结束标记之间的文档。单标记只有开始标记，例如<p>，如果使用单标记，只要在文档相应位置插入单标记即可。

属性是标记中的参数选项。大多数标记都有一些自己的属性，有些标记也有共用的一些属性，各属性之间无先后顺序，如果省略属性则采用默认值。如果要定义标记的属性则在标记名称后加空格，再书写属性名并给其赋值，属性的一般使用格式为

```
<标记名 属性1="属性值"  属性2="属性值"  …>内容</标记名>
```

例如：

```
<body text="blue"  link="red">内容</body>
```

其中，body 为标记名，text、link 为属性，"blue"和"red"为 text 和 link 属性的值。

注意： HTML 标记不区分大小写。本教材建议读者在书写标记属性值时，用""双引号将属性值引起来。

2. 文档头与文档体

HTML 文件必须由<html>标记开头，</html>标记结束，这表明该文件为 HTML 超文本文档。一个完整的 HTML 文档分为文档头和文档体两部分。文档头信息包含在<head>与</head>之间，在<head>和</head>之间可以包含有关此网页的标题、导入样式表等信息。文档体包含在<body>和</body>标记之间，是网页的主体部分。<html>和</html>标记将文档头和文档体包含在其内。

【例 2.1】 创建一个简单但较完整的 HTML 文件 ch2_1.html，文件内容如下：

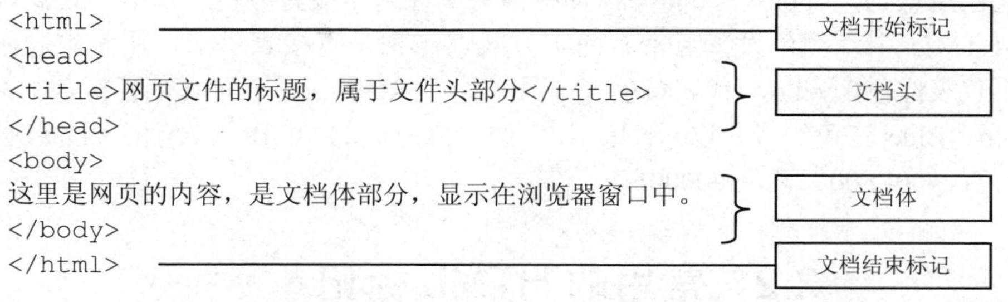

在 E:\programJsp 文件夹中创建 ch2 文件夹，修改 Tomcat 的 server.xml 配置文件，将其设置为虚拟目录，目录名称为 ch2，然后将 ch2_1.html 文件保存在该文件夹内，使用浏览器查看 ch2_1.html 效果，页面效果如图 2.1 所示。

分析例 2.1 文档 ch2_1.html，<html>和</html>在文档的最外层，它表示文档是 HTML

协议描述的。<head>与</head>是文档的头部标记，在浏览器中头部信息不显示在浏览器的正文显示区里。此标记中可以插入其他标记，用以说明文档文件的一些公共属性。<body></body>之间的文档是正文。

图 2.1　例 2.1 的页面显示效果

注意： 本章的所有例题都保存在 E:\programJsp\ch2 目录中，因为前面已经将 E:\programJsp\ch2 目录设置为 ch2，所以用户在浏览器中输入 http://127.0.0.1:8080/ch2/ch2_1.html 即可访问该页面。

3. <body></body>标记与颜色设定

<body></body>标记为文档体标记，网页上所见到的内容都是放到<body></body>之间。主体标记有影响整个页面显示方式的属性，网页的背景颜色、文本颜色、背景图片都在<body>的属性里做设置。其一般格式如下：

```
<body text="文本颜色" link="链接颜色" vlink="已访问过的链接颜色"
alink="单击时的链接颜色" bgcolor="背景颜色" background="背景图片"
leftmargin="页面的左边距" topmargin="页面的上边距"
</body>
```

其中的属性说明如下。

- text 属性为网页文本颜色。
- link 属性为超级链接文字的颜色，默认值为蓝色下划线。
- vlink 属性为已单击文本的颜色。
- bgcolor 属性为网页的背景颜色，只能设置单一颜色。
- background 属性为网页背景图片属性，属性值可指定一幅浏览器支持的图片文件名。

读者可能已注意到，<body></body>标记中很多属性都是设置颜色，那么颜色属性如何设置？在 HTML 中，颜色设置通常有两种方式，一种是直接表示法，用常数表示颜色，例如 red 代表红色，yellow 代表黄色等；另一种用 RGB 三色表示法，任何颜色都是由 Red、Green、Blue 三色组成，以#开头，用三组十六进制的 RGB 值表示，例如红色：#FF0000、绿色：#00FF00、蓝色:#0000FF。

2.2　常用的 HTML 标记

介绍了 HTML 的一些基本知识之后，本节重点介绍一些常用的 HTML 标记。熟悉这些标记的用途和使用方法，对学习 JSP 会有很大的帮助。

2.2.1　HTML 的文字标记

1. <title></title>标记

该标记在<head></head>标记之间，<title></title>标记之间的内容将显示到浏览器的标题栏上，如图 2.1 所示。

2. <p>、
和<hr>标记

1) <p>标记

<p>是一个段落的开始标记，其使用的一般格式如下：

```
<p  align="段落对齐方式">段落文字</p>
```

其中的 align 属性为段落文字的对齐方式，取值为 Left、Right、Center 等。

2)
标记

是换行标记，换行字符(CR，LF)在 HTML 文件中不起作用，必须使用
换行标记。

3) <hr>标记

<hr>用于插入一条水平线，其使用的一般格式如下：

```
<hr  size="线的粗细"  width="线的宽度">
```

其中的属性说明如下。

- size 属性用于设置线的粗细，属性取值的单位是 pixel(像素)。
- width 属性用于设置线的宽度，属性取值的单位是 pixel 或百分比(%)。

【例 2.2】演示上述标记的示例(其代码参见 ch2_2.html)。页面显示效果如图 2.2 所示。

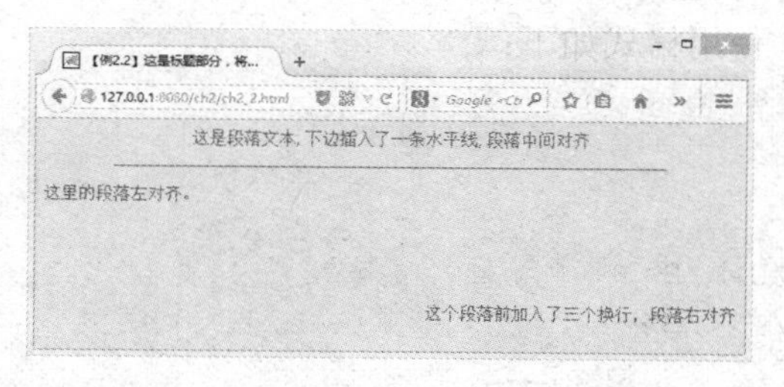

图 2.2　ch2_2.html 的页面显示效果

ch2_2.html 文件中的代码如下：

```
<html>
<head>
<title>这是标题部分，将显示在浏览器窗口标题栏</title>
</head>
<body text="red" bgcolor="yellow" >
<p align="center">这是段落文本,下边插入了一条水平线,段落中间对齐</p>
<hr size="2" width=80%>
```

```
<p align="left">这里的段落左对齐。</p>
<br/><br/><br/>
<p align="right">这个段落前加入了三个换行，段落右对齐</p>
</body>
</html>
```

3. <hn></hn>标记

该标记用来设置网页中标题的文字。标题分 6 级，使用时<hn>标记中的 n 用 1～6 之间的数字取代。<h1>是最大的标题，<h6>是最小的标题。<hn>的一个使用格式如下：

```
<hn align="对齐方式">标题内容</hn>
```

其中的 align 为标题文本对齐属性，取值为 left、right、center 等。

4. <font></font>标记

该标记用来控制文字的字体、大小和颜色，它的使用格式如下：

```
<font face="字体名称" size="字体大小" color="字体颜色">显示的文字</font>
```

其中的属性说明如下。

- face 属性的取值可以是系统所装字体的名称，例如“宋体”、“幼圆”、“楷体”、“黑体”等，默认值是“宋体”。
- size 属性的取值是从 1 到 7 之间的数字，也可以以 3 为基准设置，-1 代表 2，+1 代表 4，默认值为 3。
- color 属性的取值可以用常数表示，如 red，也可用 RGB 表示，如#00ff00。

5. <B></B><I></I><U></U>标记

<B></B>为文字粗体标记，<I></I>为文字斜体标记，<U></U>为文字加下划线的标记。它们可以组合使用，使用格式如下：

```
<B>粗体字</B>
<I>斜体字</I>
<U>加下划线字</U>
<B><U>粗体+下划线</U></B>
<B><I>粗体+斜体字</I></B>
```

6. 和标记

是文本下标标记，是文本上标标记。使用格式为：

```
<sub>下标文本</sub>
<sup>上标文本</sub>
```

【例 2.3】演示标题、文字与段落的示例(其代码参见 ch2_3.html)，页面效果如图 2.3 所示。

ch2_3.html 文件中的代码如下：

```
<html>
<head>
```

```
    <title>【例 2.3】演示标题、文字、段落的示例</title>
</head>
<body>
    <h1>Web 开发基础(标题 1)</h1>
    <h3> 创建 CSS(标题 3)</h3>
<font face="幼圆" size=4 color="red" >当读到一个样式表时，浏览器会根据它来格式
化 HTML 文档。插入样式表的方法有三种：内联样式表、内部样式表和外部样式。(字体：幼圆，
字号 4，颜色：红)</font><br>
     <b>内联样式表(粗体)</b><br>
    <i>内部样式表(斜体)</i><br>
    <u>外部样式(下划线)</u>
    <h1>高中数理化学习专题(标题 1)</h1>
    <h3>分子式学习：(标题 3)</h3>
    <p>使用下标表示水的化学分子式：
   <b><i>H<sub>2</sub>O</i></b><br>
    <h3>数学函数学习：(标题 3)</h3>
    <p>使用上标表示 X 的平方：
    <b><u>X<sup>2</sup></u></b>
</body>
</html>
```

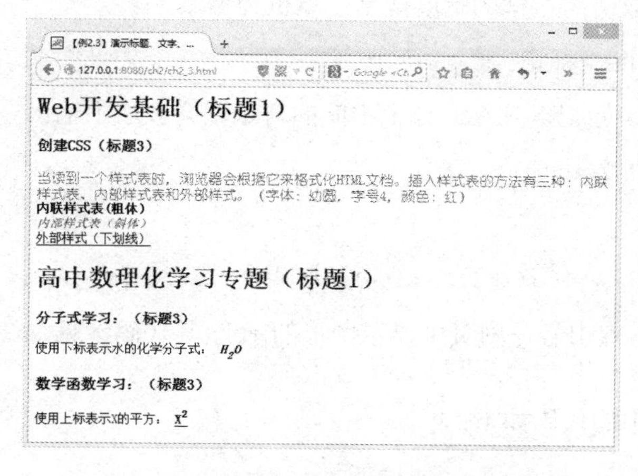

图 2.3　ch2_3.html 的页面效果

2.2.2　特殊标记和图形标记

1. 和" 标记

在 HTML 文档中，空格并不具备调整间距的功能，连续多个空格出现，仅第一个空格有效。HTML 使用( )表示空格，(")表示双引号。HTML 中还有其他特殊字符使用字符代码表示，请查阅有关手册。

2. <!-- -->标记

该标记是 HTML 注释标记，注释标记的内容不会在浏览器中显示出来，只是为了查找或记忆有关的内容。

3. <img>标记

该标记用来在网页中显示图形。图形文件与网页文件是分别存储的，通过<img>标记将图形显示在网页。该标记使用格式如下：

```
<img  src="文件名"  height="图形显示高度"  width="图形显示宽度"
  border="边框粗细"  alt="说明文字"  align="对齐方式" hspace="水平间距"
  vlign="垂直间距">
```

其中的属性说明如下。

- src 属性的取值为图形文件的 URL 路径，可以用绝对路径或相对路径表示。绝对路径是文件名的完整路径；相对路径是指相对当前网页文件名的路径。例如："http://www.sina.com/logo.gif"是绝对路径，"logo.gif"是相对路径。
- height 属性的取值为图形的高度，单位可以是像素(pixel)或浏览器窗口高度的百分比。
- width 属性的取值为图形的宽度，单位可以是像素或百分比。
- alt 属性为图形的文字说明。
- align 属性为图像与文字之间的排列属性，可取值有 baseline、top、middle、bottom、texttop、absmiddle、absbottom、left、right 等，默认为 bottom。
- border 属性取值为边框的宽度数字，单位为像素。

4. <center></center>标记

该标记为对象居中标记，凡在此标记中间的对象都会被居中输出。例如：

```
<center>文字居中</center>
```

或者：

```
<center><img src="logo.gif" alt="图形居中"  ></center>
```

【例 2.4】空格、双引号与网页中显示图形的示例(其代码参见 ch2_4.html)，页面效果如图 2.4 所示。

ch2_4.html 文件中的代码如下：

```
<html>
<head>
  <title>【例 2.4】标记示范</title>
</head>
<body>
     <center>
     <!--这是一个<img>标记和 和&quot;<center>标记的示范例子-->
     <p align="left">这是一个&lt;img&gt;标记的例子，    
         (这里有 几个 空格)显示当前目录下&quot;logo.jpg&quot;
     图形文件，显示效果如下图：
      <img src="logo.jpg" alt=" ch2_4_logo.jpg文件" width="256"
      height="192" hspace="30" vspace="20" border="10"
      align="middle" >用户登录
     </p>
      </center>
```

```
</body>
</html>
```

图 2.4　ch2_4.html 的页面效果

2.2.3　超级链接标记

超级链接是互联网的灵魂，通过超级链接标记将互联网上的资源织成了一张巨大的信息网络，极大地方便了用户的访问。作为 Web 程序设计人员必须熟练掌握 HTML 的超级链接技术。<a></a>标记为超级链接标记，一般使用格式为如下：

```
<a  href="资源地址"  target="窗口名" title="指向链接时显示的文字" >超链接显示名称</a>
```

其中的属性及参数说明如下。

- href 属性的取值为链接的目标地址。目标地址可以是绝对路径、相对路径。绝对路径是 URL 地址，如"http://www.sina.com/web/index.html"、"ftp://ftp.hebau.edu.cn/"、"mailto:yxq@163.com ? cc=dkd@163.com & subject=建议 &bcc = yxq@hebau.edu"等。相对路径是相对于当前网页文件的所在目录的路径。
- target 属性的取值为链接的目标窗名，可以是_parent、_blank(新窗口)、_self、_top(浏览器的整个窗口打开，忽略任何框架)等值，也可以是窗口名称或 id，其默认为原窗口。
- title 属性的取值为指向链接时所显示的标题文字。
- <a>与</a>之间的“超级链接显示名称”，可以是文本，也可以是<img>图形链接。

【例 2.5】超级链接的示例(其代码参见 ch2_5.html)，页面效果如图 2.5 所示。

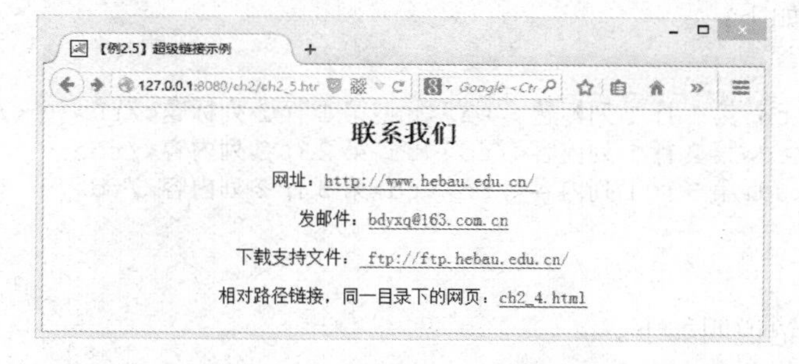

图 2.5　ch2_5.html 的页面效果

ch2_5.html 文件中的代码如下：

```
<html>
<head>
   <title>【例 2.5】超级链接示例</title>
</head>
<body>
     <h2 align="center">联系我们</h2>
     <p align="center">网址：<a href="http://www.hebau.edu.cn/"
      title="登录我们网站" target="_blank">http://www.hebau.edu.cn/</a></p>
     <p align="center">发邮件：<a title="发送邮件给我们"
      href="mailto:bdyxq@163.com.cn ?cc=yxq@hebau.edu.cn&subject=hello
      &body=杨老师你好！ &bcc=yxq@hebau.edu.cn">bdyxq@163.com.cn</a></p>
     <p align="center">下载支持文件：<a href="ftp://ftp.hebau.edu.cn"
      target="_parent" title="下载文件">
      ftp://ftp.hebau.edu.cn</a>/</p>
     <p align="center">相对路径链接，同一目录下的网页：<a href="ch2_4.html"
      title="相对链接同一目录" target="_top">ch2_4.html</a></p>
</body>
</html>
```

2.3 表　格

表格在网页中应用非常广泛，子标签灵活，属性丰富，在过去设计师经常使用 table 元素进行页面布局。现在 table 标签的作用回归到了表现数据，并推荐使用 DIV+CSS 的形式对页面进行布局。

2.3.1 定义表格的基本语法

在 HTML 中表格是通过<table></table>表格标记、<tr></tr>行标记、<th></th>列标题标记和<td></td>列内容标记等配合使用来定义的。<table></table>标记表明表格的开始和结束；在<table></table>之间的<tr></tr>标记表明“表格行”的开始和结束；在<tr></tr>之间的<th></th>表明“表格列标题”的开始和结束；在<tr></tr>之间的<td></td>表明“表格列内容”的开始和结束。简单讲就是<table></table>之间定义行，<tr></tr>标记之间定义列。一般使用格式如下：

```
<table>
        <tr><th>第 1 行 1 列标题</th><th>第 1 行 2 列标题</th> …</tr>
        <tr><td>第 2 行 1 列内容</td><td>第 2 行 2 列内容</td> … </tr>
        <tr><td>第 3 行 1 列内容</td><td>第 3 行 2 列内容</td> … </tr>
        ⋮
</table>
```

其中的有关标记说明如下。

- <table></table>内可以建立多行。

- <tr></tr>内可以建立多列。
- <th></th>定义的标题列粗体显示，也可以不用此标记。
- 一个基本表格至少要包含一行一列。

【例 2.6】基本表格示例(其代码参见 ch2_6.html)，页面效果如图 2.6 所示。

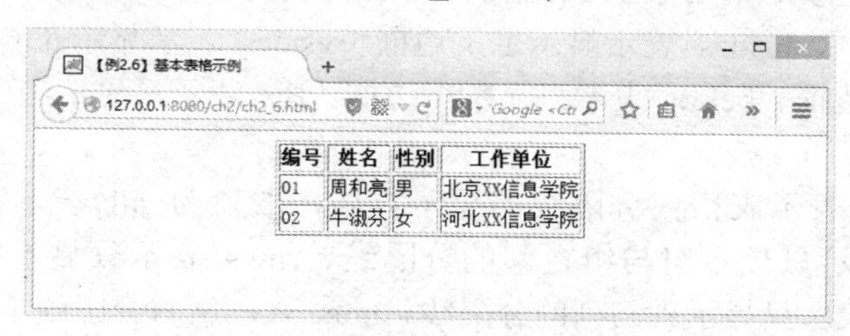

图 2.6　ch2_6.html 的页面效果

ch2_6.html 文件中的代码如下：

```
<html>
 <head>
   <title>【例 2.6】基本表格示例</title>
 </head>
 <body>
   <center>
   <table border="1">
      <tr><th>编号</th><th>姓名</th><th>性别</th><th>工作单位</th></tr>
      <tr><td>01</td><td>周和亮</td><td>男</td><td>北京 XX 信息学院
      </td></tr>
      <tr><td>02</td><td>牛淑芬</td><td>女</td><td>河北 XX 信息学院
      </td></tr>
    </table>
    </center>
</body>
</html>
```

2.3.2　表格<table>标记的属性

HTML 5 中的表格标记<table>有很多属性，可选属性的格式如下：

```
<table align="水平摆放位置" width="表格宽度" height="表格高度"
    frame="四周边框显示状态" border="四周边框宽度值"  rules="分隔线的显示状态"
    cellpadding="数值" cellspacing="数值"
    bordercolor="边框颜色值" bgcolor="背景颜色值">
    ⋮
</table>
```

其中的属性说明如下。

- align 属性的取值为表格在页面上水平摆放位置，取值为 left、center、right 值。
- border 属性的取值单位为像素，值为表格边框的宽度。
- cellpadding 属性的取值单位为像素，值为单元格内容与单元格边界之间的空白距离。

- cellspacing 属性的取值单位为像素，值为单元格之间的距离。
- width 属性为表格宽度，取值单位为像素或页面宽度的百分比。
- height 属性为表格高度，取值单位为像素或页面高度的百分比。
- frame 属性为表格四周边框的显示状态，取值为 box 表示四周都显示、void 表示不显示边框、hsides 表示显示上下边框、vsides 表示显示左右边框、above 表示只显示上边框、below 表示只显示下边框、rhs 表示只显示右边框、lhs 表示只显示左边框。
- rules 属性表示表格内分隔线的显示方式，取值为 all 表示显示所有分隔线、groups 表示只显示组与组之间的分隔线、rows 表示只显示行与行之间的分隔线、cols 表示只显示列与列的分隔线、none 表示所有分隔线不显示。

【例 2.7】显示表格外表框和单元格间距的表格(示例代码参见 ch2_7.html)，页面效果如图 2.7 所示。

ch2_7.html 文件中的代码如下：

```
<html >
<head>
<title>【例 2.7】表格示例</title>
</head>
<body>
<table frame="box" width="290" height="130" border="10" align="center"
   cellpadding="4" cellspacing="2" bordercolor="#ff0000"
   bgcolor="#DDFFDD">
<caption align="top">
我的表格
</caption>
  <tr><th>编号</th><th>姓名</th><th>性别</th><th>工作单位</th></tr>
  <tr><td>01</td><td>杨得力</td><td>男</td><td>河北 XX 大学</td></tr>
  <tr><td>02</td><td>周国正</td><td>女</td><td>北京 XX 大学</td></tr>
</table>
</body>
</html>
```

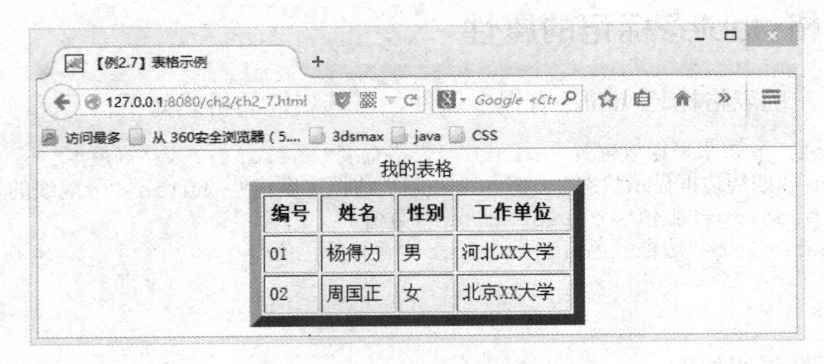

图 2.7　ch2_7.html 页面效果

【例 2.8】显示部分外边框和内部分隔线的表格示例(其代码参见 ch2_8.html)，页面效果如图 2.8～图 2.10 所示。

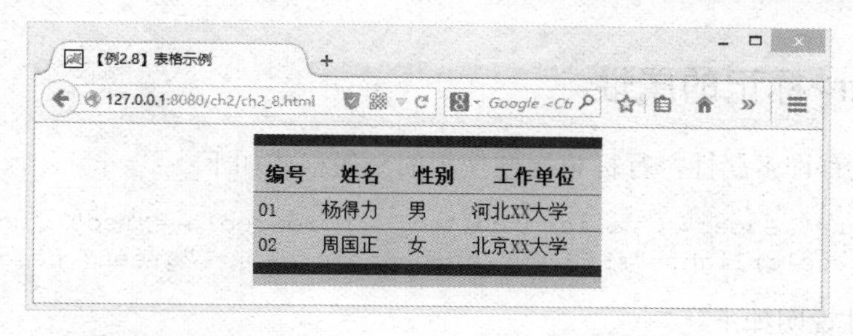

图 2.8　ch2_8.html 在火狐浏览器中的页面效果

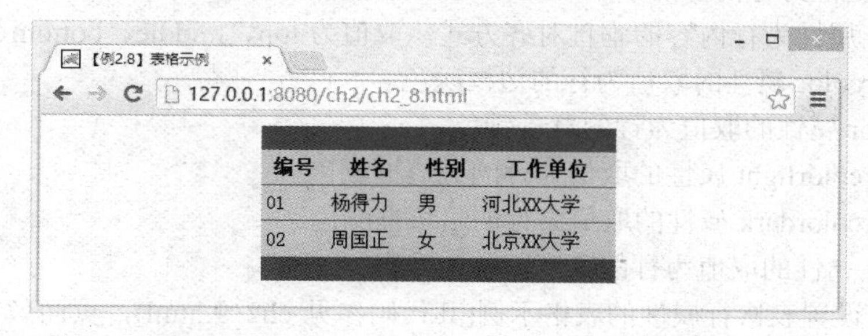

图 2.9　ch2_8.html 在 chrome 浏览器中的页面效果

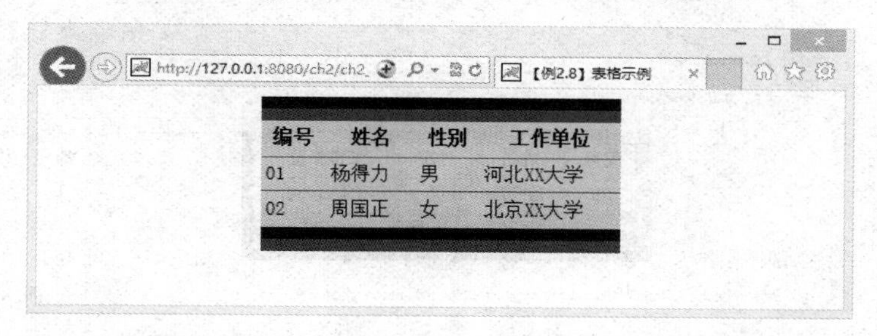

图 2.10　ch2_8.html 在 IE10 浏览器中的页面效果

ch2_8.html 文件中的代码如下：

```
<html>
<head>
<title>【例 2.8】表格示例</title>
</head>
<body>
<table frame="hsides" rules="rows" width="290" height="130"
    border="20" align="center" cellpadding="4" cellspacing="2"
    bordercolor="#ff0000" bgcolor="#55ffff">
  <tr><th>编号</th><th>姓名</th><th>性别</th><th>工作单位</th></tr>
  <tr><td>01</td><td>杨得力</td><td>男</td><td>河北 XX 大学</td></tr>
  <tr><td>02</td><td>周国正</td><td>女</td><td>北京 XX 大学</td>
  </tr>
</table>
</body>
</html>
```

2.3.3 行<tr>标记的属性

行<tr>标记有许多属性，行标记属性使用的一般格式如下：

```
<tr  align="center"  valign="bottom"  bordercolor="red"  bgcolor="blue"
    bordercolorlight="#FF0000" bordercolordark= "green" height="30" >
```

其中的属性说明如下。

- align 属性为行内容的水平对齐方式，取值为 left、right、center。
- valign 属性为行内容的垂直对齐方式，取值为 top、middle、bottom、baseline。
- bordercolor 属性的取值为行的边框颜色。
- bgcolor 属性的取值为行的背景颜色。
- bordercolorlight 属性的取值为行的亮边框颜色。
- bordercolordark 属性的取值为行的暗边框颜色。
- height 属性的取值为行的高度，单位像素。

【例 2.9】设置表格行属性的表格示例(其代码参见 ch2_9.html)，页面效果如图 2.11～图 2.13 所示。

图 2.11　ch2_9.html 在火狐浏览器中的页面效果

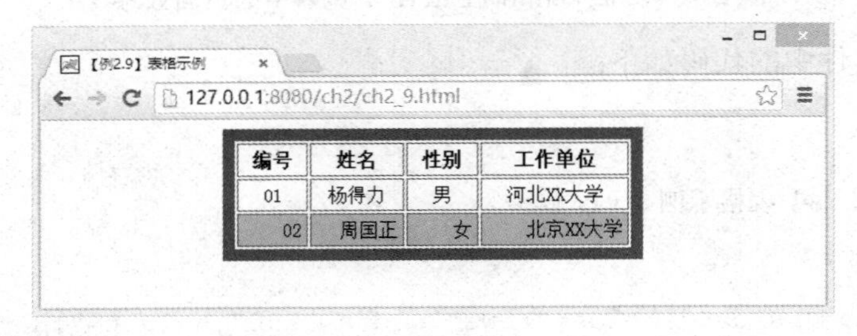

图 2.12　ch2_9.html 在 chrome 浏览器中的页面效果

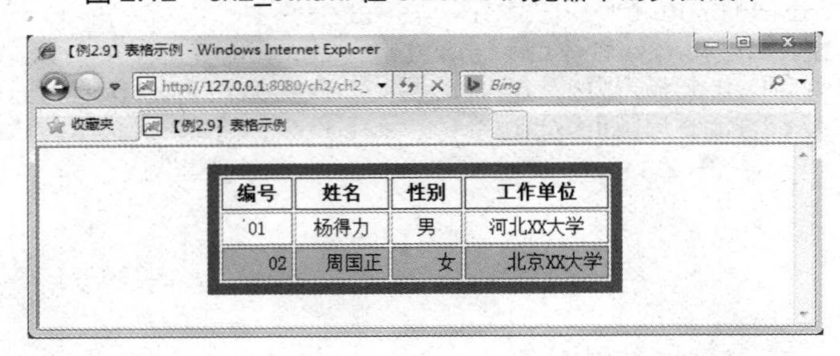

图 2.13　ch2_9.html 在 IE8 浏览器中的页面效果

ch2_9.html 文件中的代码如下：

```
<html>
<head>
<title>【例 2.9】表格示例</title>
</head>
<body>
<table width="350"  bordercolor="#FF0000" heigth="290" border="10"
    align="center" cellpadding="4" cellspacing="2">
  <tr align="center" valign="middle"  bordercolor="#0000FF">
    <th>编号</th><th>姓名</th><th>性别</th><th>工作单位</th></tr>
  <tr align="center" valign="top"  bordercolordark="#00FF00"
    bordercolorlight="#FF0000">
    <td>01</td><td>杨得力</td><td>男</td><td>河北 XX 大学</td></tr>
  <tr align="right" valign="baseline" bgcolor="#CC99FF">
    <td>02</td><td>周国正</td><td>女</td><td>北京 XX 大学</td>
  </tr>
</table>
</body>
</html>
```

2.3.4 单元格<td>和<th>标记的属性

<td>和<th>都是插入单元格的标记，标记必须嵌套在<tr></tr>标记中，且成对出现。<th>是标题标记，其中字体用粗体显示，<td>是数据标记。<td>和<th>标记属性相同，下面以<td>标记介绍单元格属性的使用方法。

```
<td width="300" height="80%" align="left"  valign="bottom"
    bordercolor="#FFFF00" bgcolor="#FF0000" bordercolordark="#FF0000"
    bordercolorlight="#FF00FF" colspan="2" rowspan="3"
    background="背景图片文件名"> 单元格内容</td>
```

其中的属性说明如下。

- width 属性为单元格宽度，取值为像素数或百分数。
- height 属性为单元格的高度，取值为像素数或百分比。
- align 属性为单元格内容的水平排列方式，取值为 left、right、center。
- valign 属性为单元格内容的垂直对齐方式，取值为 middle、top、bottom、baseline。
- bordercolor 属性为单元格边框颜色。
- bgcolor 属性为单元格背景颜色。
- bordercolorlight 属性为单元格亮边框的颜色。
- bordercolordark 属性为单元格暗边框的颜色。
- colspan 属性为单元格向右打通的列数，用于合并单元格。
- rowspan 属性为单元格向下打通的行数，用于合并单元格。

【例 2.10】设置单元格格式或列格式的表格示例(其代码参见 ch2_10.html)，页面效果如图 2.14～图 2.16 所示。

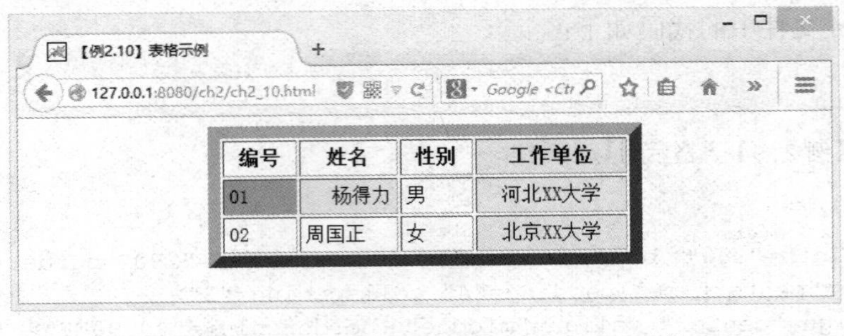

图 2.14　ch2_10.html 在火狐浏览器中的页面效果

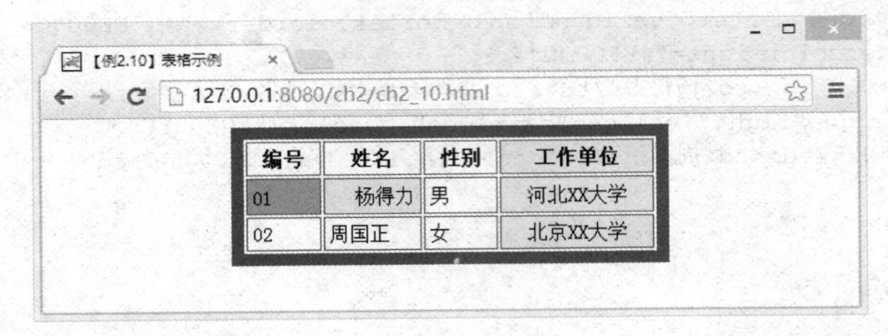

图 2.15　ch2_10.html 在 chrome 浏览器中的页面效果

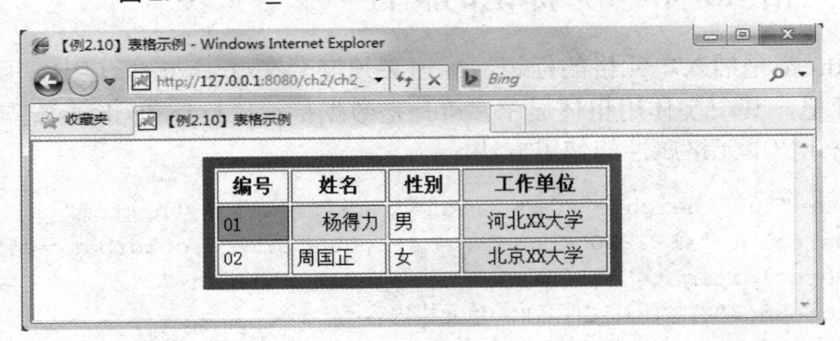

图 2.16　ch2_10.html 在 IE8 浏览器中的页面效果

ch2_10.html 文件中的代码如下：

```
<html>
<head>
<title>【例 2.10】表格示例</title>
</head>
<body>
<table width="350"  bordercolor="#FF0000" heigth="290" border="10"
    align="center" cellpadding="4" cellspacing="2">
      <tr >
         <th>编号</th><th>姓名</th><th>性别</th>
         <th align="center" valign="middle" bgcolor="#FFFF00">工作单位</th>
   </tr>
      <tr >
         <td width="50" height="30" align="left" valign="bottom"
         bordercolor="#0000FF" bgcolor="#00FF00">01</td>
         <td align="right"  bordercolorlight="#00FF00"
```

```
        bordercolordark="#333333" bgcolor="#FFFF00">杨得力</td>
        <td>男</td>
        <td align="center" valign="middle" bgcolor="#FFFF00">河北XX大学</td>
      </tr>
  <tr >
    <td>02</td>
    <td>周国正</td>
    <td>女</td>
    <td align="center" valign="middle" bgcolor="#FFFF00">北京XX大学</td>
  </tr>
</table>
</body>
</html>
```

【例 2.11】设定跨多行多列的单元格表格示例(其代码参见 ch2_11.html)，页面效果如图 2.17 所示。

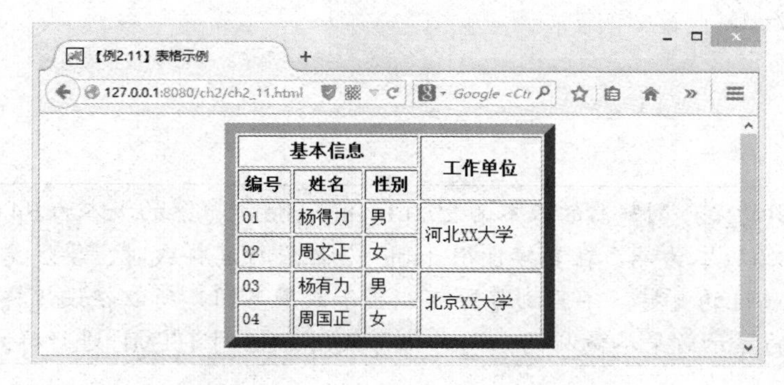

图 2.17　ch2_11.html 的页面效果

ch2_11.html 文件中的代码如下：

```
<html>
<head>
<title>【例 2.11】表格示例</title>
</head>
<body>
<table width="302"  bordercolor="#FF0000" heigth="290" border="10"
 align="center" cellpadding="4" cellspacing="2">
  <tr >
    <th colspan="3">基本信息</th>
    <th width="101" rowspan="2" >工作单位</th>
  </tr>
  <tr >
    <th width="39">编号</th>
    <th width="53">姓名</th>
    <th width="39">性别</th>
  </tr>
  <tr >
    <td>01</td>
    <td>杨得力</td>
    <td>男</td>
```

```
      <td rowspan="2">河北 XX 大学</td>
    </tr>
    <tr >
      <td>02</td>
      <td>周文正</td>
      <td>女</td>
    </tr>
    <tr >
      <td>03</td>
      <td>杨有力</td>
      <td>男</td>
      <td rowspan="2">北京 XX 大学</td>
    </tr>
    <tr >
      <td>04</td>
      <td>周国正</td>
      <td>女</td>
    </tr>
  </table>
  </body>
  </html>
```

注意： 本教材使用的火狐浏览器的版本为 29.0.1，chrome 浏览器的版本为 34.0.1847.131m，IE 浏览器的版本为 8。在直接使用 table 属性设计其样式时，要注意各个浏览器对 html 标签属性的支持，不同浏览器可能显示效果不同，所以在设计样式时要做到各大浏览器的显示效果尽量相同，同时推荐使用 CSS 对 HTML 进行样式设计。

2.4 页面布局

页面布局技术是 Web 应用程序开发的关键技术之一，DIV+CSS 页面布局模式是 W3C 标准的一个典型应用，有许多功能上的优点，下面介绍 DIV+CSS 模式的页面布局及实例。

2.4.1 CSS 简介

CSS 是 Cascading Style Sheet 的缩写，对应的中文为级联样式表。简单讲，CSS 就是用来控制一个文档中的某一区域外观的一组格式属性。CSS 有上百个控制属性，如 background-color、font 等，通过为网页元素的 CSS 样式属性赋予不同的值，来控制网页的外观。CSS 具有如下功能：如字体、颜色、布局、间距、边框等。

- CSS 可以使样式代码独立于 HTML 页面，有利于对站点的统一控制，样式表修改后，引入样式表的网页自动更新，使网页修改、维护更容易。
- CSS 可以提高开发效率，有利于分工合作，使数据设计人员只需要用基本的标记呈现数据即可，而页面的外观显示则由样式设计人员设计。
- CSS 样式可以使网页的打开速度加快，因为相同的样式表不用浏览器重复打开。
- CSS 与层结合，可以精确地实现页面元素的布局。

CSS 样式通过特定的机制与特定的文档相联系，并对此文档进行表现控制。那么 CSS 样式和网页元素是如何具体结合，来实现对网页元素外观控制的？

1. 创建 CSS 样式

当读到一个样式表时，浏览器会根据它来格式化 HTML 文档。根据插入样式表的不同方法可将样式表分为三种：内联样式表、内部样式表和外部样式表。

1) 内联样式表

要使用内联样式表，需要在相关的标签内使用样式(style)属性。style 属性可以包含任何 CSS 属性。例如：

```
<table style="border-collapse:collapse">
```

代码中 style="border-collapse:collapse"控制表格的边框显示为不折叠。这样可以实现单像素边框表格，默认的表格边框为双线的立体效果。

这种 CSS 样式与 HTML 标记书写在一起，简单直观并且能够单独控制个别元素的外观，和传统的外观控制方式没有本质区别，样式代码分布在整个文档中，使样式的修改很困难，样式需要重复加载，运行效率较低。由于要将表现和内容混杂在一起，内联样式会损失掉样式表的许多优势。请慎用这种方法。

【例 2.12】内联样式示例(其代码参见 ch2_12.html)，页面显示效果如图 2.18 所示。

图 2.18　ch2_12.html 的页面效果

ch2_12.html 文件中的代码如下：

```
<html>
<head>
   <title>内联样式示例</title>
</head>
<body>
 <table style= "border-collapse:collapse;
  background-color:#66CCFF;
  border-color:#FF0066;"
  border="3" align="center">
  <tr><th>编号</th><th>姓名</th><th>性别</th> <th>工作单位</th></tr>
  <tr><td>01</td><td>杨得力</td><td>男</td><td>河北 XX 大学</td></tr>
  <tr><td>02</td><td>周国正</td><td>女</td><td>北京 XX 大学</td></tr>
</table>
</body>
</html>
```

2) 内部样式表

内部样式表是使用<style>标记将一段 CSS 代码嵌入 HTML 文档中。一般是使用<style>标记将一段 CSS 代码插入 HTML 文档头部，也就是<head></head>标记之间。例如：

```
<head>
   <style type="text/css">
   <!--
      table {   border-collapse:collapse;
                background-color:#00FF00;
                border-color:#FF0000;
            }
     -->
   </style>
</head>
```

上述代码中：

```
<style type="text/css">
<!-- css 语句-->
</style>
```

是 CSS 语句的容器，<!-- -->是为了使不兼容 CSS 的浏览器忽略这段内容，避免将其显示在页面上。

【例 2.13】内部样式示例(其代码参见 ch2_13.html)，页面显示效果如图 2.19 所示。

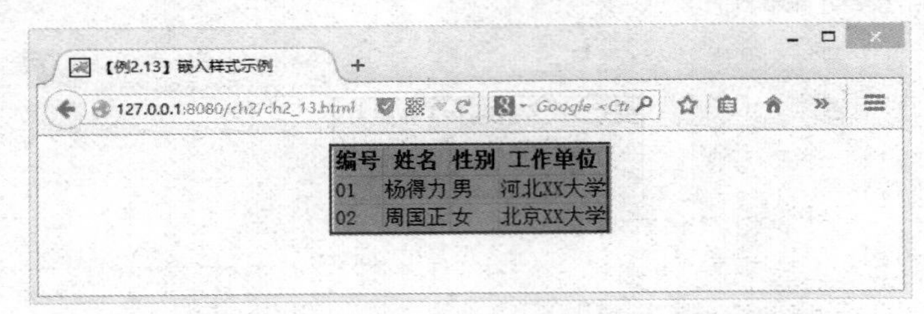

图 2.19　ch2_13.html 的页面效果

ch2_13.html 文件中的代码如下：

```
<html>
<head>
   <title>【例 2.13】内部式示例</title>
   <style type="text/css">
   <!--
      table {   border-collapse:collapse;
                background-color:#00FF00;
                border-color:#FF0000;
            }
     -->
   </style>
</head>
<body>
 <table border="3" align="center">
  <tr><th>编号</th><th>姓名</th><th>性别</th> <th>工作单位</th></tr>
```

```
  <tr><td>01</td><td>杨得力</td><td>男</td><td>河北XX大学</td></tr>
  <tr><td>02</td><td>周国正</td><td>女</td><td>北京XX大学</td></tr>
 </table>
</body>
</html>
```

从上例可以看出，CSS集中写入其语句容器中，使用和维护比内联样式方便。内部样式一般用于在整体文档CSS样式控制的基础上，对文档的局部细节进行修饰的地方。当单个文档需要特殊的样式时，就应该使用内部样式表。

3) 外部样式表

外部样式表(又称外联样式表)是通过外部样式文件对文档进行外观控制，CSS样式语句存储在一个独立的文本文件(样式文件)中，通过在文档中指定样式文件来控制文档外观。当样式需要应用于很多页面时，外部样式表将是理想的选择。在使用外部样式表的情况下，可以通过改变一个文件来改变整个站点的外观。每个页面使用<link>标签链接到样式表，一般使用格式如下：

```
<link  rel="stylesheet" href="css/stylesheet1.css" type="text/css"/>
```

该标记一般都放在文档的<head></head>标记之间。href属性指定了样式文件的路径，rel和type属性表明这是一个样式文件。

【例2.14】外联样式示例(其代码参见ch2_14.css和ch2_14.html)，页面效果如图2.20所示。

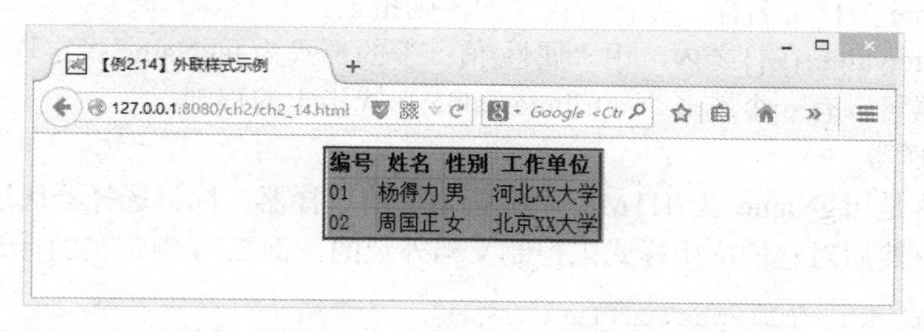

图2.20 ch2_14.html的页面效果

ch2_14.css文件中的代码如下：

```
table { border-collapse:collapse;
        background-color:#00FF00;
        border-color:#FF0000;
      }
```

ch2_14.html文件中的代码如下：

```
<html>
<head>
   <title>外联样式示例</title>
   <link rel="stylesheet" href="ch_14.css" type="text/css"/>
</head>
<body>
 <table border="3" align="center">
  <tr><th>编号</th><th>姓名</th><th>性别</th> <th>工作单位</th></tr>
```

```
  <tr><td>01</td><td>杨得力</td><td>男</td><td>河北 XX 大学</td></tr>
  <tr><td>02</td><td>周国正</td><td>女</td><td>北京 XX 大学</td></tr>
 </table>
</body>
</html>
```

2. CSS 选择器

在上面的 CSS 例子中，反复地使用了下面的代码：

```
table { border-collapse:collapse;
        background-color:#00FF00;
        border-color:#FF0000;
      }
```

这一段完整的代码在 CSS 中称为选择器。选择器是样式表中使用一定规则指定的一个或一组标记，CSS 通过被规则指定的标记，对文档中使用该标记的内容进行统一的外观控制。定义选择器的一般格式如下：

```
tagName  {attribute:value;
          attribute:value;
       ⋮
       }
```

其中的参数说明如下。

- tagName 为标记名称，被{}所包含的规则指定。
- attribute:value;在{}之内，用“属性:值；”的形式为 tagName 指定规则。
- 选择器可以有一个或多个 attribute:value;形式指定的规则。

1) 标记选择器

标记选择器是 tagName 为 HTML 中标记名称的选择器。标记选择器通过选择所有指定标记的节点，然后对它们应用样式来控制文档外观的。例 2.14 中定义的标记选择器：

```
table { border-collapse:collapse;
        background-color:#00FF00;
        border-color:#FF0000;
      }
```

其作用范围为文档中所有使用<table>标记的表格。标记选择器可同时指定多个标记，如：h1，h2，h3{...}。对于层次关系的标记，使用 ul lt{...}这种格式，样式表会按照标记的层次关系控制外观。

2) 类型选择器

标记选择器便于统一控制节点样式，对页面中所有这类标记都有效。标记选择器具有通用性。有时候用户为了缩小标记选择器的作用范围，也使用类型选择器。类型选择器就是定义选择器时，为 tagName 标记名称指定 className 属性。类型选择器一般定义格式：

```
tagName.className  {attribute:value;
                    attribute:value;
                    ⋮
                   }
```

其中 tagName 可以省略，但是“.className”不能省略，这样对于所有指定 class 并且名称与 className 相同的标记起作用。在 W3C 中所有标记的 id 属性不能重名，class 名称可以重名。

3) id 选择器

和类型选择器一样，在定义选择器时，在 tagName 名称后加上 idName，“.”用“#”替代，则称为 id 选择器。其一般定义格式如下：

```
tagName#idName  {attribute:value;
                 attribute:value;
                 ⋮
                }
```

其中 tagName 可以省略，但是“idName”不能省略，这样对于所有指定 id 并且名称与 idName 相同的标记起作用。在 W3C 中所有标记的 id 属性不能重名。

4) 派生选择器

通过依据元素在其位置的上下文关系来定义样式，可以使标记更加简洁。

在 CSS1 中，通过这种方式来应用规则的选择器被称为上下文选择器(contextual selectors)，这是由于它们依赖于上下文关系来应用或者避免某项规则。在 CSS2 中，它们称为派生选择器，但是无论如何称呼它们，它们的作用都是相同的。

派生选择器允许用户根据文档的上下文关系来确定某个标签的样式。通过合理地使用派生选择器，可以使 HTML 代码变得更加整洁。

例如，希望列表中的 strong 元素变为斜体字，而不是通常的粗体字，可以这样定义一个派生选择器：

```
li strong{
    font-style:italic;
    font-weight:normal;
}
```

请注意标记<li>中的标记 <strong> 的上下文关系：

```
<p><strong>我是粗体字，不是斜体字，因为我不在列表当中，所以这个规则对我不起作用
</strong></p>
<ol>
<li><strong>我是斜体字。这是因为 strong 元素位于 li 元素内。</strong></li>
<li>我是正常的字体。</li>
</ol>
```

在上面的例子中，只有 li 元素中的 strong 元素的样式为斜体字，无须为 strong 元素定义特别的 class 或 id，代码更加简洁。

5) 属性选择器

对带有指定属性的 HTML 元素设置样式。可以为拥有指定属性的 HTML 元素设置样式，而不仅限于 class 和 id 属性。

注释：只有在规定了 !DOCTYPE 时，IE7 和 IE8 才支持属性选择器。在 IE6 及更低的版本中，不支持属性选择器。

下面的例子为 title="mytitle" 的所有元素设置样式：

```
[title=mytitle]{
    Border:5px solid blue;
}
```

注意： 有的 CSS 样式可以继承，如字体、文本的定义，另外一些样式是不可继承的，如边框、间距、布局、定位等。当样式定义重复出现的时候，最后定义的样式起作用，也就是 CSS 的就近原则。

3. CSS 的盒子模型

盒子模型是 CSS 的重要概念，它是所有布局控制的基础。将所有的 HTML 元素都放置到一个盒子中，通过控制盒子的外观来实现整个页面外观的控制，这个模型就是 CSS 的盒子模型。

在 CSS 中一个盒子模型包括 4 个区，分别是内容(Content Container)、内边距(padding)、边框(border)、外边距(margin)、内容高度(height)、内容宽度(width)等。W3C 的盒子模型如图 2.21 所示。

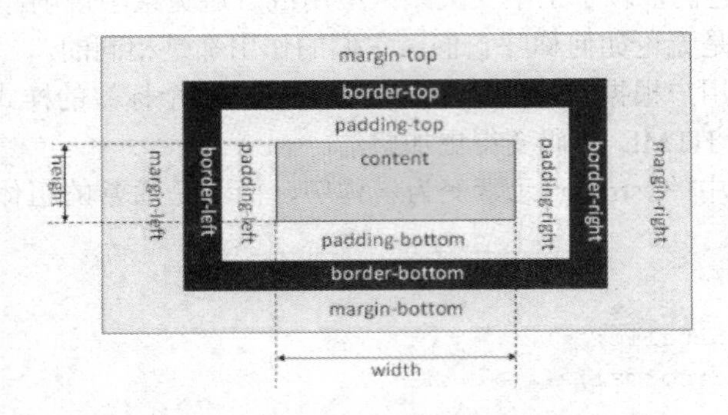

图 2.21　W3C 的盒子模型

在指定一个元素大小时，就是根据盒子模型的参数指定的，例如分层标记 div 的 id 选择器定义格式如下：

```
div#header{
            width:700px;
            height:100px;
            border-width:2px;
            padding:10px;
            margin:10px;
            background-color: #ccc;
            }
```

其中的标记及属性说明如下。

- div 为标记名称。
- header 为标记的 id 号，标记名与 id 号之间用#分隔。
- width 属性指的是内容的宽度，border-width 指的是边框线的宽度，其他参照模型，取值为像素数。

- background-color 取颜色值，十六进制码的第 1、2 位和第 3、4 位以及第 5、6 位分别相同时，可以省写为 3 位，“#”不能省略。
- margin 属性可以简写成"margin:10px;"，表示四个方向值相同；当值不一样时，可以四个方向分别定义，如“margin-top:1px;margin-right:2px;margin-bottom:3px; margin-left:4px;”。也可以简写成：“margin:1px 2px 3px 4px;”，方向顺序为“上、右、下、左”。也可以单独定义某个方向，而其他方向是默认值，例如：“border-left:1px solid #333;”
- padding 属性与 margin 属性取值规则相同。
- 盒子的宽度为 width+2*padding+2*margin+2*border-width=744，这是根据 CSS 的定义计算得到的。

注意：IE 和 W3C 分别有一套盒子模型，火狐浏览器中采用 W3C 标准模型，而 IE 中则采用 Microsoft 自己的标准。IE 的盒子模型如图 2.22 所示。

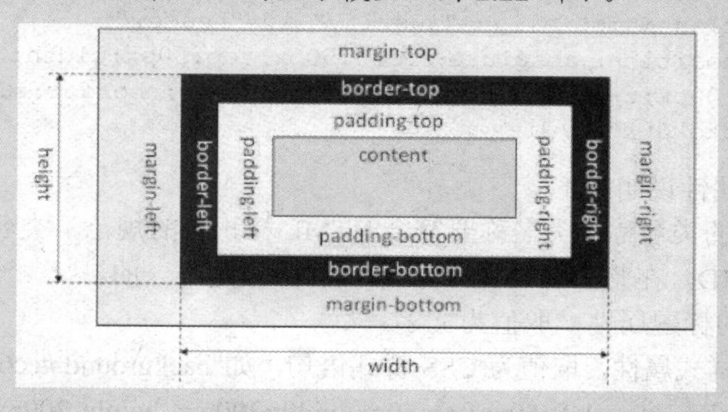

图 2.22　IE 的盒子模型

很明显，W3C 认为，content 的宽度即 CSS 中设置的 width 值的大小，而 IE 标准认为 content 的宽度为“width+padding+border”，高度同理。

例如当使用 DIV 的 CSS 样式如下：

```
div{
  border:1px solid black;
  padding:1px;
  margin:1px;
  width:100px;
  height:50px;
}
```

则 IE 认为 content 的宽度实际为：1+1+100+1+1;

W3C 标准认为 content 的宽度实际为：100。

2.4.2　DIV 层

1. DIV 标记

层又称为 Block-Level 区块或 Div Element 或 CSS-Layer，中文一般都称之为层。DIV

标记是为 HTML 文档定义层的标记。<div></div>标记定义的层可以方便地放在页面的任何一个位置，除了平面上的并行定位，层还增加了三维空间的定位 z-index，因为 z-index 定义了堆叠的顺序，类似于图形设计中使用的图层，所以拥有了 z-index 属性的元素被形象地称为层。层的主要功能如下：

- 可以利用层准确定位网页元素。在层中可以存放 HTML 文档所包含的元素，例如文本、图像、其他层等，并精确定位到页面，这极大地方便了页面的布局。
- 可以产生重叠效果。因为层可以重叠，将网页元素放入层中，就可以产生许多重叠效果。
- 实现网页上的下拉菜单。层可以隐藏和显示，因此用此特性可以实现级联下拉菜单。
- 将层和时间轴配合，可以实现网页动画。

在 HTML 中使用<div></div>定义层，<div>标记使用内联样式定义层的格式如下：

```
<div class="className" align="justify" id="header"
     style="position: absolute; left:50px;top:50px;width:200px;
     height:200px;z-index:1;background:#CC6666; border:#CC99CC"
     title="dkdkdkk"></div>
```

其中的参数及属性说明如下。

- class 为层的类名称，类名称要符合 HTML 标识符的规范，类名称可以重名。
- id 为层的 ID，在整个 HTML 文档中，id 必须是唯一的。
- title 为层的标题属性，取值为文字。
- style 为层样式属性，取值为 CSS 样式语句，如"background:#cc6666;border:#cc99cc; position：absolute；left:50px;top:50px;width:200px; height:200px;z-index:1;"，样式语句采用“属性:属性值;”形式。

下面通过一个层定义的例子，演示<div>标记的使用方法，体会层定义中各参数的含义。读者在学习此例时，可以通过调整参数值并观察实际显示效果，注意在 IE 和 Firefox 浏览器中显示的区别。

【例 2.15】<div>标记使用示例(其代码参见 ch2_15.html)，页面效果如图 2.23 所示。整个页面由四个层组成，容器层 id 为 main，在 main 层中有 left、center、right 层，分别给不同的层设置了不同的背景色，便于读者观察、理解。

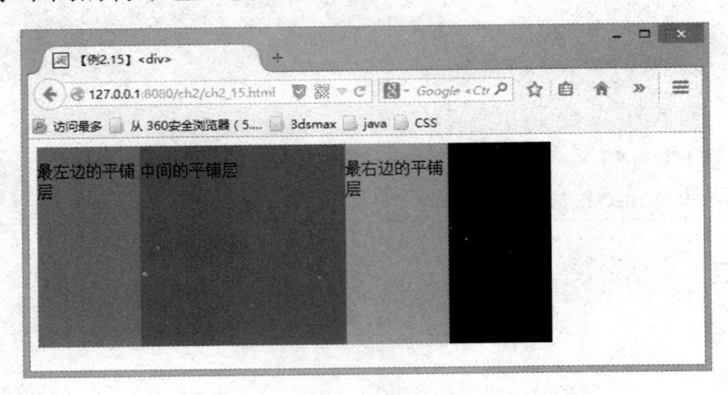

图 2.23　ch2_15.html 的页面效果

ch2_15.html 文件中的代码如下：

```
<html>
<head>
  <title>&lt;div&gt;</title>
</head>
<body>
<div id="main"  style="background-color:#000000; width:500px;
    height:200px;">
    <div id="left" style="width:100px; height:200px; float:left;
          background-color:#ff00ff;">
          <p>最左边的平铺层</p></div>
    <div id="center" style="width:200px; height:200px; float:left;
          background-color:#ff0000;">
          <p>中间的平铺层</p></div>
    <div id="right" style="width:100px; height:200px; float:left;
           background-color:#00ff00;">
     <p>最右边的平铺层</p></div>
 </div>
 </body>
</html>
```

2. CSS 中的文档流模型

所有的块元素在 HTML 文档中是按照它们出现在文档中的先后顺序排列的(当然，嵌套不在此列)，每一个块都会另起一行。

【例 2.16】 CSS 中的文档流模型示例(其代码参见 ch2_16.html 和 ch2_16.css)，页面效果如图 2.24 所示。

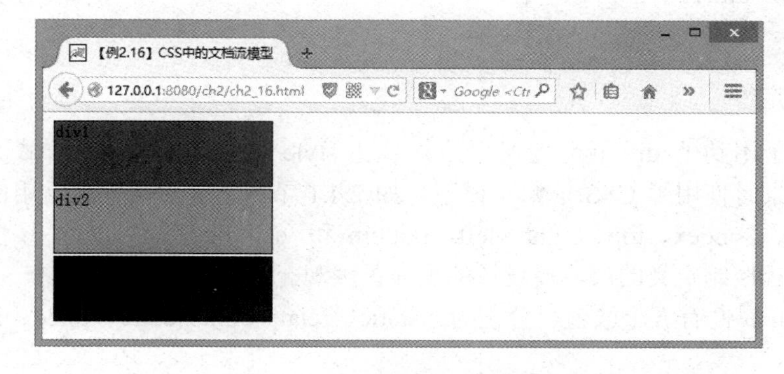

图 2.24　ch2_16.html 页面效果

ch2_16.html 文件中的代码如下：

```
<html>
<head>
    <title>【例 2.16】CSS 中的文档流模型</title>
     <link href="ch2_16.css" rel="stylesheet" type="text/css" />
  </head>
<body>
     <div id="div1">div1</div>
```

```
    <div id="div2">div2</div>
    <div id="div3">div3</div>
</body>
</html>
```

ch2_16.css 样式表文件中的代码如下：

```
#div1 {
    border: 1px solid #000099;
    height: 60px;
    width: 200px;
    margin:2px;
    background: #FF0000;
}
#div2 {
    border: 1px solid #000099;
    height: 60px;
    width: 200px;
    margin:2px;
    background: #00FF00;
}
#div3 {
    border: 1px solid #000099;
    height: 60px;
    width: 200px;
    margin:2px;
    background: #0000FF;
}
```

3. DIV 层定位

前面的例子用到了 div 标记定义层，并通过 style 属性为层使用了样式。样式在控制层的大小、背景时使用了 CSS 的盒子模型，如 2.4.1 节所述。盒子模型常用的定位属性有 position、float、z-index、top、right、left、bottom 和 visibility 等，对这些属性的理解，会极大地帮助读者控制定义的层，使设计的页面更美观。

(1) position 属性有五个取值，分别为：static、relative、absolute、fixed、inherit。

说明如下。

- 取值为 static 或者没有设置 position 属性的时候，元素的位置按正常方式定位，也即按文本流的正常顺序，从上至下或者从左到右顺序定位。static 为默认值，也可以理解为“无定位”。
- 取值为 relative 时，层的显示方式按相对于元素“原来的位置”来定位，此时需要指定 left、top 等属性，指定偏移量。
- 取值为 absolute 时，使用 top、left、bottom 和 right 属性，可以将元素按照父容器指定的坐标定位在页面的任意位置。使用绝对定位的层前面的或者后面的“层”会认为这个“层”并不存在，也就是在 z 方向上，它是相对独立出来的，丝毫不影响到其他 z 方向的“层”。

- 取值为 fixed 值时，使用 top、left、bottom 和 right 属性，相对于浏览器窗口定位层的位置，层固定漂浮在浏览器的某个位置，不会随页面内容的移动而移动。注意，在 IE 浏览器中会出现异常，在 Firefox 浏览器中显示正常。
- 取值为 inherit 值时，层继承父容器的 position 属性值。

(2) top、right、bottom、left 属性的取值既可以是像素数，也可以是百分比，并且只有父容器的 position 取值为非 static 值时才有效。

(3) float 属性用于控制文本流的显示方向，可以取 left、right、none 和 inherit。float 属性取值为 left 或 right 后，会影响到后面其他的层，要清除这些影响，可以设置 clear:both; 属性。需要注意，当 div 设置 position 取值为 absolute 值后，float 属性无效。

(4) z-index 属性用于控制层在 z 轴上的排列顺序，值为整数，值越大层越靠上面。理解 z-index 属性，可以将盒子模型想象成三维结构。该属性在设置了 position 并取值为 absolute 或 relative 时有效。

(5) visibility 属性用于控制层的显示与隐藏。可以取 visible、hidden、collapse 和 inherit 等值。取 visible 值，元素可见，取 hidden 时，元素隐藏。关于更多属性，请参考 CSS 手册。

【例 2.17】相对定位。

在文档流中，每个块元素都会被安排到流中的一个位置，可以通过 CSS 中的定位属性来重新安排它的位置。定位分为相对定位和绝对定位，相对定位是相对于该块元素在文档流中的位置的，比如，可以使用相对定位把 div2 放到 div1 的右侧，效果如图 2.25 所示。其代码参见 ch2_17.html 和 ch2_17.css。

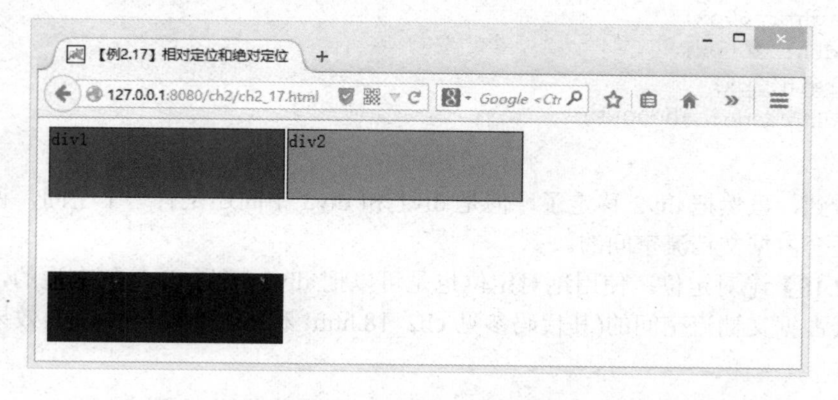

图 2.25　ch2_17.html 的页面效果

ch2_17.html 文件中的代码如下：

```
<html>
<head>
   <title>【例 2.17】相对定位</title>
   <link href="ch2_17.css" rel="stylesheet" type="text/css" />
 </head>
 <body>
   <div id="div1">div1</div>
```

```
    <div id="div2">div2</div>
    <div id="div3">div3</div>
</body>
</html>
```

ch2_17.css 样式表文件中的代码如下：

```
#div1 {
    border: 1px solid #000099;
    height: 60px;
    width: 200px;
    margin:2px;
    background: #FF0000;
}
#div2 {
    border: 1px solid #000099;
    height: 60px;
    width: 200px;
    margin:2px;
    position: relative;
    top: -64px;
    left: 204px;
    background: #00FF00;
}
#div3 {
    border: 1px solid #000099;
    height: 60px;
    width: 200px;
    margin:2px;
    background: #0000FF;
}
```

可以看到，虽然把 div2 移走了，但是 div1 和 div3 中间还是有一个空间，说明相对定位的元素是会占据文档流空间的。

【例 2.18】绝对定位。使用绝对定位也是可以把 div2 摆到 div1 的右边的，而且绝对定位是不会占据文档流空间的(其代码参见 ch2_18.html 和 ch2_18.css)，页面效果如图 2.26 所示。

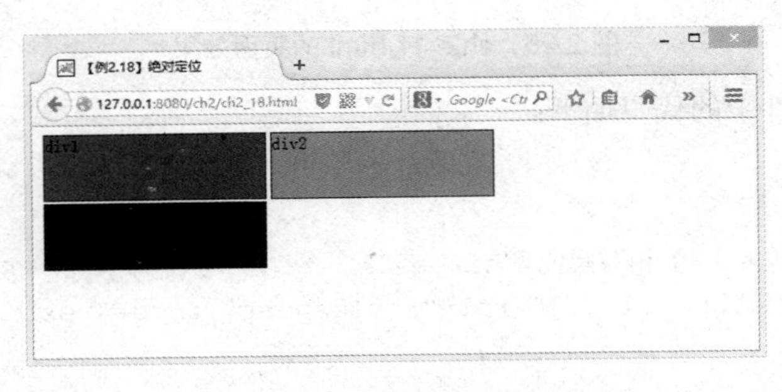

图 2.26　ch2_18.html 的页面效果

ch2_18.html 文件中的代码如下：

```
<html>
<head>
    <title>【例 2.18】绝对定位</title>
    <link href="ch2_18.css" rel="stylesheet" type="text/css" />
  </head>
  <body>
    <div id="div1">div1</div>
    <div id="div2">div2</div>
    <div id="div3">div3</div>
</body>
</html>
```

ch2_18.css 样式表文件中的代码如下：

```
#div1 {
   border: 1px solid #000099;
   height: 60px;
   width: 200px;
   margin:2px;
   background: #FF0000;
}
#div2 {
   border: 1px solid #000099;
   height: 60px;
   width: 200px;
   margin:2px;
   position: absolute;
   top: 6px;
   left: 214px;
   background: #00FF00;
}
#div3 {
   border: 1px solid #000099;
   height: 60px;
   width: 200px;
   margin:2px;
   background: #0000FF;
}
```

【例 2.19】综合实例，在盒子模型指导下，采用外部样式定义层并设置层的 position 等属性的示例(其代码参见 ch2_19.html 和 ch2_19.css)，页面效果如图 2.27 所示。

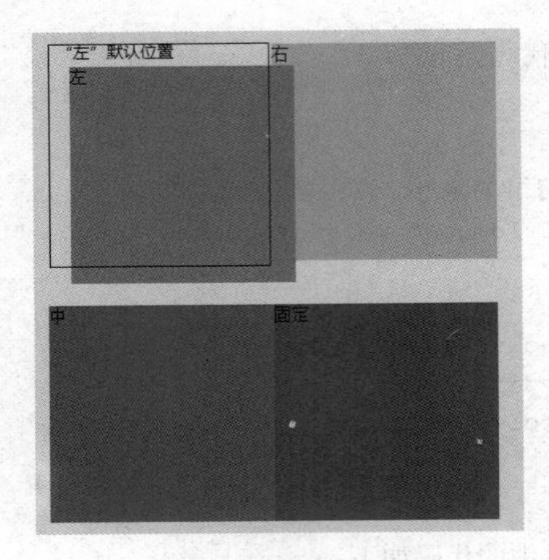

图 2.27　ch2_19.html 的页面效果

ch2_19.html 文件中的代码如下：

```
<html>
<head>
  <link rel="stylesheet" href=" ch2_19.html" type="text/css" />
</head>
<body>
 <div id="main">
    <div id="left">左</div>
    <div id="center">中</div>
    <div id="right">右</div>
    <div id="fixed">固定</div>
 </div>
</body>
</html>
```

ch2_19.css 样式表文件中的代码如下：

```
#main {
    width:600px;
     height:500px;
     margin:30px;
     padding:30px;
     z-index:1;
     background-color:#CCCCCC;
}
#left {
    position:relative;
     float:left;
     left:20px;
     top:20px;
     width:200px;
     height:200px;
```

```
    background-color:#FF00FF;
    z-index:2;
}
#center {
    position:absolute;
    left:68px;
    top:300px;
    width:200px;
    height:200px;
    background-color:#FF0000;
    z-index:4;
}
#right {
    float:left;
    width:200px;
    height:200px;
    background-color:#00FF00;
    z-index:3;
}
#fixed {position:fixed;
        left:268px;
         top:300px;
         width:200px;
         height:200px;
         background-color:#0033FF;
         z-index:6;
         }
```

注意：在此例中，分别写了 DIV 部分(结构)以及 CSS 部分(表现)，然后通过 DIV 的 ID 属性，使用到了相对应的 CSS 样式：<div id="main"></div>。请读者仔细体会各层的位置是如何定位的，可以通过修改属性参数，比较显示效果来加深理解。

2.4.3　DIV+CSS 页面布局

页面布局一直是 Web 应用程序界面设计的一个重要内容，使用框架技术和表格技术实现页面布局，是 DIV+CSS 模式出现之前的主流方式。在 W3C 标准中，DIV+CSS 实现页面布局是一个典型的应用。

1. DIV+CSS 页面布局的优点

在 W3C 标准中，网页主要由三部分组成：结构(Structure)、表现(Presentation)和行为(Behavior)等。结构主要包括 DIV 在内的一系列的 XHTML 标记，表现主要包括 CSS 层叠样式表，行为主要包括对象模型(如 W3C DOM)、ECMAScript 等。利用这种模式开发的网页是符合 W3C 设计标准的，有下列优点。

- 网页开发与维护变得更简单、容易。因为使用了更具有语义和结构化的 XHTML，让程序员更加容易、快速地理解网页代码。

- 网页下载、读取速度变得更快。使用 DIV+CSS 模式开发网页，网页 HTML 代码减少，下载速度更快；因为网页使用相同的 CSS 样式表，不用重复下载和加载，使显示速度加快。
- 网页可访问性和适应性提高。语义化的 HTML 使结构和表现相分离，通过使用不同的 CSS 样式表，可以很方便地让读屏器、掌上电脑、智能电话等访问。

2. DIV+CSS 页面布局工作流程

在学习了 DIV 和 CSS 相关知识后，读者要问，如何使用 DIV+CSS 实现网页的布局？下面就用本教材第 11 章的网上报名系统页面布局为例，介绍页面布局的工作流程。

页面布局的工作流程，没有统一的规定，是人们在网站开发中不断总结出来的，不同的开发团队使用的流程也有区别，这里给读者一个大体的建议。

【例 2.20】网站页面布局示例，ch2_20.html 页面效果如图 2.28 所示。

(1) 使用图形图像制作工具(如 Photoshop)或手工绘制出网站页面的效果图，以像素为单位给出页面布局中各板块元素的大小及颜色的 RGB 值，如图 2.29 所示。

(2) 对照效果图，在页面制作工具(如 Dreamweaver、EditPlus 等)用 DIV+CSS 代码绘制页面框架。在容器层 div#container 中，定义了四个子层：div#header、div#sidebar1、div#mainContent、div#footer，它们在父容器中“按文本流”方式排列，即默认从上到下排列。可用通过层的 float:left；属性值来让其从左到右排列。

图 2.28　ch2_20.html 的页面效果

ch2_20.html 文件中的 div 代码如下：

```
<!DOCTYPE HTML PUBLIC "-//W3C//DTD HTML 4.01 Transitional//EN">
<html>
    <head>
        <title>【例 2.20】职称计算机考试报名</title>
    <link href="mystylesheet.css" rel="stylesheet" type="text/css" />
  </head>
  <body class="twoColHybLtHdr">
    <div id="container">
      <div id="header">
        <h1>XX 省职称计算机考试报名系统</h1>
```

```
      </div>
      <div id="sidebar1">
        <h3>左侧菜单</h3>
        <p>考生报名</p>
        <p>考生登录</p>
        <p>浏览信息</p>
        <p>修改信息</p>
        <p>注销考试</p>
        <p>修改密码</p>
        <p>上传照片</p>
        <p>退出登录</p>
        <p>返回主页</p>
        <h3>左侧栏高</h3>
        <p>栏高自动调整，指定 height 属性，可固定栏高。</p>
      </div>
      <div id="mainContent">
          <center>
              <img src=" ch2_20_welcome.jpg" width="100%" ></img>
          </center>
      </div>
      <br class="clearfloat" />
      <div id="footer">
        <p>清华大学出版社《JSP 编程技术教程》教材编写组 2014 年 5 月 11 日</p>
      </div>
    </div>
    </body>
</html>
```

图 2.29　页面布局效果图

与 ch2_20.html 外联的 CSS 样式表文件 mystylesheet.css 中的代码如下：

```
@charset "gb2312";
body  {
    font: 100% "宋体","方正姚体","华文隶书", "隶书",  "新宋体", "幼圆";
    background: #666666;
```

```
    margin: 0;
    padding: 0;
    text-align: center;
    color: #000000;
}

.twoColHybLtHdr #container {
    width: 80%;
    background: #FFFFFF;
    margin: 0 auto; /*页边空白*/
    border: 1px solid #000000;
    text-align: left;
}
.twoColHybLtHdr #header {
    background: #DDDDDD;
    padding: 0 10px;
    height:70px;
}
.twoColHybLtHdr #header h1 {
    margin: 0;
    text-align:center;
    padding: 10px 0;
}
.twoColHybLtHdr #sidebar1 {
    float: left;
    width: 8em;
    background: #EBEBEB;
    padding: 15px 0;
}
.twoColHybLtHdr #sidebar1 h3, .twoColHybLtHdr #sidebar1 p {
    margin-left: 20px;
    margin-right: 20px;
}
.twoColHybLtHdr #mainContent {
    margin: 0 20px 0 9em;
}
.twoColHybLtHdr img {
    height:545px;
}
.twoColHybLtHdr #footer {
    padding: 0 10px;
    background:#DDDDDD;
    text-align:center;
}
.twoColHybLtHdr #footer p {
    margin: 0;
    padding: 10px 0;
```

```
    }
    .fltrt {
        float: right;
        margin-left: 8px;
    }
    .fltlft {
        float: left;
        margin-right: 8px;
    }
    .clearfloat {
        clear:both;
       height:0;
       font-size: 1px;
       line-height: 0px;
    }
```

(3) 细化每个板块，填充内容信息。将每个板块的文本、图形、链接等添加到板块中，细化网站的页面。

> 注意：读者在实现此例时，要使用网页制作工具的代码视图，这样有利于对标记、CSS样式属性的记忆和理解，对今后的实际开发大有益处。

2.5 上机实训

实训目的

- 理解HTML文件组成、URL、标记与属性、CSS样式、CSS选择器、层、页面布局等概念。
- 掌握HTML常用标记的使用。
- 掌握网页文字与段落、图形图像、超级链接、表格等内容的创建与表现方法。
- 掌握DIV+CSS页面布局方法。

实训内容

完成“XX省职称计算机考试报名系统——查看报考信息”页面，页面效果如图2.30所示。

要求：

(1) 参照例2.29完成页面布局。

(2) 参照图2.30完成页面内容区“职称计算机考试报名信息”表格的设计制作。

(3) 完成表格信息填写，完成照片显示、邮件链接、毕业学校网站链接。

(4) 样式表文件、图形文件和页面文件都存储在ch2虚拟目录中，照片名称为photo1.jpg，页面存成HTML文件类型，表格中的数据均采用静态数据直接输入。

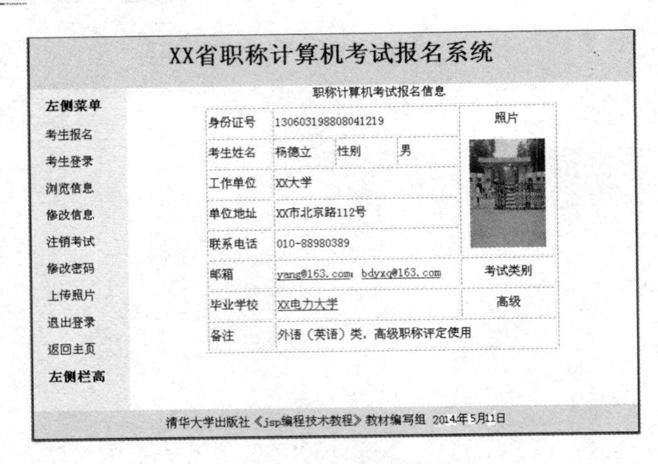

图 2.30　实训页面效果

实训总结

通过本章的上机实训，学员应该能够理解 DIV+CSS 页面布局的基本概念和原理；掌握 HTML 常用标记的使用；掌握网页文字、表格、图形、链接的制作；掌握 DIV+CSS 模式布局的步骤和方法。

2.6　本 章 习 题

思考题

(1) 什么是 HTML/XHTML?

(2) 什么是 CSS? 与 HTML/XHTML 是什么关系。

(3) 什么是 CSS 的选择器、盒子模型?

(4) DIV 层如何定位?

(5) DIV+CSS 模式的页面布局的工作流程?

(6) 异形表格如何实现?

拓展实践题

(1) 上网查看 3 个以上著名网站，分析网站页面布局采用的技术。

(2) 使用网页制作工具，代码模式下设计“网上职称计算机报名系统”的静态页面。

第 3 章　JSP 语法基础

学习目的与要求：

JSP 语法是 Web 应用编程的基础。本章主要学习 JSP 页面的构成、表达式和变量的声明、Java 程序片、表达式、指令标记和动作标记等内容。通过本章的学习，读者要掌握页面成员变量和方法的声明格式、程序片的编写、表达式的使用技术，重点掌握 page 和 include 指令标记的使用方法和技巧，掌握 jsp:include、jsp:forward 和 jsp:param 动作标记的使用方法。JSP 的语法规则是学习后续章节的基础，读者要仔细研究打好基础。

3.1　JSP 页面的基本结构

前面已经介绍 HTML 页面是由 HTML 标记和文档数据构成。在 HTML 页面文件中加入 JSP 脚本元素、JSP 标记等就构成了一个 JSP 页面。为了简单清楚，读者可以认为：一个完整的 JSP 页面由 7 种元素组成。

(1) 普通的 HTML 标记。

(2) JSP 指令标记。

(3) JSP 动作标记，以“<jsp:”开头，以“/>或</jsp:动作>”结束的标记。

(4) 变量声明与方法声明(Declaration)。

(5) 程序片(Scriptlet)。

(6) 表达式(Exception)。

(7) 注释(Comment)。

下面的例子中就包含了其中的 6 种元素。为了调试运行 JSP 页面，读者要为本章的例题创建一个名称为 ch3 的 Web 虚拟目录，本教材中 ch3 对应物理路径为 E:\programJsp\ch3，除非特别说明本章的例子都保存在该目录中。

【例 3.1】演示 JSP 页面基本构成，程序功能：获得并显示系统时间；获得并显示客户端 IP 地址；显示客户访问页面的次数。页面效果如图 3.1 所示。

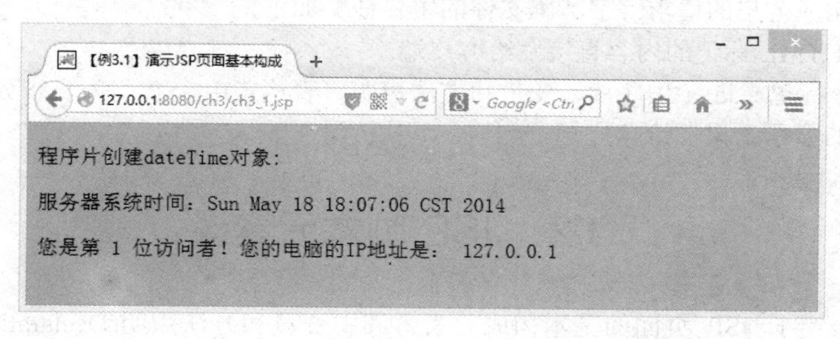

图 3.1　ch3_1.jsp 页面效果

ch3_1.jsp 文件内容：

```
1  <%@ page contentType="text/html;charset=GB2312" %> <!--jsp 指令标记-->
2  <%@ page import="java.util.Date" %>           <!--jsp 指令标记-->
3  <%! Date dateTime;                            //变量声明
4     int countNum;
5     public void setCount(){                    //方法声明
6       countNum++;
7    }
8  %>
9  <html>                                         <!--HTML 标记-->
10  <body  bgcolor="cyan">
11    <font size=4><p>程序片创建 dateTime 对象:
12      <%  dateTime=new Date();                     //java 程序片
13          out.println("<p>服务器系统时间: "+dateTime);
14           setCount();
15           String str=request.getRemoteAddr();
16        %></p>
17   <p>您是第
18        <%= countNum %>                          <!--表达式-->
19        位访问者! 您的电脑的 IP 地址是:
20        <%= str%></p>                              <!--表达式-->
21    </font>
22  </body>                                         <!--HTML 标记-->
23  </html>
```

上述代码中第 1、2 行为 JSP 指令标记，以“<%@”开头，以“%>”结束；第 3～7 行为变量和方法的声明，以“<%!”开头，以“%>”结束，第 3～4 行声明了两个变量；第 5～7 行声明 setCount 方法；第 12～16 行为 Java 程序片，以“<%”开头，以“%>”结束；第 18、20 行为 Java 表达式，以“<%=”开头，以“%>”结束；第 1、2 等行的行末，以“<!--”开始，“以-->”结束的内容为注释；第 3、5 等行的行末以“//”开头的内容为 Java 注释。本例中 JSP 动作标记没有演示，后面将详细介绍。

当服务器上的 JSP 页面被首次请求时，JSP 页面被转译称 Java 文件，并由 Java 编译器编译成字节码文件，然后由 JSP 引擎执行该字节码文件响应客户请求。当多个客户同时请求同一页面时，JSP 引擎为每一个客户建立一个线程，该线程负责将常住内存的字节码文件执行来响应客户的请求。字节码文件的主要任务如下。

- 把 HTML 标记的内容直接发给客户端。
- JSP 标记、Java 程序片、变量和方法声明、表达式由 JSP 引擎负责处理，并将需要的显示结果发给客户端。

3.2 JSP 脚本元素

3.1 节介绍了 JSP 页面的基本构成元素，其中变量和方法声明(Declaration)、表达式(Expression)和 Java 程序片(Scriptlet)统称为 JSP 脚本元素；指令标记、JSP 动作标记统称为 JSP 标记。本节介绍脚本元素的基本知识。

3.2.1　变量与方法的声明

变量和方法在“<%!”和“%>”标记之间声明，声明变量和方法的语法和格式同 Java 语言。声明的语法格式如下：

```
<%! declarations %>
```

1. 变量声明

声明变量就是在“<%!”和“%>”标记之间放置 Java 的变量声明语句。变量的数据类型可以是 Java 的任何数据类型。例如：

```
<%! Date dateTime;
    int countNum;
%>
```

dateTime 和 countNum 就是声明的变量。在“<%!”和“%>”之间声明的变量又称为页面成员变量，其作用范围为整个 JSP 页面，与书写位置无关，但一般放在页面的前面；JSP 引擎转译页面时，将“<%!”和“%>”标记之间声明的变量作为类成员变量来处理，变量占用的内存空间直到 JSP 引擎关闭时才释放；当多个客户访问同一个页面时，JSP 引擎为客户创建的线程之间共享页面成员变量，每个客户线程对页面成员变量的操作都会影响它的值。

例 3.1 就利用了页面成员变量被多客户线程共享的特性，实现了一个简单的计数器，页面每刷新一次，计数 countNum 变量就会加 1。程序的效果如图 3.1 所示。

2. 方法声明

声明方法就是在“<%!”和“%>”标记之间放置 Java 的方法声明。“<%!”和“%>”之间声明的方法在整个页面内有效，称为页面的成员方法；页面成员方法在 Java 程序片中被调用，在方法内声明的变量称为局部变量，只在方法内有效，方法调用为其分配空间，调用完毕释放变量空间。

【例 3.2】方法声明示例，程序功能：求连续的多个数之和并显示。程序效果如图 3.2 所示。

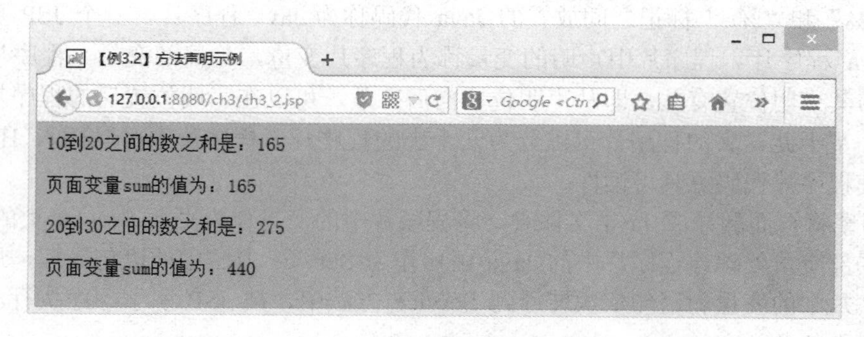

图 3.2　ch3_2.jsp 页面效果

ch3_2.jsp 文件内容：

```
<%@ page contentType="text/html;charset=GB2312" %>
<%@ page import="java.uitl.*" %>
<%! int sum=0;
    public int sumMethod(int begin,int end){
        int subsum=0;
        for(;begin<=end;begin++){
            subsum+=begin;
        }
        sum+=subsum;
        return subsum;
    }
%>
<html>
<head>
<title>【例 3.2】方法声明示例</title>
</head>
<body bgcolor=cyan>
<% //调用 sumMethod 方法
   out.println("<p>10 到 20 之间的数之和是: "+sumMethod(10,20));
   out.println("<p>页面变量 sum 的值为: "+sum);
   //调用 sumMethod 方法
   out.println("<p>20 到 30 之间的数之和是: "+sumMethod(20,30));
   out.println("<p>页面变量 sum 的值为: "+sum);
%>
</body>
</html>
```

注意：变量 sum 是页面成员变量，页面每刷新一次其值会累计增加；subsum 是方法内的局部变量，只在方法内有效，方法每次调用结果相同；页面成员方法可以在页面的 Java 程序片中调用。

3.2.2 程序片

在“<%”和“%>”标记之间放置的 Java 代码称为 Java 程序片。一个 JSP 页面可以有多个 Java 程序片。程序片中声明的变量称为程序片变量，是局部变量。程序片变量的有效范围与其声明位置有关，即从声明位置向后有效，可以在声明位置后的程序片、表达式中使用；基于此，大的程序片可以分为几个小的程序片，程序片中可以插入 HTML 标记，以便使程序片代码更具可读性。

JSP 引擎将页面翻译成 Java 文件时，将程序片中的变量作为页面 Servlet 类的 Service 方法中的局部变量处理，程序片中的 Java 语句作为 Service 方法中的语句处理，最终将页面所有程序片中的变量和语句依次转译到 Service 方法中，被 JSP 引擎顺序执行。程序片可以完成以下任务。

- 程序片可以操作页面的成员变量，页面成员变量在客户线程之间共享。

- 程序片可以调用方法，调用的方法必须是页面成员方法或内置对象的方法。
- 声明和操作程序片变量。

如果多个客户访问一个 JSP 页面，JSP 页面的程序片就会被执行多次，分别运行在不同的线程中。程序片变量不同于在“<%!”和“%>”之间声明的页面成员变量，不能在不同客户访问页面的线程之间共享，也就是运行在不同客户线程的 Java 程序片变量互不干扰；当一个客户线程执行完程序片后，该线程的程序片变量内存空间随即释放。

【例 3.3】抽奖程序示例。程序功能：每个访问者随机抽取幸运数，如果幸运数等于总访问次数与 10 的余数，则为幸运访问者。页面效果如图 3.3 所示。

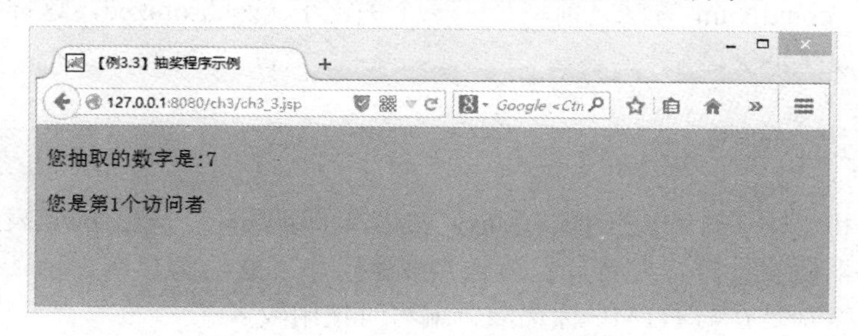

图 3.3　ch3_3.jsp 页面效果

ch3_3.jsp 文件代码：

```
<%@ page contentType="text/html;charset=GB2312" %>  <!--jsp 指令标记-->
<%! int lucknum=0,count=0;
    public synchronized void countNum(){
          count++;
          lucknum=count%10;
    }
%>
<html>
<head>
<title>【例 3.3】抽奖程序示例</title>
</head>                                                <!--HTML 标记-->
<body  bgcolor="cyan">
   <font size=4>
     <%  countNum();                                      //程序片
          int num=(int)(Math.random()*10)+1;
           if (num==lucknum)
           {
     %>      <p>您访问的幸运数是:<%= num %>
      <%  }
           else
           {
      %>
              <p>您抽取的数字是:<%= num %>
               <p>您是第<%= count %>个访问者
      <%
```

```
            }
        %>
</font>
</body>
</html>
```

该程序第 2 行声明了一个计数器变量和一个幸运数变量，计数器变量和幸运数变量在客户线程之间共享，直至服务器关闭。当一个客户线程在执行 synchronized 方法时，其他客户必须等待，直到这个客户线程调用执行完毕该方法后，其他客户线程才能执行。因此，在第 3 行的 countNum 方法之前增加了一个关键字 synchronized，这样可以防止客户线程之间调用它发生冲突。

3.2.3 表达式

在“<%=”和“%>”标记之间放置 Java 表达式，可以直接输出 Java 表达式的值。表达式的值由服务器负责计算，并将计算值转换成字符串发送给客户端显示。表达式在 JSP 编程中较常用，特别是在与 HTML 标记混合编写时使用较多。

下面通过一个例演示表达式与 HTML 标记混合编码的使用。

【例 3.4】用不同颜色输出 26 个英文字母。页面效果如图 3.4 所示。

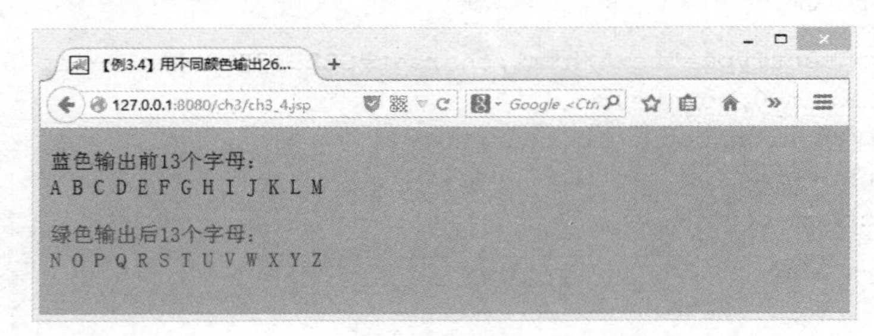

图 3.4　ch3_4.jsp 页面效果

ch3_4.jsp 文件代码：

```
<%@ page contentType="text/html;charset=GB2312" %>  <!--jsp 指令标记-->
<html>
<head>
<title>【例 3.4】用不同颜色输出 26 个英文字母</title>
</head>                                                <!--HTML 标记-->
<body  bgcolor="cyan">
   <font size=4>
   <% char begin='A';                                  /*Java 程序片*/
      int  ix=13;
   %>
   <font color="blue"><p>蓝色输出前 13 个字母：<br>   <!--插入 Html 标记-->
   <%                                                  /*Java 程序片*/
      for(;begin<'A'+ix;begin++){
   %>
        <%= begin %>                                   <!--Java 表达式-->
```

```
    <%
        }
    %>
    </font>
    <font color="green"><p>绿色输出后 13 个字母：<br>  <!--插入 Html 标记-->
    <%
       for(;begin<'N'+ix;begin++){
    %>
          <%= begin %>                              <!--Java 表达式-->
    <%
       }
    %>
    </font>
    </font>
</body>
</html>
```

3.3　注　　释

适当的注释可以增强程序的可读性，它可以方便程序的调试和维护。程序员在程序中书写注释是程序员一个良好的习惯，读者在学习时应该注意这些习惯的养成，为成为一个优秀的程序员做好准备。

3.3.1　输出型注释

输出型注释是指会被 JSP 引擎发送给客户端浏览器的注释，这种注释可以在浏览器的源码中看到，浏览器将其作为 HTML 的注释处理。输出型注释的内容写在“<!--”和“-->”之间，格式如下：

```
<!-- 注释内容 [<%= 表达式 %>] -->
```

例如：

```
<!-- 下面是 Java 的程序片 -->
```

在客户端的 HTML 源码中为“<!-- 下面是 Java 的程序片 -->”。

输出型注释还可以在注释内容中加入表达式，例如：

```
<! -- 页面加载时间：<%= (new java.util.Date()).toLocaleString()%> -->
```

在客户端的 HTML 源码中为：“<! -- 页面加载时间：20:08:00 -->”

3.3.2　隐藏型注释

在标记“<%--”和“--%>”之间加入的内容称为隐藏型注释，它们会被 JSP 引擎忽略，不会发送的客户端浏览器中，所以称为隐藏型注释。其使用格式如下：

```
<%-- 注释内容 --%>
```

隐藏型注释一般写在 Java 程序片的前面，对程序片做出说明。

读者需要注意，在 Java 程序片中可以使用 Java 语言的注释方法，例如：

//注释内容

/*注释内容*/

/**注释内容*/

这些注释内容都会被 JSP 引擎忽略，不会发送到客户端浏览器中。

【例 3.5】注释示例，程序功能：巴西世界杯倒计时，页面效果如图 3.5 所示。

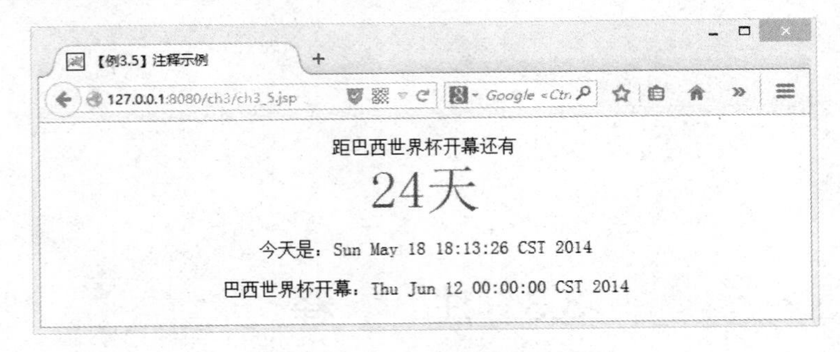

图 3.5　ch3_5.jsp 页面效果

(1) ch3_5.jsp 文件内容：

```
<%@ page contentType="text/html;charset=GB2312" %>
<%@ page import="java.util.*" %>
<html>
<head>
<title>【例 3.5】注释示例</title>
</head>
<body>
<!--程序加载时间：
    <%= (new java.util.Date()).toLocaleString()%>
    输出型注释
-->
<%--展示时间数据的 Java 程序片(隐藏注释)--%>
<%  //创建日期型对象
    Date dateNow=new Date();
    /*相对于 1900 年
        设置世界杯开幕时间*/
    Date dateTemp=new Date(114,5,12);
    /**求两个时间相对于 1900 年 1 月 1 日
     *的毫秒数
     */
    double  secondnum=(double)(dateTemp.getTime()-dateNow.getTime());
    long  daynum=(long)(secondnum/1000/60/60/24);
%>
<center>
```

```
  <p>距巴西世界杯开幕还有<br>
    <font color="red" size="8"><%=daynum%>天</font>
  <p>今天是：<%= dateNow.toString()%>
  <p>巴西世界杯开幕：<%= dateTemp.toString()%>
</center>
</body>
</html>
```

(2) 客户端 HTML 文件内容：

```
<html>
<body>
<!--程序加载时间:
    2014-5-18 16:20:43
    输出型注释
-->
<center>
  <p>距巴西世界杯开幕还有<br>
    <font color="red" size="8">24 天</font>
  <p>今天是：Sun May 18 16:20:43 CST 2014
  <p>巴西世界杯开幕：Thu Jun 12 00:00:00 CST 2014
</center>
</body>
</html>
```

注意：输出注释一般放在 JSP 页面的文档头中或文档体的开头，如果放在 Java 程序片之间，JSP 引擎将会忽略<! -- -->类型的注释，不发送到客户端。

3.4 JSP 指令标记

JSP 页面第一次被请求时，会被 JSP 引擎转译成 Servlet 的 Java 文件，然后再被编译成字节码文件执行。JSP 指令标记为 JSP 页面转译提供整个页面的相关信息。JSP 指令标记的使用格式：

```
<%@ directive {attribute="value"}* %>
```

"<%@"字符串作为指令的起始标记，“<”、“%”和“@”之间不能加空格，作为一个整体使用。JSP 指令标记有三种：page、include 和 taglib，taglib 允许用户自定义标签库，下面介绍 Page 和 Include 其具体使用方法。

3.4.1 page 指令标记

page 指令用来指定整个 JSP 页面的一些属性的属性值，属性值用双引号括起来。可以用一个 page 指令指定多个属性的取值。使用格式如下：

```
<%@ page 属性 1="属性值" 属性 2="属性值" --- %>
```

或

```
<%@ page 属性 1="属性值"%>
<%@ page 属性 2="属性值"%>
⋮
<%@ page 属性 n="属性值"%>
```

page 指令的作用对整个页面有效，一般写在页面的最前面。page 指令可以设置的属性有 contentType、import、language、session、buffer、autoFlash、isThreadSafe、info 等。

(1) contentType 属性。

该属性用来设置 JSP 页面的 MIME(Multipurpose Internet Mail Extention)类型和字符编码集，取值格式为"MIME 类型"或"MIME 类型;charset=字符编码集"。下面的设置：

```
<%@ page contentType="text/html;charset=GB2312" %>
```

其含义是告诉浏览器启用 HTML 解析器来解析执行所收到的信息，页面字符集为 GB2312 字符集，也就将客户端将收到的信息当作一般的网页来处理。本教材的 JSP 页面一般都做此设置。下面的设置：

```
<%@ page contentType="application/msword" %>
```

其含义是告诉浏览器启用本地的 MS-Word 应用程序来解析收到的信息。可以设置的常用 MIME 类型有"application/vnd.ms-powerpoint"、"application/vnd.ms-excel"等。如果没有用 page 指令标记设置 contentType 属性，JSP 引擎会默认 contentType 值为"text/html; charset=ISO-8859-1"。page 指令只能为 contentType 属性指定一个值，不能为它多次指定不同的值。

【例 3.6】用 Excel 解析收到的信息。效果如图 3.6 所示。

```
<%@ page contentType="application/vnd.ms-excel" %>
<html>
<head>
<title>【例 3.6】用 Excel 解析收到的信息</title>
</head>
<body bgcolor="cyan">
<font size="4" color="blue">
<table width="350" bordercolor="#FF0000" heigth="290" border="10"
    align="center" cellpadding="4" cellspacing="2">
  <tr align="center" height="30" valign="middle"
bordercolor="#0000FF">
    <th>编号</th>
    <th>姓名</th>
    <th>性别</th>
    <th>工作单位</th>
  </tr>
  <tr align="center" valign="top" bordercolordark="#00FF00"
    bordercolorlight="#FF0000">
    <td>01</td>
    <td>杨得力</td>
```

```
      <td>男</td>
      <td>河北 XX 大学</td>
    </tr>
    <tr align="right" valign="baseline" bgcolor="#CC99FF">
      <td>02</tds>
      <td>周国正</td>
      <td>女</td>
      <td>北京 XX 大学</td>
    </tr>
  </table>
  </font>
  </body>
  </html>
```

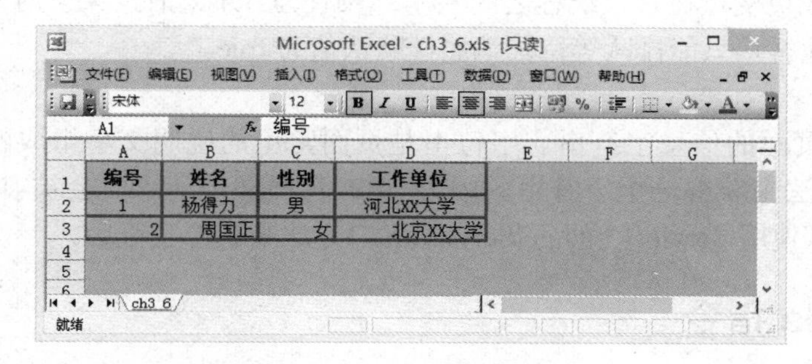

图 3.6 Excel 打开网页效果

提示： MIME 的中文名为多功能网络邮件扩展标准，读者可以简单理解为：MIME 是服务器通知客户端所接收文件类型的一种手段。不同 MIME 类型的文件，客户端会指定不同程序或插件打开，例如 Flash 动画指定其播放插件来播放。

(2) import 属性。

该属性用来导入页面中要用到的包或类，导入的包或类可以是 Java 环境的核心类，也可以是用户自己编写的包或类。可以为该属性指定多个值，例如：

```
<%@ page import="java.io.*","java.util.Date" %>
```

默认情况下，JSP 页面 import 属性已有如下值：

```
java.lang.*、javax.servlet.*、javax.servlet.jsp.*、javax.servlet.http.*
```

这些值不必再进行导入，这里*的意思是该包下所有的类。导入包或类后，在 JSP 页面中的程序片、变量和方法声明、表达式中就可以使用导入的类。

(3) language 属性。

该属性定义 JSP 页面使用的脚本语言，默认是 Java，目前只能支持 Java，所以一般不对该属性设置值。

(4) session 属性。

该属性设置是否在 JSP 页面中使用默认的 session 对象。设置为 false 时不允许使用，设置为 true 时允许使用，默认为 true。例如：

```
<%@ page session="false" %>
```

(5) buffer 属性。

该属性用来设定 out 对象输出缓存处理的缓冲区的大小，可以取的值有 none、8kb 或给定的 kb 值，none 为没有缓存直接输出。默认值为 8kb。

(6) autoFlash 属性。

该属性设置缓冲区填满时，是否自动刷新，默认值为 true。当设置为 false 值时，如果缓冲区填满时，就会出现溢出异常。autoFlash 属性与 buffer 属性有联系，如果 buffer 属性值为 none，此属性不能设置为 false。

(7) isThreadSafe 属性。

该属性用来设置是否可以多线程访问。当设置为 true 时可以允许多个客户同时访问一个页面，页面的成员变量在客户线程之间共享；设置为 false 时，某一时刻只能有一个用户访问页面，其他客户只能排队等待。该属性默认值为 true。

(8) info 属性。

该属性设置页面的信息字符串。字符串是页面常用的说明文字组成的串，可以通过 getServletInfo()方法来获得这个字符串。因为 JSP 页面在被请求时会被 JSP 引擎转译成 Servlet，所以可以使用 Servlet 类的 getServletInfo()方法来获得字符串。

3.4.2 include 指令

include 指令用于在 JSP 页面静态插入一个文件，被插入的文件可以是 JSP 页面、HTML 网页、文本文件或一段 Java 代码。使用了 include 指令的 JSP 页面在转换成 Java 文件时，将被插入的文件在当前 JSP 页面出现该指令的位置做整体插入，合并成一个新的 JSP 页面，然后 JSP 引擎再将这个新的 JSP 页面转译成 Java 文件。因此，必须保证插入文件后形成的新的 JSP 页面符合 JSP 语法和逻辑规则。include 指令的使用格式为：

```
<%@ page include file="文件的 url" %>
```

插入文件可以使用绝对路径和相对路径，一般要使用相对路径。相对路径是相对于当前网页所在目录的路径，例如，当前网页所在目录是 Web 服务根目录，被插入的文件在根目录下的 text 文件中，文件名为 head.txt，相对路径的使用格式如下：

```
<%@ page include file="text/head.txt" %>
```

当被插入文件被修改后，JSP 引擎会重新将当前的 JSP 页面和修改后的被嵌入文件合并成一个新的 JSP 页面，然后在转译、编译成新的 Java 字节文件供客户访问。

使用 include 指令可以实现代码的复用，提高代码的使用效率。例如，页面一般都需要一个标题栏、导航栏、页脚等，用户可以将标题栏等写成单独的文件，每一个 JSP 页面都可以在适当的位置用 include 指令插入标题栏、导航栏、页脚等文件，这样就极大地提高了代码的复用和效率，使代码更便于修改。

下面的例子中，两个 JSP 页面嵌入了相同的标题栏 head.txt、导航栏 left.txt、页脚 footer.txt 等文本文件，这些文本文件内容是多个页面共有的元素。

【例 3.7】 include 指令标记示例。编写 head.txt、left.txt 和 footer.txt 文本文件，制作

ch3_7.jsp 和 ch3_7_1.jsp 两个页面，在适当的位置使用 include 指令将文本文件插入页面中，页面使用 mystylesheet.css 样式表。页面效果如图 3.7、图 3.8 所示。

head.txt 文件保存在 ch3 目录中，内容如下：

```
<%@ page contentType="text/html;charset=GB2312" %>
<h1>XX 省职称计算机考试报名系统</h1>
```

left.txt 文件保存在 ch3 目录中，内容如下：

```
<%@ page contentType="text/html;charset=GB2312" %>
   <h3>左侧菜单</h3>
  <p><a href="ch3_7_1.jsp">考生报名</a></p>
  <p><a href="ch3_7.jsp">返回主页</a></p>
```

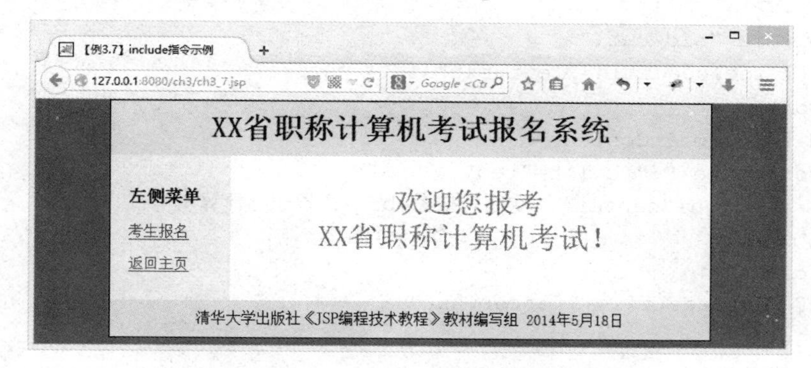

图 3.7　ch3_7.jsp 页面效果

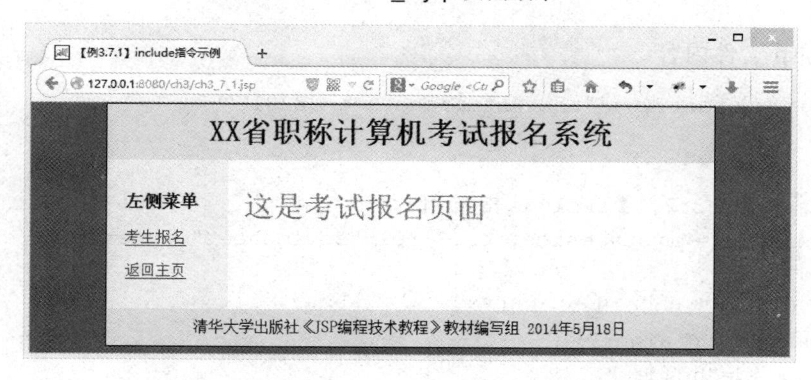

图 3.8　ch3_7_1.jsp 页面效果

footer.txt 文件保存在 ch3 目录中，内容如下：

```
<%@ page contentType="text/html;charset=GB2312" %>
   <br class="clearfloat" />
   <div id="footer">
   <p>清华大学出版社《JSP 编程技术教程》教材编写组 2014 年 5 月 18 日</p>
   </div>
```

mystylesheet.css 样式文件保存在 ch3 目录，内容参见 2.4.3 节例 2-20mystylesheet.css 样式文件内容，这里省略。

ch3_7.jsp 文件内容：

```
<%@  page contentType="text/html;charset=GB2312" %>
<html>
  <head>
    <title>【例 3.7】include 指令示例</title>
     <link href="mystylesheet.css" rel="stylesheet" type="text/css" />
  </head>
  <body class="twoColHybLtHdr">
    <div id="container">
      <div id="header">
      <%@ include file="head.txt" %>
      </div>
      <div id="sidebar1">
       <%@ include file="left.txt" %>
      </div>
      <div id="mainContent">
      <font size="6" color="red" >
         <p align="center" >欢迎您报考<br>XX 省职称计算机考试！</p>
       </font>
      </div>
      <%@  include file="footer.txt" %>
</div>
</body>
</html>
```

ch3_7_1.jsp 文件内容：

```
<%@  page contentType="text/html;charset=GB2312" %>
<html>
<head>
    <title>【例 3.7.1】include 指令示例</title>
     <link href="mystylesheet.css" rel="stylesheet" type="text/css" />
  </head>
  <body class="twoColHybLtHdr">
    <div id="container">
      <div id="header">
      <%@ include file="head.txt" %>
      </div>
      <div id="sidebar1">
       <%@ include file="left.txt" %>
      </div>
      <div id="mainContent">
       <font size="6" color="red">
      <p>这是考试报名页面</p>
       </font>
      </div>
      <%@  include file="footer.txt" %>
```

```
</div>
</body>
</html>
```

注意：Tomcat 7.0 允许在被嵌入文件中指定 contentType 属性，但要与当前页面的属性值相同，例如 head.txt、left.txt、footer.txt 文本文件，第一行内容为<%@ page contentType="text/html;charset=GB2312" %>。读者在练习此例时，可以将 2.4.3 节的例 2.20 样式表复制到 ch3 目录中来。关于页面中的 div 层标记，请参见第 2 章。

3.5　JSP 动作标记

与 JSP 指令标记不同，JSP 动作标记影响 JSP 运行时功能，可以将代码处理程序与特殊的 JSP 标记关联在一起。JSP 动作标记在 JSP 程序设计中经常用到，本节做详细介绍。

3.5.1　jsp:include 动作标记

include 动作标记用来在 JSP 页面中动态包含一个文件，包含页面程序与被包含页面程序是彼此独立的，互不影响。jsp:include 标记的一般使用格式：

```
<jsp:include page="文件的url" />
```

或

```
<jsp:include page="文件的url">
    <jsp:param 子标记/>
</jsp:include>
```

jsp:include 动作标记与 include 指令标记包含文件的处理时间和方式不同。include 指令标记插入的文件在页面转译时就合并到一起了，被包含文件与当前页面组合而成的新页面必须符合 JSP 的语法和逻辑规则，由于是提前合并编译，所以执行速度快；而 jsp:include 动作标记被包含的文件语法和逻辑独立于当前页面，单独被 JSP 引擎编译，当前页面执行时再将被包含文件的运行结果传送给客户端，由于是执行页面时处理包含文件，所以执行速度较慢，但可以利用 param 子标记传递参数，使用更灵活。

jsp:include 动作标记可以包含动态文件，也可以包含静态文件，动态文件被 JSP 引擎执行后将结果发送给客户端，静态文件则直接发给客户端。如果包含动作不使用参数，则可以使用<jsp:include page="文件的 url"/>格式，如果包含动作需要传递参数，则使用下面的格式：

```
<jsp:include page="">
    <jsp:param 子标记/>
    <jsp:param 子标记/>
</jsp:include>
```

注意：在书写“<jsp:include”动作标记时，字符和字符之间没有空格，“<jsp:include”必须作为一个整体的字符串来使用。

3.5.2 jsp:param 动作标记

jsp:param 动作标记不能单独使用，必须作为 jsp:include、jsp:forward、jsp:plugin 标记的子标记使用，并为它们提供参数。jsp:param 动作标记的使用格式如下：

```
<jsp:param  name="参数名字" value="指定的参数值" >
```

其中：name 是参数名字、value 是参数的值，当该标记作为 jsp:include 的子标记时，“name—value”为页面之间提供参数传递。被加载的 JSP 文件可以使用 request 对象获取 jsp:include 动作标记的 param 子标记中 name 属性提供的值。获取 jsp:param 子标记中参数的方法如下：

```
request.getParameter("name");
```

下面的例子使用了 include 动作标记和 param 子标记。ch3_8.jsp 文件使用 include 动作标记加载 JSP 文件 ch3_8_1.jsp，两个文件都保存在 ch3 目录中。ch3_8_1.jsp 页面可以比较两个数的大小并输出字符串，被加载时使用 request 对象的 getParameter 方法获取 param 子标记提供的两个数，程序片中调用比较方法输出结果。

【例 3.8】 include 与 param 子标记使用示例。页面效果如图 3.9 所示。

ch3_8.jsp 文件内容如下：

```
<%@ page contentType="text/html;charset=GB2312" %>
<html>
<head>
<title>【例 3.8】这是 Include 动作标记示例</title>
</head>
<body bgcolor="cyan">
<% double dx=3.14,dy=4.3;
%>
<p>主页面信息：加载 ch3_8_1.jsp 文件，求两个数的最大值:下面开始加载。
  <jsp:include page="ch3_8_1.jsp" >
      <jsp:param name="dx" value="<%=dx%>" />
       <jsp:param name="dy" value="<%=dy%>" />
  </jsp:include>
<p>主页面信息：现在已经加载完毕。</p>
</body>
</html>
```

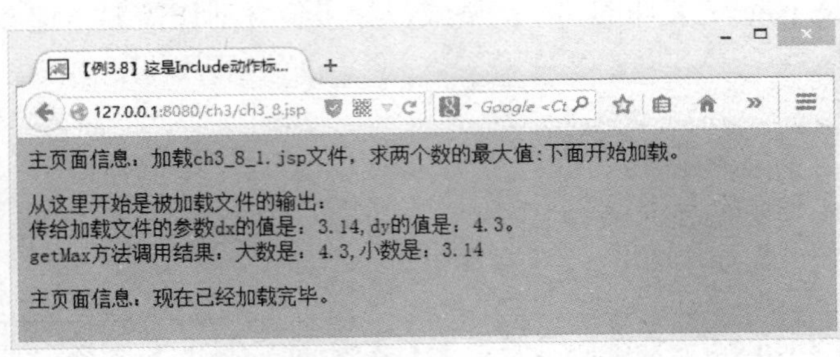

图 3.9 用 param 子标记传递参数页面效果

ch3_8_1.jsp 文件内容：

```
<%@ page contentType="text/html;charset=GB2312" %>
<%! public String getMax(double x,double y){
    if (x>y){
        double temp;
        temp=x;
        x=y;
        y=x;
    }
    return ("大数是："+y+",小数是："+x);
}
%>
<% String dx=request.getParameter("dx");
   String dy=request.getParameter("dy");
   double x=Double.parseDouble(dx);
   double y=Double.parseDouble(dy);
%>
<html>
<body>
<p>从这里开始是被加载文件的输出：<br>
传给加载文件的参数 dx 的值是：<%=dx%>,dy 的值是：<%=dy%>。<br>
getMax 方法调用结果：<%=getMax(x,y)%>
</body>
</html>
```

3.5.3　jsp:forward 动作标记

jsp:forward 动作允许将用户请求定位到其他页面，基本的语法格式是：

```
<jsp:forward page={"要转向的页面 url"|"<%=表达式%>"} />
```

或者

```
<jsp:forward page={"要转向的页面 url"|"<%=表达式%>"} >
   <jsp:param name="属性名" value="属性值" />
</jsp:forward>
```

该动作标记的作用是从该动作标记处停止当前页面的执行，转到 page 属性指定的页面执行。page 属性值是相对的 URL 地址，可以是静态的 URL，也可以是由表达式计算等到的 URL 地址。如果需要向转向的页面传送数据信息，需要使用第二种格式，通过 jsp:param 子标记传递参数。被转向的页面可以使用 request 对象的 getParameter 方法获取传递的参数值。不需要传递参数值时，使用第一种格式。

在下面的例子中，ch3_9.jsp 页面中使用了 jsp:forward 动作标记转向 ampm.jsp 页面。在 ch3_9.jsp 页面中判断系统时间是上午还是下午，根据判断结果传给 ampm.jsp 页面不同的参数值，向访问网页的用户问好。ch3_9.jsp、ampm.jsp 两个文件都保存在 ch3 目录中。

【例 3.9】 jsp:forward 动作标记示例。页面效果如图 3.10 所示。

图 3.10　ch3_9.jsp 页面效果

ch3_9.jsp 文件内容:

```
<%@ page contentType="text/html;charset=GB2312" %>
<%@ page import="java.util.*" %>
<html>
<head>
   <title>【例 3.9】JSP:FORWARD 使用示例</title>
</head>
<body>
  <p>这是第一个页面的输出</p>
  <%
  if (Calendar.HOUR>Calendar.AM){
  %>
     <jsp:forward page="ampm.jsp">
        <jsp:param name="hello" value="Good p.m.!"/>
     </jsp:forward>
  <%} else {
  %>
     <jsp:forward page="ampm.jsp">
        <jsp:param name="hello" value="Good a.m.!"/>
     </jsp:forward>
  <%}
  %>
</body>
</html>
```

ampm.jsp 文件内容:

```
<%@ page contentType="text/html;charset=GB2312" %>
<%@ page import="java.util.*" %>
<html>
<head>
   <title>【例 3.9】JSP:FORWARD 转向的页面</title>
</head>
<body background="cyan" >
  <%
   String str=request.getParameter("hello");
  %>
  <font size="5" color="red">
```

```
      <%=str%>
   </font>
</body>
</html>
```

注意：在浏览器中看到的页面地址是 ch3_9.jsp，但实际显示的内容是 ampm.jsp 页面的内容。在 ch3_9.jsp 页面输出到客户端的内容用户看不到。

3.5.4 jsp:plugin 动作标记

jsp:plugin 动作标记用来在客户端浏览器中执行一个 Bean 或者显示一个 Applet，而这种执行、显示方式往往需要浏览器 Java 插件。jsp:plugin 动作标记指示 JSP 页面加载 Java plugin，插件的下载由用户负责，插件下载后就可执行所指定的 Bean 或 Applet 程序；插件下载安装一次即可，不用每次运行页面都重复安装。jsp:plugin 动作标记的一般使用格式如下：

```
<jsp:plugin type="bean|applet" code="类字节文件名"
    codebase="类字节文件所在路径" jreversion="Java版本1.2" --->
    <jsp:fallback>
        客户端是否支持插件下载的信息
    </jsp:fallback>
</jsp:plugin>
```

例如，运行 applet 小程序的格式：

```
<jsp:plugin type="applet" code="应用程序.class" codebase="应用程序的路径"
    jreversion="小程序的虚拟机版本号" width="小程序窗口宽度"
    height="小程序窗口高度">
    <jsp:fallback>
       applet载入错误提示信息
    </jsp:fallback>
</jsp:plugin>
```

下面演示 Tomcat 7.0 自带的一个<jsp:plugin>动作标记使用例子。

【例 3.10】 jsp:plugin 动作标记示例。

在客户端浏览器地址栏中输入地址：

```
http://127.0.0.1:8080/examples/jsp/plugin/plugin.jsp
```

页面显示效果如图 3.11 所示。

plugin.jsp 文件内容：

```
<html>
<title> Plugin example </title>
<body bgcolor="white">
<h3> Current time is : </h3>
<jsp:plugin type="applet" code="Clock2.class" codebase="applet"
        jreversion="1.2" width="160" height="150" >
```

```
    <jsp:fallback>
        Plugin tag OBJECT or EMBED not supported by browser.
    </jsp:fallback>
</jsp:plugin>
<p>
<h4>
<font color=red>
The above applet is loaded using the Java Plugin from a jsp page using
the
plugin tag.
</font>
</h4>
</body>
</html>
```

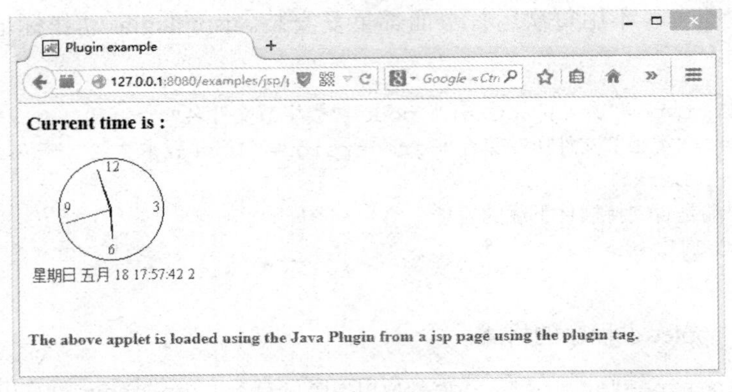

图 3.11　plugin.jsp 页面效果

注意：如果客户端浏览器不支持该 Applet 的运行，则会弹出一个警示对话框，选择“是”就会自动下载支持运行 Applet 的插件，然后根据提示安装，安装完毕后浏览器开始使用下载的 Java 虚拟机，即可运行该 Applet，以后就不必再重复安装。若果客户端提示“应用程序已被阻止”，此时在电脑上找到“控制面板→程序→Java→安全”，将安全级别调到“中”即可。

3.5.5　jsp:useBean 相关动作标记

JSP+JavaBean+Servlet 是实际工程中使用较多的 MVC 模式开发，这种模式可以较大程度地实现页面静态内容与动态内容的分离。jsp:useBean 动作标记就是用来在 JSP 页面中创建并使用一个 JavaBean 组件的指令，它让 HTML 完成 JSP 页面的静态内容，JavaBean 组件完成 JSP 页面的动态内容，真正实现了页面静态和动态的分离。jsp:useBean 动作标记的使用格式：

```
<jsp:useBean id="bean的名称" scope="bean的有效范围" class="包名.类名">
</jsp:useBean>
```

其中：

- id 属性值是 Bean 在这个 JSP 页面中的名称，在其指定的有效范围内，都可以使用此 id 来访问这个 Bean 的实例。
- scope 属性值是 Bean 的有效范围，可以取 page、request、session、application 等值，默认为 page。
- class 属性值是 Bean 的包名和类名。

jsp:useBean 动作标记是一个非重要且常用的标记，与其配合使用的相关动作标记还有 jsp:setProperty 和 jsp:getProperty。jsp:setProperty 动作标记用来设置 Bean 的属性值，jsp:getProperty 动作标记用来得到 Bean 的属性值。详细内容将在第 5 章介绍，这里不再赘述。

3.5.6 特殊字符

在 JSP 页面中字符'<'、'>'等用作系统标识符，有特定的意义，如果用户想将这些字符作为一般的文本字符数在页面中显示，就需要做特殊的处理，否则就会和系统的标识符产生冲突，因为 JSP 编译器不知道它是命令标记还是普通字符。JSP 采用一些替代字符来解决这些特殊字符的显示。表 3.1 就是 JSP 特殊字符的替代对照表。

表 3.1 JSP 特殊字符替代表

需要替代的字符	替代字符	需要替代的字符	替代字符
®	®	<	<
×	×	>	>
©	©	&	&
"	"	空格	

下面的例子使用了一些特殊的字符<、>等。

【例 3.11】特殊字符替代示例。页面效果如图 3.12 所示。

ch3_11.jsp 文件内容：

```
<%@ page contentType="text/html;charset=GB2312" %>
<html>
<head><title>【例 3.11】特殊字符使用示例&lt;html&&gt;</title>
</head>
<body>
这在 HTML 标记中的示例：<p>
         &lt;html&gt;
         &lt;body&gt;
         &amp;=&amp;amp;
         &quot;=&amp;quot;
<p>这是在程序片中的示例：
<p>
<%
   out.println("\"&lt;jsp&gt;\"");
%>
```

```
</body>
</html>
```

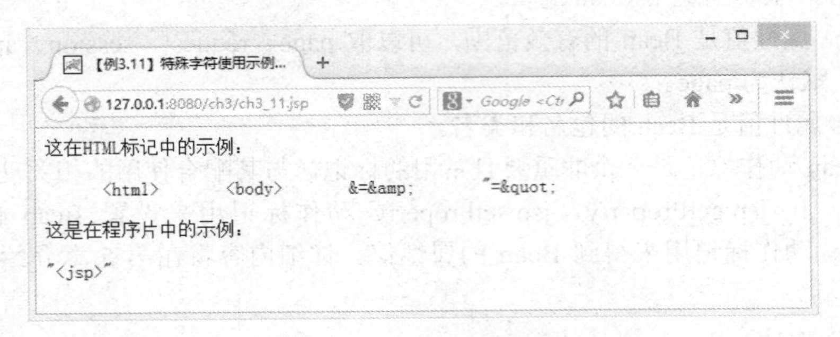

图 3.12　ch3_11.jsp 页面效果

注意： 上例中在 HTML 标记内使用的特殊替代字符与 HTML 特殊字符相同；在程序片中也可以使用特殊字符，Java 语言中的转义字符"\"也可以使用。

3.6 上 机 实 训

实训目的

- 了解 JSP 页面的构成，熟记构成 JSP 页面的 7 种元素的标记。
- 理解页面成员变量和程序片局部变量之间的区别。
- 掌握页面成员变量、成员方法，程序片局部变量和方法内局部变量声明与使用。
- 掌握 JSP 页面注释的两种方法。
- 掌握 Include 指令标记和 jsp:include 动作标记的使用方法和区别。
- 掌握 page 指令标记中 contentType、import 属性的设置方法。

实训内容

实训 1　编写一个对 1 到 100 之间的整数求和的 JSP 程序。

要求：

(1) 在程序中对语句进行说明。

(2) 说明采用两种注释方式。

实训 2　完成“XX 省职称计算机考试报名系统”中的首页及其他任意 2 个页面，页面效果参见图 2.28 或第 11 章“网上报名系统案例”。

要求：

(1) 使用层和文件包含设计首页 index.jsp 页面，要求使用中文显示。

(2) 首页和其他页面由 4 部分组成：head.txt 标题、left.txt 左侧导航栏、footer.txt 页脚和 maincontent，标题、导航栏、页脚采用 include 文件包含方式完成。

(3) 文件 head.txt、left.txt、footer.txt 文件内容参见第 11 章。

(4) 做好三个页面之间的链接。

实训 3　在实训 2 的主页面设置一个超级链接，使用 jsp:forward 动作标记将链接指向

clock.jsp 页面。

要求：

(1) clock.jsp 页面插入一个显示时钟的 Applet 小程序。

(2) 显示时钟的小程序 Applet 在 Tomcat 7.0 自带的例子中，参见例 3.10。

实训总结

通过本章的上机实训，学员应该能够理解 JSP 页面的构成要素及其作用；掌握页面成员变量和成员方法的声明和使用；掌握 JSP 页面程序片的编写技术和表达式的使用方法；掌握常用的 JSP 指令标记和动作标记的使用方法。

3.7 本章习题

思考题

(1) include 标记与 include 动作标记有什么区别？

(2) 如何保证页面跳转时当前页面与跳转页面之间的联系？

(3) 如果有两个用户访问一个 JSP 页面，该页面的程序片将被执行几次？

(4) 在<%!和%>之间声明的变量和在<%和%>之间声明的变量有何区别？

(5) 是否允许一个 JSP 页面为 contentType 设置两次不同的值？

(6) JSP 的特殊字符与 Java 语言的转义字符关系？

拓展实践题

完成“XX 省职称计算机考试报名系统”的所有页面、导航栏、标题栏、页脚文件的设计，并使用 include 指令标记包含在页面中。

第 4 章　JSP 内建对象

学习目的与要求：

为了方便用户开发，JSP 内建了很多对象，这些对象是 JSP 编程的基本手段，使用这些对象配合适当的作用范围可以实现很多页面处理功能。本章主要学习 JSP 内建的 out、request、response、session 和 application 等对象。通过本章的学习，学员要了解 HTTP 的请求消息结构和响应消息结构以及请求响应模型；了解表单验证的知识；掌握 request 对象获得表单数据的方法，out 输出各种格式数据的方法，response 对象响应等方法，session 对象的生命周期和保存、获取数据的方法。掌握 application 对象数据的保存与获取方法。

4.1　内建对象概述

JSP 内建对象是指 JSP 页面系统中已经默认设置的 Java 对象，这些对象无须显式声明，直接就可以在 Java 程序片和表达式中使用。JSP 的内置对象有 request、response、session、application、out、config、exception、page 和 pageContext 对象，这些对象分别完成不同的任务。request、response、session、application 和 out 对象是 Web 程序中常用的对象，读者要熟练掌握它们的使用方法和技巧。在介绍它们之前先要了解一下 HTTP，这对学习本章内容有很大帮助。

4.1.1　什么是 HTTP

什么是 HTTP 呢？HTTP(Hyper Text Transfer Protocol)是超文本传输协议的缩写，它用于传送 WWW 方式的数据。当用户在浏览器地址栏中输入某个网址的时候，浏览器会向该网址的服务器发送请求，这个请求(消息)使用 HTTP，其中包括请求的主机名、HTTP 版本号等信息。服务器在收到“请求”后响应该请求，使用 HTTP 将回复的“消息”发回给客户端浏览器。客户端的浏览器在接收到服务器传回的信息后，将其解释并显示在浏览器的窗口中。这个过程采用了 HTTP 的请求/响应模型。

在“请求/响应”模型中，HTTP 消息是服务器和客户端通信的方式。理解消息的格式和内容是理解协议的重点。通常 HTTP 消息包括客户机向服务器的请求消息和服务器向客户机的响应消息。这两种类型的消息由一个起始行，一个或者多个头域，一个指示头域结束的空行和可选的消息体组成。HTTP 的头域包括通用头，请求头，响应头和实体头四个部分。每个头域由一个域名，冒号(:)和域值三部分组成。域名是大小写无关的，头域可以被扩展为多行，在每行开始处，使用至少一个空格或制表符，每行结束处是回车符和换行符。

一个典型的 HTTP 请求消息包括请求行、多个请求头和信息体，请求行包括请求方法、协议和资源，请求头包含请求的信息和信息体的附加信息，一个请求还可能包括信息体。下面是一个典型的 HTTP 请求消息：

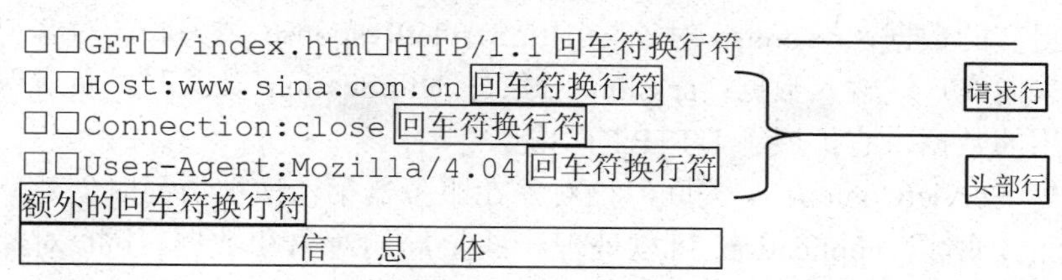

该消息的首行是请求行，规定了请求的方法是 GET，请求的资源是 index.htm，使用的协议版本为 HTTP/1.1；第 2～4 行是头，Host 头规定了主机地址 www.sina.com.cn；头部 Connection:close 告知服务器不使用持久连接，服务器发出所请求的对象后应关闭连接；User-agent 头部行指定用户代理，也就是产生当前请求的浏览器的类型；接下来是一个头域的空行(额外的回车和换行)；头域空行下面是可选的信息体，信息体中一般包含表单使用 POST 或 GET 方法发送的内容。请求的方法除了 GET 以外，还有 POST、HEAD、DELETE、TRACE 及 PUT 等方法。

一个 HTTP 响应消息都由状态行开始，可以包括多个响应头和可能的多个信息体。响应状态行指明了使用的协议版本、结果代码和结果描述；响应头包括服务器的一些信息和信息体的一些附加信息；信息体就是客户请求的网页运行结果，对于 JSP 来讲就是网页的静态信息。HTTP 响应消息由服务器发送给客户端浏览器。下面是一个典型的 HTTP 响应消息：

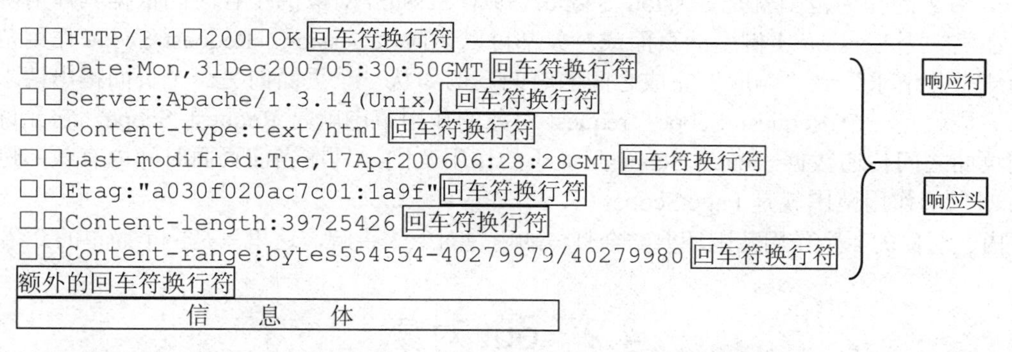

该 HTTP 响应信息的第 1 行是响应行，HTTP/1.1 表示支持的本版，200 是一个三个数字的结果代码，表示处理成功，OK 是对结果代码 200 的一个简单描述；结果代码可能能取 5 个不同的值：1xx、2xx、3xx、4xx、5xx，HTTP/1.1 200 OK 表示成功响应，HTTP/1.1 404 Object Not Found 表示客户端请求出错。响应头 Date 域指明了响应的时间，Content-type 域指明了信息体的编码方式，Content-length 指明了信息体的长度等；最后是响应消息的信息体。

4.1.2　内建对象

1. 重要的 5 个对象

request、response 和 session 对象是 JSP 内建对象中较重要的三个对象。它们提供了 B/S 模式服务器和客户端“请求/响应”模式的通信控制。request 对象与 HttpServletRequest

类关联，是 javax.servlet.ServletRequest 的一个子类，控制客户浏览器的请求，用 request 对象可以获得客户端提交的数据；response 对象与 HttpServletResponse 类关联，对客户浏览器进行响应，可用来向客户端输入数据；session 对象与 HttpSession 类关联，负责保存这个会话期间需要使用的数据，它扩展了 HTTP 的功能。

application 对象与 ServletContext 类关联，该对象在服务器第一次被访问时创建，直到服务器关闭服务为止。所以 application 可以处理一些全局的对象和数据。out 对象是 PrintWriter 类的一个实例，用来向客户端输出数据。config 对象是 ServletConfig 类的一个对象，是 JSP 配置处理程序的句柄，在 JSP 页面范围内有效。pageContext 对象用来管理属于 JSP 中特殊可见部分中已命名的对象。

2. 内建对象的 4 种作用范围

在详细介绍这些对象之前，需要对它们的作用范围有一些了解。所谓内建对象的作用范围(Scope)是指内建对象实例在多长时间和多大范围内可以被有效地访问，也就是它的生存时间和范围。在 JSP 中为这些对象定义了 4 种作用范围，即 Application Scope、Session Scope、Page Scope 和 Request Scope。

Application Scope 作用范围是指从服务器运行到服务器关闭之间的时间范围，Application 对象的作用范围是 Application Scope，任何页面或对象都可以访问 Application 对象。Session Scope 作用范围是指客户端和服务器会话开始到会话结束之间的时间；session 对象的作用范围就是 Session Scope，每个 Session 对象的生存时间依据具体情况而定，在同一个 Session 实例中的页面或对象都可以访问该 Session。Request Scope 作用范围是指一个“请求”到“响应”完成之间的时间。例如从一个页面向另一个页面提出请求到请求完成就是一个 Request Scope；request 对象的作用范围就是 Request Scope，它允许在不同页面之间传递数据。Page Scope 是最小的作用范围，只在当前页面时间内有效，例如 page 对象的作用范围就是 Page Scope。

内建对象的 4 个作用范围还要结合具体的内建对象来讲解，这里只是一个简单的介绍。

4.2　out 对象

out 对象是一个输出流，它实现了 javax.servlet.JspWriter 接口，用来向客户端输出数据。out 对象可以调用很多方法，用于向客户端输出各种数据。out 对象常用的方法如下。

- out.print()：输出各种类型数据。例如：out.print(boolean)，out.print(double)，out.print(String)，out.print(char)，out.print(float)，out.print(long)等。
- out.println()：输出各种类型数据并换行。例如 out.println(boolean)，out.println(double)，out.println(String)，out.println(char)，out.println(float)，out.println(long)等。
- out.newLine()：输出一个换行符。
- out.flush()：输出缓冲区中的内容。
- out.close()：关闭输出流。

这里强调 out.print()方法和 out.println()方法的区别，out.println()在输出数据的最后自动加入一个换行，而 out.print()只输出数据不加入换行。如果使用 out.print()输出数据，可使用 out.print("<br>")来控制换行。

下面的例子使用 out 对象向客户端发送数据。本章使用虚拟目录 ch4，ch4 映射的物理目录为 E:\programJsp\ch4。在 Tomcat 7.0 服务器上建立虚拟目录的具体方法，参见第 1 章 1.3 节的有关内容。

【例 4.1】 out 对象输出示例(其代码参见 ch4_1.jsp)，页面效果如图 4.1 所示。

图 4.1　ch4_1.jsp 页面效果

ch4_1.jsp 文件内容：

```
<%@ page contentType="text/html;charset=GB2312" %>
<%@ page import="java.util.*" %>
<%! public String getWeek(int n){
    String strweek[]=
     {"星期日","星期一","星期二","星期三","星期四","星期五","星期六"};
    return strweek[n];
  }
%>
<html>
<body  bgcolor="#99CCCC">
<!--程序加载时间：
   <%= (new java.util.Date()).toLocaleString()%>
   输出型注释
-->
<%--使用 out 对象输出表格和各种数据--%>
<%  /*创建一个世界杯对象，相对于 1900 年
        设置世界杯揭幕战日历对象*/
   Calendar calendar1=Calendar.getInstance();
    calendar1.setTime(new Date(114,5,13,4,0));
    String strYear=String.valueOf(calendar1.get(Calendar.YEAR)),
          strMonth=String.valueOf(calendar1.get(Calendar.MONTH)+1),
           strDay=String.valueOf(calendar1.get(Calendar.DAY_OF_MONTH)),
           strWeek=getWeek(calendar1.get(Calendar.DAY_OF_WEEK)-1);
   int hour=calendar1.get(Calendar.HOUR_OF_DAY),
          minute=calendar1.get(Calendar.MINUTE),
           second=calendar1.get(Calendar.SECOND);
   out.println("<table border=\"2\" align=\"center\">");
    out.print("<tr><td>2014 巴西世界杯揭幕战北京时间</td>");
    out.print("<td>"+strYear+"年</td>");
    out.print("<td>"+strMonth+"月</td>");
```

```
        out.print("<td>"+strDay+"日</td>");
        out.print("<td>"+strWeek+"</td>");
        out.print("<td>"+hour+"时</td>");
        out.print("<td>"+minute+"分</td>");
        out.print("<td>"+second+"秒</td>");
        out.println("</tr>");
        out.println("</table>");
%>
</body>
</html>
```

该例创建了一个 Calendar 日历对象，使用 Date 类为其赋值，调用日历对象的 get()方法获得日历的年、月、日和时间部分值。使用 out 对象的 print()方法输出了一系列的表格元素，并将日历对象中的年月日输出在表格中。out 对象不仅可以输出表格元素，还可以输出任意的 HTML 元素、JavaScript 脚本代码和数据库中的数据信息。使用 out 对象输出时，要注意 JSP 的特殊字符和 Java 语言中的转义字符等。

4.3 request 对象

如前所述，HTTP 是客户服务器之间一种请求(request)与响应(reponse)的通信协议。在 JSP 中，内建 request 对象与 HttpServletRequest 类关联，它使用 HTTP 封装了用户提交的请求信息，用户可以通过 request 对象的方法获取用户浏览器所提交请求中的各项参数和选项。

4.3.1 获取客户信息

在 Web 程序设计中，用户一般是通过 HTML 表单提交信息的。表单的一般格式是：

```
<form  method=get|post  action="目的页面或 Servlet"  name="表单名字" >
    提交手段部分
</form>
```

其中：

- <form>是表单标记。
- method 属性：表单数据的提交方式，取值为 get 或 post。get 方式提交的表单数据会在浏览器的地址栏中看到，而 post 方式提交的数据在地址栏看不到。
- action 属性：表单数据提交的目的 JSP 页面或者 Servlet 对象。
- name 属性：表单的名称，可以在 JavaScript 代码中使用它。
- 提交手段是指文本框、列表等表单元素。

request 对象可以使用 getParameter(String s)方法获得该表单提交的信息。例如：

```
<form  method=post  action="directpage.jsp" name="form1" >
    <input type="text" name="examineeName" value="yang" >
    <input type="submit" value="提交" name="submit1">
</form>
```

该表单使用了 post 方法提交数据，表单的名称为“form1”，提交信息的手段是 text 文本框，单击页面上的“提交”按钮后，向服务器的 directpage.jsp 页面提交数据。用户在 directpage.jsp 页面可以使用 request.getParameter("examineeName")方法获得表单提交的数据。

下面通过一个例子演示表单提交数据，目的页面使用 request 的 getParameter(String s)方法获得数据。

【例 4.2】获得简单表单数据(其代码参见 ch4_2.jsp 和 ch4_2_directpage.jsp)，该例中有两个页面，一个页面 ch4_2.jsp 负责提供数据输入界面，另一个页面 directpage.jsp 负责接受 ch4_2.jsp 页面提交的两个数据并按数值大小输出。页面效果如图 4.2、图 4.3 所示。

图 4.2　ch4_2.jsp 使用表单提交数据

图 4.3　ch4_2_directpage.jsp 页面返回处理结果

ch4_2.jsp 文件内容：

```
<%@ page contentType="text/html; charset=gb2312" language="java"%>
<html>
<head><title>【例 4.2】获取简单表单信息</title></head>
<body  bgcolor="#99CCCC">
<font size="3">
<form method="post" action="directpage.jsp" name="form1" >
   <p>输入第一个数：
   <input type="text" name="ix" value="8" size="6">
      输入第二个数：
   <input type="text" name="iy" value="6" size="6">
   <input type="submit" name="submit1" value="提交" size="4">
</form>
</font>
</body>
</html>
```

ch4_2_directpage.jsp 文件内容：

```
<%@ page contentType="text/html; charset=gb2312" language="java"%>
<html>
```

```
<head><title>【例 4.2】获取简单表单信息</title></head>
<body  bgcolor="#99CCCC">
<font size="3">
<%
   String strNumx=request.getParameter("ix");
   String strNumy=request.getParameter("iy");
   try{
     double ix=Double.parseDouble(strNumx);
      double iy=Double.parseDouble(strNumy);
      if(ix>iy){
      out.print("<p>您输入的两个数中大数是:");
      out.print((int)ix);
      out.print("  小数是: ");
      out.print((int)iy);
      }
      else{
      out.print("<p>您输入的两个数中大数是:");
      out.print((int)iy);
      out.print("  小数是: ");
      out.print((int)ix);
      }
    }
  catch(NumberFormatException ee){
     out.println("<br>请输入数字字符");
  }
%>
</font>
<body>
</body>
</html>
```

提示：directpage.jsp 页面使用 request 的 getParameter 方法获得 ch4_2.jsp 页面提交的数据“ix”和“iy”，并将它们放在字符串 strNumx 和 strNumy 中；程序片中使用 Double 的 parseDouble 方法将字符串转化为 double 类型数据，然后对两个数进行比较，按要求格式输出。

下面的例子在同一个页面中接受表单提交的数据。

【例 4.3】获取当前页面提交信息示例(其代码参见 ch4_3.jsp)，页面效果如图 4.4 所示。

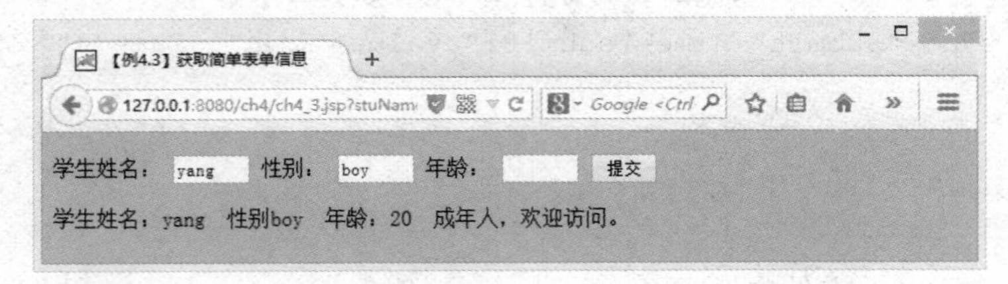

图 4.4　获取当前表单信息

ch4_3.jsp 文件中的代码如下：

```
<%@ page contentType="text/html; charset=gb2312" language="java"%>
<html>
<head><title>【例 4.3】获取简单表单信息</title></head>
<body  bgcolor="#99CCCC">
<font size="3">
<form method="get" action="" name="form1" >
   <p>学生姓名：
   <input type="text" name="stuName" value="yang" size="6">
     性别：
   <input type="text" name="stuSex" value="boy" size="6">
     年龄：
   <input type="text" name="stuAge"  size="6">
   <input type="submit" name="submit1" value="提交" size="4">
</form>
<%
   String  strName=request.getParameter("stuName"),
           strSex=request.getParameter("stuSex"),
            strAge=request.getParameter("stuAge");
    double  dAge=0;
    if(strAge==null){
      strAge="";
    }
    try{  dAge=Double.parseDouble(strAge);
    if(dAge>=18){
    out.print("<p>学生姓名："+strName+"  性别"+strSex);
    out.print("  年龄："+strAge+"  成年人，欢迎访问。");
    }
    else{
    out.print("<p>学生姓名："+strName+"  性别"+strSex);
    out.print("  年龄："+strAge+"  未成年人，谢绝访问！
");
    }
    }
    catch(NumberFormatException ee){
          out.print("在年龄字段请输入数字");
    }
%>
</font>
<body>
</body>
</html>
```

使用 request 对象获取当前数据时要注意空对象的问题，如果使用了空对象页面在编译时会出现空对象错误 java.lang.NullPointerException。注释例子中的下面语句：

```
/*if (strAge= =null){
   strAge="";}
```

```
*/
```

然后再使用浏览器访问 ch4_3.jsp 页面，就会出现如下编译错误：

```
org.apache.jasper.JasperException: An exception occurred processing JSP
page /ch4_3.jsp at line 23
20: /*if(strAge==null){
21:    strAge="";
22: }*/
23: try{  dAge=Double.parseDouble(strAge);
24:        if(dAge>=18){
25:         out.print("<br><br><p>学生姓名: "+strName+"  性别
"+strSex);
26:         out.print("  年龄: "+strAge+"  成年人，欢迎访
问。");
root cause
java.lang.NullPointerException
```

原因是用户第一次请求该页面时，并没有为年龄提交数据，strAge 对象还没有创建，语句 Double.parseDouble(starAge)就使用了空对象，编译器就会报 NullPointerException 错误。实际编程时为了避免使用空对象，一般都使用该例示范的技巧来处理类似问题。

注意：在例 4.3 中，如果用户在学生姓名文本框中输入汉字提交后，姓名显示会出现乱码，关于汉字处理的问题，下一节将具体介绍。

4.3.2 处理汉字

request 对象在获取客户提交的汉字字符时，会出现乱码问题。如例 4.3，在学生姓名字段输入汉字“王嗥”并提交，名字显示的信息为乱码。解决 request 对象获取提交汉字乱码问题，可以采用两种方法：一种方法是将 request 对象获取的字符串用 ISO-8859-1 进行编码，并将编码放到一个字节数组中，然后再用这个字节数组创建 String 对象即可，具体代码如下所示：

```
String strName=request.getParameter("stuName");
byte b[]=strName.getBytes("ISO-8859-1");
strName=new String(b);
```

另一种方法是将页面的 contentType 属性指定为下面的值：

```
<%@ page contentType="text/html;Charset=GB2312" %>
```

也就是将 Charset 属性的首字母大写。

用户可以将汉字特殊处理代码编写成一个函数，保存在一个文本文件中，需要的时候将其包含到用户的 JSP 文件中即可。

下面通过一个例子演示汉字的处理。

【例 4.4】显示汉字(其代码参见 ch4_4.jps 和 ch4_4_show.jsp)，ch4_4.jsp 是一个表单页面，负责输入注册名称和密码的汉字信息，并提交给页面 ch4_4_show.jsp。

ch4_4_show.jsp 负责用户提交信息的显示。汉字处理的方法保存在单独的文件 chString.txt 中。页面效果如图 4.5 和图 4.6 所示。

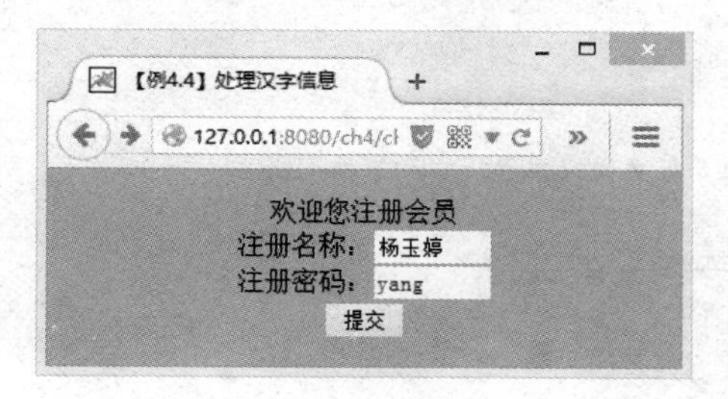

图 4.5　汉字信息表单

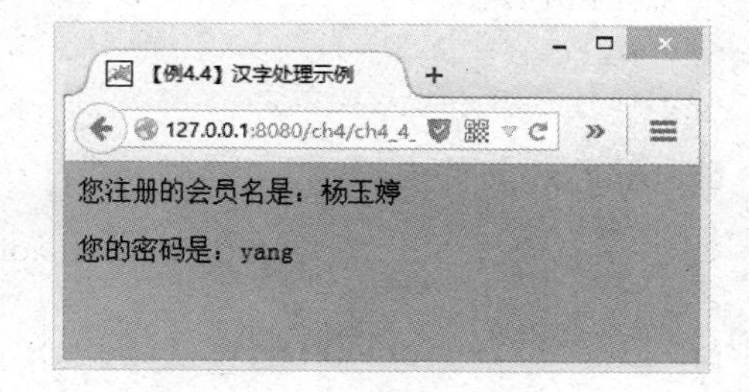

图 4.6　显示汉字

ch4_4.jsp 文件中的代码如下：

```
<%@ page contentType="text/html;charset=GB2312" %>
<html>
<head>
<title>【例 4.4】处理汉字信息</title>
</head>
<body bgcolor="cyan">
<form method="get" action="ch4_4_show.jsp" name="form1">
   <p align="center">欢迎您注册会员
   <br>注册名称：<input type="text" name="strName" size="8" >
   <br>注册密码：<input type="text" name="strPassword" size="8">
   <br> <input type="submit" name="submit" value="提交">
</form>
</body>
</html>
```

ch4_4_show.jsp 文件内容：

```
<%@ page contentType="text/html;charset=GB2312" %>
<%@ include file="chString.txt" %>
<html>
<head>
<title>【例 4.4】汉字处理示例</title>
</head>
<body bgcolor="cyan">
<% String strName=request.getParameter("strName");
   String strPassword=request.getParameter("strPassword");
   if (strName==null)
   {  strName="";
   }
   if(strPassword==null)
   {  strPassword="";
   }
   strName=chString(strName);
   strPassword=chString(strPassword);
```

```
%>
<p>您注册的会员名是：<%= strName %>
<p>您的密码是：<%= strPassword %>
</body>
</html>
```

chString.txt 文件内容：

```
<%@ page contentType="text/html;charset=GB2312" %>
<%!
public String chString(String s)
{
   String str=s;
   try
   {
     byte b[]=str.getBytes("ISO-8859-1");
     str=new String(b);
     return str;
   }
   catch(Exception e)
   {
     return str;
   }
}
%>
```

4.3.3 处理表单子标记

表单是客户提交数据的重要手段，表单一般格式如下：

```
<form  method="post|get" name="表单名称" action="目的页面或 Servlet">
   提交手段部分
</form>
```

其中提交手段是表单子标记定义的一些文本框、列表、文本区、单选按钮或复选框等页面元素，这些元素都是以图形界面的形式提供数据输入接口，方便用户输入数据。前面已经对表单的一般格式和简单文本框表单子标记<input>做了一些介绍，常用的表单子标记还有：

```
<input …>
<select …></select>
<option …></option>
<textarea …></textarea>
```

接下来对这些表单子标记做详细的介绍。

1. <input>标记

在表单中<input>子标记非常重要，它能够将浏览器中的控件加载到页面中，为页面提供图形界面的数据输入方式以及表单的提交键等。<input>子标记的通用使用格式：

```
<input type="控件类型" name="控件名称" 属性="属性值">
```

其中：

- type 属性：指定控件的类型，取值为 text(文本框)、radio(单选按钮)、checkbox(复选框)、password(密码框)、hidden(隐藏字段)、submit(提交按钮)、reset(重置)等。
- name 属性：指定控件对象名称，通过该属性，使用 request.getParameter 方法获得用户提交的数据。

(1) 文本框。

当该标记的 type 属性指定为 text 时，控件显示为文本框。文本框的定义格式一般为：

```
<input type="text" name="文本框名称" value="默认值" size="文本框大小"
   algin="对齐方式" maxlength="可输入的最多字符数">
```

其中：

- size 属性：文本框的长度，单位为字符宽度，取值为数字。
- value 属性：文本框默认文本值。
- algin 属性：文本框在浏览器窗体中的对齐方式，可取值 top、left、middle 等值。
- maxlength 属性：文本框输入的最多字符数目。
- name 属性：文本框的名称。通过 request.getParameter 方法获得文本框的值，如果用户没有在文本框中输入任何值，getParameter 方法获得的字符串长度为 0，即空串""。

例如：

```
<input type="text" name="stuName" value="杨得力" size="10" align="middle"
maxlength="10">
```

(2) 单选按钮。

当该标记的 type 属性指定为 radio 值时，控件显示为单选按钮。单选按钮定义格式一般为：

```
<input type="radio" name="单选按钮名称" value="单选按钮值" align="对齐方式"
   checked="默认值">
```

其中：

- value 属性：指定单选按钮的值。
- align 属性：单选按钮在浏览器窗口的对齐方式。
- checked 属性：取值为非空字符串，则该单选按钮为默认选中按钮。
- name 属性：单选按钮的名称。名称相同的多个单选按钮为一组，同组当中的按钮只能有一个被选中。request.getParameter 方法获得的值是选中按钮的 value 属性的值。

例如：

```
<form method="post" action="" name="">
   <input type="radio" name="sex" value="女" align="top" >
```

```
    <input type="radio" name="sex" value="男" align="top" checked="男">
</form>
```

(3) 复选框。

当该标记的 type 属性取值为 checkbox 时，控件显示为复选框。复选框一般定义格式为：

```
<input type="checkbox" name="复选框名称" value="复选框值" align="对齐方式"
    checked="默认值">
```

其中：

- value 属性、align 属性、checked 属性与单选按钮含义相同。
- name 属性：复选框的名称。名称相同的多个复选框为一组，同组中的按钮可以多选。可以通过 request.getParameterValues 方法获得多个被选中的 value 属性的值。

例如：

```
<form method="post" action="" name="">
    <input type="checkbox" name="course" value="高数" align="top" >
    <input type="checkbox" name="course" value="英语" align="top"
     checked="英语">
</form>
```

(4) 密码框。

该标记的 type 属性取值为 password 时，控件显示为密码框。密码框只能输入字母、数字和一些特殊字符等，用户输入字符后输入的信息用“*”回显，起到保密的作用。密码框定义格式类参考文本框，这里不再赘述。

(5) 隐藏框 hidden。

该标记的 type 属性取值为 hidden 时，控件不在页面中显示，也就是没有可视的输入界面。它的属性类似于文本框，给定的默认数据直接提交到服务器。一般使用格式如下：

```
<input type="hidden" name="hinfo" value="admin user">
```

其中的属性与文本框类似，可以调用 request.getParameter 方法，通过控件的 name 属性指定的名字得到提交的信息。

(6) 提交按钮。

当该标记的 type 属性取值为 submit 时，控件显示为“提交”按钮。一个表单必须有一个提交按钮，才能将表单的数据提交到服务器。提交按钮的定义格式为：

```
<input type="submit" name="ok"  value="提交" size="12">
```

单击提交按钮后，request 对象就可以获得表单提交的数据信息，包括提交按钮的 value 属性的值。

(7) 重置按钮。

当该标记的 type 属性取值为 reset 时，控件显示为“重置”按钮。单击重置按钮清空表单中数据，以便重新输入。

下面通过一个例子，来演示<input>子标记的使用方法。例子中的 ch4_5.jsp 页面使用了<input>的各种类型控件实例，负责提交数据。ch5_6_show.jsp 页面负责提交数据的显示。以网上报名系统页面为例演示。

【例 4.5】 <input>子标记的使用示例。页面显示效果如图 4.7 和图 4.8 所示。

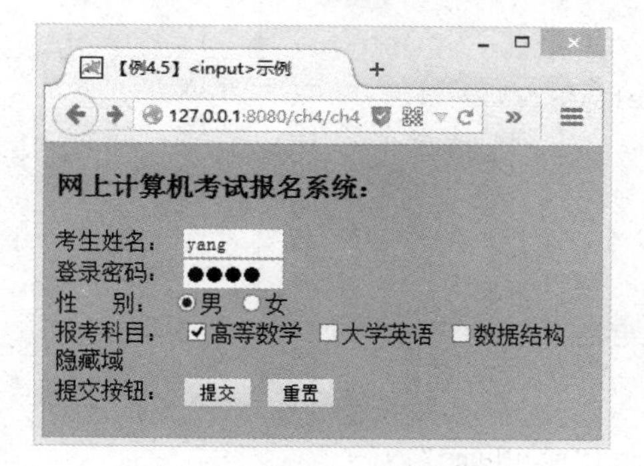

图 4.7　表单提交数据

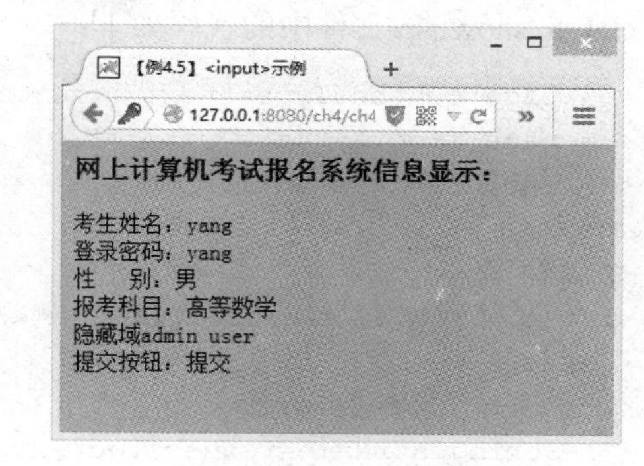

图 4.8　显示提交的数据

ch4_5.jsp 文件的代码如下：

```
<%@ page contentType="text/html;charset=GB2312" %>
<%@ page import="java.util.*"%>
<html>
<head>
<title>【例 4.5】&lt;input&gt;示例</title>
</head>
<body bgcolor="cyan">
<form method="post" action="ch4_5_show.jsp" name="form1">
    <h3>网上计算机考试报名系统：</h3>
    <p>考生姓名：
    <input type="text" name="stuName" value="yang" size="8"
align="left">
    <br>登录密码：
    <input type="password" name="stuPassword" value="yang"
        size="8" align="left">
    <br>性   别：
    <input type="radio" name="stuSex" value="男" checked="男"
        align="left">男
    <input type="radio" name="stuSex" value="女" align="left">女
    <br>报考科目：
    <input type="checkbox" name="course" value="高等数学"
         checked="高等数学" >高等数学
    <input type="checkbox" name="course" value="大学英语" >大学英语
    <input type="checkbox" name="course" value="数据结构" >数据结构
    <br>隐藏域
    <input type="hidden" name="hinfo"  value="admin user" >
    <br>提交按钮：
    <input type="submit"  name="submit" value="提交" size="6">
    <input type="reset" name="reset" value="重置" size="6">
```

```
</form>
</body>
</html>
```

ch4_5_show.jsp 文件中的代码如下：

```
<%@ page contentType="text/html;charset=GB2312" %>
<%@ page import="java.util.*"%>
<%@ include file="chString.txt" %>
<html>
<head>
<title>【例 4.5】&lt;input&gt;示例</title>
</head>
<body bgcolor="cyan">
<%  String strName=request.getParameter("stuName");
   String strPassword=request.getParameter("stuPassword");
    String strSex=request.getParameter("stuSex");
    String strCourse[]=request.getParameterValues("course");
    String strHinfo=request.getParameter("hinfo");
    String strSubmit=request.getParameter("submit");
    strName=chString(strName);
    strSex=chString(strSex);
    strHinfo=chString(strHinfo);
    strSubmit=chString(strSubmit);
     out.print("<h3>网上计算机考试报名系统信息显示：</h3>");
     out.print("<p>考生姓名：");
     out.print(strName);
     out.print("<br>登录密码：");
     out.print(strPassword);
     out.print("<br>性   别：");
     out.print(strSex);
     out.print("<br>报考科目：");
     if(strCourse==null)
        out.print("没有选课");
      else
      {
         for(int k=0;k<strCourse.length;k++){
             out.print(chString(strCourse[k])+"  ");
          }
      }
      out.print("<br>隐藏域");
      out.print(strHinfo);
      out.print("<br>提交按钮：");
      out.print(strSubmit+" ");
%>
</body>
</html>
```

ch4_5_show.jsp 文件中包含的 chString.txt 文件代码参见例 4.4。

2. <select>和<option>

<select>和<option>子标记用来定义下拉列表框和滚动列表框控件，为用户提供选择输入的图形界面。<select>作为<form>的子标记，<option>又作为<select>的子标记，定义下拉列表框的格式如下：

```
<form method="post|get" action="目的页面" name="表单名称">
    <select name="下拉列表名称">
        <option value="选项值 1" selected="true">选项值 1</option>
        <option value="选项值 2">选项值 2</option>
            ...
        <option value="选项值 n">选项值 n</option>
    </select>
</form>
```

定义滚动列表框的使用格式如下：

```
<form method="post|get" action="目的页面" name="表单名称">
    <select name="滚动列表名称"  size="6" >
        <option value="选项值 1" selected="true">选项值 1</option>
        <option value="选项值 2">选项值 2</option>
            ...
        <option value="选项值 n">选项值 n</option>
    </select>
</form>
```

selected 属性值字符串不为空，则该选项为默认选项。相对于下拉列表框来讲，滚动列表框的定义格式只是增加了 size 属性，size 属性值指定了滚动列表可见选项的行数。下拉列表和滚动列表的选项值为选中 option 的值。调用 request 的 getParameter 方法获得选中项的值。

3. <textarea>标记

<textarea>子标记用来定义一个多行文本区。多行文本区不同于文本框，它可以输入或显示多行文本。一般用于备注型等数据的输入。定义多行文本区的格式如下：

```
<form method="post|get" action="目的页面" name="表单名称">
    <textarea name="文本区名称" rows="文本可见行数" cols="文本可见列数">
    </textarea>
</form>
```

下面通过一个例子来介绍<select>和<textarea>标记的使用方法。例子是网上计算机报名系统的报名界面，滚动列表选择报名考试的类别，下拉列表选择学历。

【例 4.6】<select>和<textarea>子标记示例(其代码参见 ch4_6.jsp 和 ch4_6_show.jsp)，页面 ch4_6.jsp 提交数据，ch4_6_show.jsp 页面负责显示数据，页面效果如图 4.9 和图 4.10 所示。

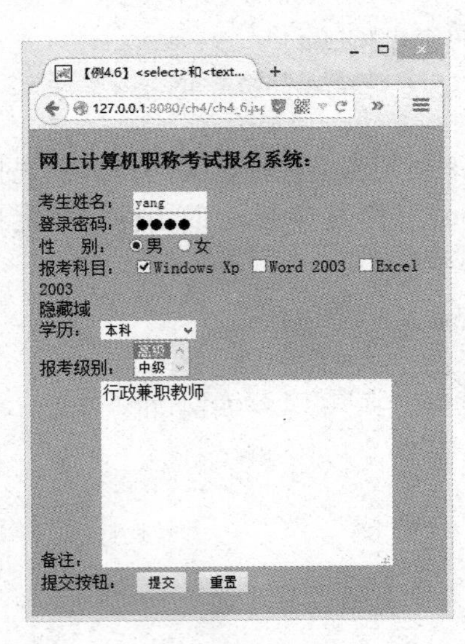

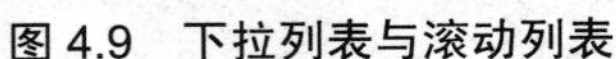

图 4.9　下拉列表与滚动列表

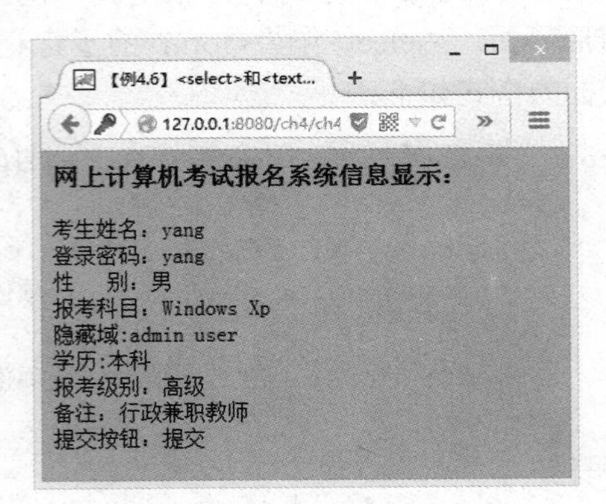

图 4.10　列表数据显示

ch4_6.jsp 文件中的代码如下：

```
<%@ page contentType="text/html;charset=GB2312" %>
<%@ page import="java.util.*"%>
<html>
<head>
<title>【例 4.6】&lt;select&gt;和&lt;textarea&gt;示例</title>
</head>
<body bgcolor="cyan">
<form method="post" action="ch4_6_show.jsp" name="form1">
      <h3>网上计算机职称考试报名系统：</h3>
      <p>考生姓名：
      <input type="text" name="stuName" value="yang" size="8"
align="left">
      <br>登录密码：
      <input type="password" name="stuPassword" value="yang" size="8"
            align="left">
      <br>性   别：
      <input type="radio" name="stuSex" value="男" checked="男"
            align="left">男
      <input type="radio" name="stuSex" value="女" align="left">女
      <br>报考科目：
      <input type="checkbox" name="course" value="Windows Xp"
            checked="true" >Windows Xp
      <input type="checkbox" name="course" value="Word 2003" >Word 2003
      <input type="checkbox" name="course" value="Excel 2003" >Excel 2003
      <br>隐藏域
      <input type="hidden" name="hinfo"  value="admin user" >
      <br>学历：
      <select name="level"  >
```

```
            <option value="专科">专科</option>
            <option value="本科" selected="true">本科</option>
            <option value="硕士研究生">硕士研究生</option>
            <option value="博士研究生">博士研究生</option>
        </select>
        <br>报考级别:
        <select name="exmaclass" size="2">
            <option value="高级" selected="true">高级</option>
            <option value="中级">中级</option>
        </select>
        <br>备注:
        <textarea name="memo" rows="8" cols="30">在这里输入您报考的备注信息
        </textarea>
        <br>提交按钮:
        <input type="submit" name="submit" value="提交" size="6">
        <input type="reset" name="reset" value="重置" size="6">
</form>
</body>
</html>
```

ch4_6_show.jsp 文件中的代码如下：

```
<%@ page contentType="text/html;charset=GB2312" %>
<%@ page import="java.util.*"%>
<%@ include file="chString.txt" %>
<html>
<head>
<title>【例 4.6】&lt;select&gt;和&lt;textaera&gt;示例</title>
</head>
<body bgcolor="cyan">
<%  String strName=request.getParameter("stuName");
    String strPassword=request.getParameter("stuPassword");
    String strSex=request.getParameter("stuSex");
    String strCourse[]=request.getParameterValues("course");
    String strHinfo=request.getParameter("hinfo");
    String strSubmit=request.getParameter("submit");
    String strLevel=request.getParameter("level");
    String strExamclass=request.getParameter("exmaclass");
    String strMemo=request.getParameter("memo");
    strName=chString(strName);
    strSex=chString(strSex);
    strHinfo=chString(strHinfo);
    strSubmit=chString(strSubmit);
    strLevel=chString(strLevel);
    strExamclass=chString(strExamclass);
    strMemo=chString(strMemo);
     out.print("<h3>网上计算机考试报名系统信息显示：</h3>");
     out.print("<p>考生姓名：");
```

```
        out.print(strName);
        out.print("<br>登录密码: ");
        out.print(strPassword);
        out.print("<br>性   别: ");
        out.print(strSex);
        out.print("<br>报考科目: ");
        if(strCourse==null)
           out.print("没有选课");
        else
        {
            for(int k=0;k<strCourse.length;k++){
                out.print(chString(strCourse[k])+"  ");
            }
        }
        out.print("<br>隐藏域:");
        out.print(strHinfo);
        out.print("<br>学历:");
        out.print(strLevel);
        out.print("<br>报考级别: ");
        out.print(strExamclass);
        out.print("<br>备注: ");
        out.print(strMemo);
        out.print("<br>提交按钮: ");
        out.print(strSubmit+" ");
%>
</body>
</html>
```

4.3.4 表单验证

表单验证是指在客户端的浏览器对用户输入的信息进行合法性的验证，如用户名或密码是否为空，电子邮件是否含有“@”和“.”等字符，身份证号码是否数字或字母等。如果客户输入的信息不符合规则要求或不合法，浏览器则提示用户进行修改，而不会直接发送给服务器，这样使得客户端信息验证能够有效地减轻服务器的负担和网络通信的压力。表单(Form)及表单元素为表单验证提供了方法、事件和属性，用户使用 JavaScript 语言来编写验证的代码，也就是处理表单事件的函数或方法，表单事件和 JavaScript 代码在客户端浏览器的运行环境中触发和执行。本节简单介绍 JavaScript、表单编程和正则表达式，以及使用它们实现表单验证的实用技术。

1. JavaScript 简介

JavaScript 是一种脚本语言，结构简单，使用方便，其代码可以直接嵌入 HTML 代码中去，也可以单独保存到 JS 文件中，再在 HTML 文件中引用。JavaScript 代码可以直接在浏览器中执行，可以捕获事件，并对事件做出相应的处理。

用户可以直接将 JavaScript 代码嵌入 HTML 文件中，使用的一般格式如下:

```
<script language="JavaScript" type="text/javascript" >
   JavaScript 代码
</script>
```

用户也可以将 JavaScript 代码单独保存在扩展名为 js 的文件中，然后在 HTML 文件中使用下面的语句引入 JavaScript 代码。

```
<script type="text/javascript" src="xxx.js"
   language="JavaScript" charset="GB2312" ></script>
```

JavaScript 使用关键字 var 声明变量，变量名区分大小写，变量没有数据类型，变量的数据类型由所包含的数据类型决定，称为相应数据类型；数据类型之间可以自动转换；运算符和流程控制语句与 Java 语言相似。

JavaScript 语言是面向对象的语言，可以用面向对象的语言来编写程序。Date 对象对日期数据进行操作，提供了 get 和 set 方法。Math 对象是一个全局对象，可以完成各种类型的数学运算。Array 对象是一个数组对象，提供了对数组操作的方法。String 对象是字符串对象，String 的 match 方法从字符串中搜索出匹配 regExp 正则表达式的所有字串，search 方法从字符串中搜索匹配 regExp 正则表达式的第一个子字符串，返回其索引值。location 对象控制浏览器，history 对象控制页面的前进和后退。document 对象控制文档的 forms、images、links 属性，用 write 方法向文档写入数据。Function 对象是 JavaScript 的特色，它不同于其他语言，函数是作为对象来处理的，下面是定义函数的一种基本形式：

```
function myFunction(a,b){
   return a+b;
}
```

函数通过函数名来调用。

JavaScript 对象的详细内容请参考 JavaScript 手册。

2. 表单对象及表单域对象

在 JavaScript 中，document 对象对应 HTML 文档，form 对象对应 HTML 表单，此时每对<form></form>标记被解析为一个 form 对象，<form>标记中的属性、事件对应 form 对象中的属性和事件，这样使用 form 对象就可以很方便地访问表单中的数据，完成表单验证。

表 4.1 是 JavaScript 中表单对象的重要事件、方法和属性。获得一个 form 对象的方法如下：

```
var myform=document.forms["myForm"];
```

或者

```
var myform=document.forms[0];
```

表 4.1　form 对象的属性、方法、事件

名　称	类　型	描　述
name	属性	返回或设置表单名称
elements	属性	该表单中所有表单域的集合
length	属性	表单域的数量
submit	方法	相当于单击“提交”按钮
reset	方法	相当于单击“重置”按钮
onsubmit	事件	表单提交前触发
onreset	事件	表单重置前触发

表单元素(也称表单域)是指用于接收输入或操作的一些页面元素，例如文本框、按钮、单选按钮、复选框等，它们被包含在 HTML 文档中的<form>和</form>标记之间。在 JavaScript 中表单域标记，例如<input>等被解析为 form 对象的 element 对象。表单域的属性、方法和事件如表 4.2 所示。获得表单域的方法如下：

```
var element=theForm.elements[index];或
var element=theForm.elements["eleementName"];
```

表 4.2　表单域的通用属性、方法和事件

名　称	类　型	描　述
disabled	属性	创建只读表单域
name	属性	获取或设置表单域的名称
form	属性	获取表单域所在表单
value	属性	获取或设置表单域的值
focus	方法	让表单域获得焦点
blur	方法	让表单域失去焦点
onfocus 或 onblur	事件	获得或失去焦点时触发
onclick、onkeydown	事件	鼠标单击或按键按下触发
onmouseover、 onmouseout	事件	鼠标动作对应的事件
onchange	事件	表单域的值发生改变时发生

执行表单验证一般都发生在用户单击“提交”按钮后，数据提交服务器之前，这时如果用户数据的数据不符合验证规则，则取消提交，同时提示用户重新输入。例如 onValidate()函数是用于检验表单有效性的，返回值为 true 或 false。程序中有两种方法可以调用 onValidate()函数。

方法一，将 onValidate()函数绑定到提交按钮的单击事件，如果返回 false 则会终止单击，不会提交表单。格式如下：

```
<input type="submit" onclick="onValidate()"/>
```

方法二，将 onValidate()函数绑定到表单的提交事件，提交事件发生在提交之前，如

果返回 false，则不会向服务器提交。格式如下：

```
<form action="sompage.jsp" onsubmit="return onValidate()"/>
```

onValidate 函数一般完成文本域是否为空、下拉列表框是否为空、判断数据有效性等验证工作，判断数据有效性一般用正则表达式来实现。onValidate 函数代码如下：

```
function onValidate(){
    if(form1.stuName.value.length= =0){  //stuName 为文本框的名字
        alert("用户名不得为空！");
        return false;
    }
    if(form1.stuLevel.selectedIndex==0){ //stuLevel 为列表框名字
        alert("请选择报考级别");
        return false;
    }
    var str=form1.stuMail.value;
    if(!str.search(regExp){
        alert("请正确输入邮件地址！");
        return false;
    }
    else{
        return true;
    }
}
```

下面通过一个示例演示表单验证编程，该例是在例 4.6 基础上编写的，增加了 JavaScript 验证代码。JavaScript 代码单独保存在 check.js 文件中。

【例 4.7】表单验证示例(其代码参见 check.js 和 ch4_7.jsp)，姓名为空时的报告对话框如图 4.11 所示。

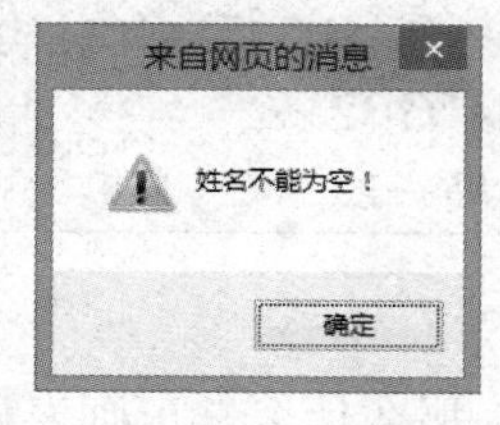

图 4.11　姓名为空时的警告对话框

check.js 文件代码如下：

```
function check(){
        var myForm=document.forms["form1"];//获得 form 对象
        var elementName=myForm.elements["stuName"];//获得表单域对象
        var elementPassword=myForm.elements["stuPassword"];//获得表单域对象
        if(elementName.value.length<=0){
            alert("姓名不能为空！");
            elementName.focus();//设置焦点
            return false;//返回 false，中断提交
```

```
        }
        if(elementPassword.value.length<=0){
            alert("密码不能为空！");
            elementPassword.focus();//设置焦点
            return false;//返回 false 中断提交
        }
}
```

ch4_7.jsp 文件代码如下：

```
<%@ page contentType="text/html;charset=GB2312" %>
<%@ page import="java.util.*"%>
<html>
<head>
<title>【例 4.7】&lt;select&gt;和&lt;textarea&gt;示例</title>
<script type="text/javascript" src="check.js"
   language="JavaScript" charset="GB2312" >
</script>
</head>
<body bgcolor="cyan">
<form method="post" action="ch4_6_show.jsp" name="form1"
          onSubmit="return check()">
     <h3>网上计算机职称考试报名系统：</h3>
     ...
     此处代码省略，请参见【例 4.6】
</form>
</body>
</html>
```

技巧：例子中，通过在 HTML 表单标记<from></form>中设置 onsubmit 属性值为"return check()"，将事件与处理函数绑定，通过<script type="text/javascript" src="check.js" language="JavaScript" charset="GB2312" >将 JS 文件引入 HTML 文档中，charset="GB2312"确保文件编码一致。

3. 正则表达式

前面已经多次提到正则表达式，那么什么是正则表达式？正则表达式就是用来描述模式匹配的规则，模式匹配是指字符串的匹配问题。模式则是指与元字符和基本字符描述格式匹配的字符串，可以用正则表达式来表示。

正则表达式由两种字符组成，一种是元字符，指“\ | ( ) { } - ^ $ * ? . +”这些字符，用来限定一定的格式；另一种是除元字符以外的其他字符称为基本字符。

在正则表达式里基本字符可以直接出现，元字符不能直接作为模式的一部分，必须转义，也就是在其前面加上转义字符“\”。元字符的含义如表 4.3 所示。正则表达式的预定义词如表 4.4 所示。一些常用的模式正则表达式如表 4.5 所示。

一些特殊控制字符的转义同 Java 语言的转义字符。例如\t，\r、\n 等。

表 4.3　元字符含义

元字符	描　述
.	表示任意一个除换行以外的字符，例如“b.b"表示 bab，bdb，bcb 等
|	表示或者的意思。例如“a|b”表示 a，或者 b
[]	结合起来表示某特定类型的字符。例如“[abc]”表示可以是 abc 中任意一个
{}	结合起来表示匹配的次数。例如“ab{2}”可匹配 abb。“ab{2,}”表示至少 2 次，可以是 abb，abbb，abbbb 等。例如“ab{2,4}”表示至少 2 次，最多 4 次，可以是 abb，abbb，abbbb
$	表示模式必须出现在目标串的结尾。例如“xy$”，表示以 xy 结尾的串
[^]	表示除了指定类型以外的字符。例如[^a-z]，表示除小写字母以外的任意一个字符
()	结合使用表示一个字模式
-	与[]结合使用，表示一段字符的范围。例如[a-z]，表示任意一个小写字母
?	表示 0 个或一个。例如“a?b”，表示 b，ab 等
*	表示 0 个或多个。例如“ab*”，表示 a，ab，abb，abbb 等
+	表示 1 个或多个。例如“ab+”，表示 ab，abb，abbb，abbbb 等
^	表示模式必须出现在目标串的开始。例如“^b”，表示 b，bb，baa 等
\f	换页匹配
\n	换行匹配
\r	匹配一个回车
\t	匹配制表符
\un	匹配 n，其中 n 是以四位十六进制表示的 Unicode 字符。例如\u00a9 匹配版权符号(©)
\xn	匹配 n，其中 n 是一个十六进制转义字码。例如\x41 匹配“A”

表 4.4　预定义词

元字符	描　述
\d	表示一个数字，同[0-9]
\D	表示一个非数字字符，同[^0-9]
\s	表示一个白字符，同[\t\n\x0b\f\r]
\S	表示一个非白字符，同[^\t\n\x0b\f\r]
\w	一个字符，可以是字母、数字或下划线，同[a-zA-Z_0-9]
\W	一个字符，不能是字母、数字或下划线，同[^a-zA-Z_0-9]

表 4.5　一些常用模式的正则表达式

正则表达式	模式含义
^\d+(\.\d+)*$	数字
^[0-9]*[1-9][0-9]*$	正整数
^(-?\d+)(\.\d+)?$	浮点数
[\u4e00-\u9fa5]	中文字符
[^\x00-\xff]	双字节字符包括中文

续表

正则表达式	模式含义
^[A-Za-z0-9]+$	由数字和 26 个英文字母组成的字符串
\n[\s|]*\r	空行
^[\w-](\.[\w-]+)*@[\w-]+(\.[\w-]+)+$	EMail 地址
\d{3}-\d{8}|\d{4}-\d{7}	国内电话号码，例如：010-75775588
\d{15}|\d{18}	身份证号

【例 4.8】正则表达式验证数据示例(其代码参见 ch4_8.jsp)。

ch4_8.jsp 文件代码如下：

```
<%@ page contentType="text/html;charset=GB2312" %>
<%@ page import="java.util.*"%>
<script language="javascript" type="text/javascript" >
<!--
  function check(){
    var strNum=document.forms[0].elements[0].value;
    var strID=document.forms[0].elements[1].value;
    var strEmail=document.forms[0].elements[2].value;
    var strCnstring=document.forms[0].elements[3].value;
    if(strNum.search("^\\d+(\\.\\d+)*$")!=0){
      alert("请输入一个数字！");
      document.forms[0].elements[0].focus();
      return false;
    }
    if(strID.search("\\d{15}|\\d{18}")!=0){
      alert("请输入一个正确的身份证号！15 位或 18 位");
      document.forms[0].elements[1].focus();
      return false;
    }
    if (strEmail.search("^[\\w-]+(\\.[\\w-]+)*@[\\w-]+(\\.[\\w-
]+)+$")!=0){
      alert("请输入一个正确的电子邮件地址！");
      document.forms[0].elements[2].focus();
      return false;
    }
    if(strCnstring.search("[\\u4e00-\\u9fa5]")!=0){
      alert("请输入中文字符！");
      document.forms[0].elements[3].focus();
      return false;
    }
  }
-->
</script>
<html>
<head>
```

```
<title>【例 4.8】正则表达式示例</title>
</head>
<body bgcolor="cyan">
<form  name="form1" method="post" action="" onSubmit="return check()">
  <br>输入一个数字：<input name="num" type="text">
  <br>输入身份证号：<input name="id" type="text">
  <br>输入邮件地址：<input name="email" type="text">
  <br>输入中文字符：<input name="cnstring" type="text">
  <br>提交和重置按钮：
  <input name="submit" value="提交" type="submit" >
  <input name="reset" value="重置" type="reset" >
</form>
</body>
</html>
```

技巧：学习正则表达式要通过例子来学习。首先通过例子将元字符的含义理解，掌握预定义词，然后阅读常用正则表达式的含义。多思考多实践就会很快掌握正则表达式。

4.3.5　常用方法举例

当浏览器访问一个页面时，会向服务器提交一个 HTTP 请求给 JSP 引擎。HTTP 请求包括一个请求行、多个请求头和信息体，请求消息结构请参见 4.1.1 节内容。通过表单提交的信息就封装在 HTTP 请求消息的信息体部分，用户使用 request 对象的 getParameter 方法可以得到通过表单提交的信息。除了 getParameter 方法外，request 对象还提供了许多访问请求消息各个域的方法。

request 对象常用方法和功能如下。

(1) getParameter(String s)：获取表单提交的信息。

(2) getProtocol()：获取客户向服务器提交信息所使用的协议。

(3) getServletPath()：获取用户请求的 JSP 页面文件的目录。

(4) getContentLength()：获取用户提交的整个信息的长度。

(5) getMethod()：获取用户提交信息的方式，例如 post、get 等。

(6) getHeader(String s)：获取 HTTP 消息中由参数 s 指定的头域名字的域值，一般 s 可取的头域名有：accept、referer、accept-language、content-type、accept-encoding、user-agent、host、content-length、connection、cookie 等，例如，s 取值为 host 将获取用户请求的服务器的主机名。

(7) getHeaderNames()：获取 HTTP 消息头域名字的一个枚举。

(8) getHeaders()：获取 HTTP 消息中指定头域名字的全部值的一个枚举。

(9) getRemoteAddr()：获取客户机的 IP 地址。

(10) getRemoteHost()：获取客户机的名称或 IP 地址。

(11) getServerName()：获取服务器名称。

(12) getServerPort()：获取服务器的服务端口号。

(13) getParameterNames()：获取客户提交的信息体部分中 name 参数值的一个枚举。

(14) getAttribute(String s)：获取由 s 指定的属性的值，不存在 s 属性则返回 null。

(15) setAttribute(String s, Object obj)：设置名字为 s 的 request 参数的值为 obj。

(16) getSession(boolean create)：返回和请求相关的 session 对象。

下面的例子示范了 request 对象的常用方法及其使用技巧。

【例 4.9】获取请求消息行、头和信息体的相关信息(其代码参见 ch4_9.jsp)，页面效果如图 4.12 所示。

ch4_9.jsp 页面中包含的 chString.txt 文件可参照例 4.4 中的内容。

ch4_9.jsp 文件中的代码如下：

```
<%@ page contentType="text/html;charset=GB2312" %>
<%@ page import="java.util.*"%>
<%@ include file="chString.txt" %>
<html><head><title>【例 4.9】request 对象常用方法示例</title></head>
<body bgcolor="cyan">
<form method="post" action="" name="form1">
     <p>输入学生姓名:
     <input type="text" name="stuName" value="yang" size="8">
     <p>输入学生密码:
     <input type="text" name="stuPassword" value="yang" size="8">
     <br>
     <input type="submit" name="submit" value="提交">
</form>
<% String strMethod=request.getMethod();
   String strProtocol=request.getProtocol();
   String strServerPath=request.getServletPath();
   String strClientIP=request.getRemoteAddr();
   String strServerName=request.getServerName();
   int iServerPort=request.getServerPort();
   out.print("<p><font color=red>□□</font>");
   out.print("请求行:");
   out.print(strMethod);
   out.print(strServerPath+"/");
   out.print(strProtocol);
   out.print("<p>客户端 ip 地址: ");
   out.print(strClientIP+"<br>");
   out.print("服务器名称: ");
   out.print(strServerName+"<br>");
   out.print("端口号: ");
   out.print(iServerPort);
   out.print("<p>下面显示的是头名字及其值，显示格式“□□头名:头值”: <br>");
   Enumeration enumHeaderNames=request.getHeaderNames();
   while(enumHeaderNames.hasMoreElements())
   {
      String strn=(String)enumHeaderNames.nextElement();
      String strv=request.getHeader(strn);
      out.print("<font color=red>□□</font>");
```

```
        out.println(strn+":"+strv);
      }
      Enumeration enumFormParaNames=request.getParameterNames();
      out.print("<p>显示表单所有参数及其值，显示格式“□□参数名:参数值”的格式。
   <br>");
      while(enumFormParaNames.hasMoreElements())
      {
         String strn=(String)enumFormParaNames.nextElement();
          String strv=chString(request.getParameter(strn));
          out.print("<font color=red>□□</font>");
          out.print(strn+":" +strv+"  ");
      }
   %>
   </body>
   </html>
```

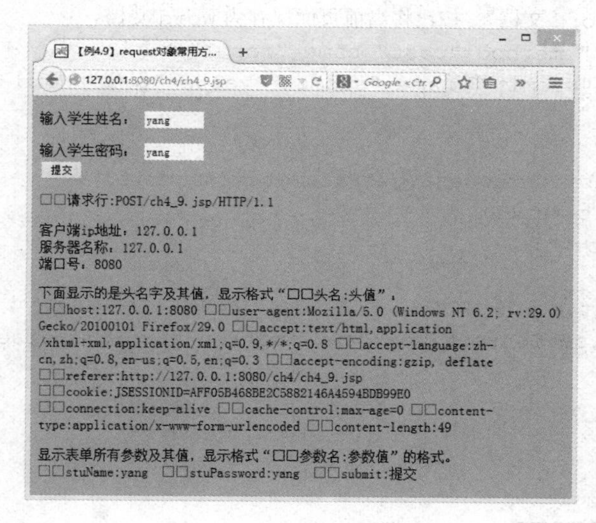

图 4.12　request 获取信息

4.4　response 对象

如前所述，HTTP“请求/响应”模型的请求消息与 request 对象对应，那么响应消息与哪个内建对象对应呢？在 JSP 中，内建 response 是 HttpServletResponse 类的一个对象，它与 request 对象对应，封装了 HTTP 响应消息，用来对客户的请求做出响应，向客户端发送数据。用户可以通过 response 对象的方法修改响应消息中的各项参数和选项，下面介绍 response 对象的几个典型应用。

4.4.1　修改 ContentType 属性

contentType 属性用来设置 JSP 页面的 MIME 类型和字符编码集，取值格式为"MIME 类型"或"MIME 类型;charset=字符编码集"。JSP 引擎根据 contentType 属性，对用户的请求做出响应。客户端收到响应后，根据 contentType 的值对信息做对应的处理。page 指令可以静态指定 contentType 的值，而 response 对象可以调用 setContentType(String s)方法动态修改 contentType 属性值，JSP 引擎会按照修改后的 MIME 类型来响应客户浏览器，参数 s 的取值可参考 3.4.1 节的内容。

下面的例子，通过单击页面的 submit 按钮，动态改变 contentType 的值，将当前页面保存为一个 Word 文档。

【例 4.10】改变 contentType 属性值示例(其代码参见 ch4_10.jsp)。

ch4_10.jsp 文件代码如下：

```
<%@ page contentType="text/html;charset=GB2312" %>
<html>
<head><title>【例 4.10】contentType 属性示例</title>
</head>
<body>
<p>单击“保存为 Word 文档” 按钮将当前页面保存为 Word 文档
<form  action="" method="post" name="form1" >
   <input type="submit"  name="submit" value="保存为 Word 文档" >
</form>
<%
   String strSubmit=request.getParameter("submit");
   if (strSubmit==null){
      strSubmit="";
   }
   if(!strSubmit.equals("")){
      response.setContentType("application/msword;charset=GB2312");
   }
%>
</body>
</html>
```

4.4.2 定时刷新页面

HTTP 响应消息主要由响应行、响应头和信息体组成，请参照 4.1.1 内容。response 对象的 addHeader(String head,String value)方法或者 setHeader(String head,String value)方法可以动态添加新的响应头和头值，如果这个头事先已经存在，则原来的头被覆盖。

下面是一个使用 response 对象的 setHeader 方法来添加一个响应头 refresh。如果为 refresh 属性指定了秒数，例如 3 秒，客户端的页面会每隔 3 秒刷新一次。

【例 4.11】为响应消息增加 refresh 头示例(其代码参见 ch4_11.jsp)。

ch4_11.jsp 文件内容如下：

```
<%@ page contentType="text/html;charset=GB2312" %>
<%@ page import="java.util.Date" %>
<html>
<head><title>【例 4.11】增加 refresh 头示例</title>
</head>
```

```
<body>
<%
   Date date=new Date();
   response.setHeader("refresh","3");
%>
<p>页面将每隔 3 秒钟刷新一次。
<p>现在的时间是：<%= date%>
</body>
</html>
```

4.4.3 重定向

当服务器响应客户端请求时，将客户端请求重新引导到另一页面，称之为重定向。例如，如果用户填写的表单信息不正确，就将客户请求重新引导到表单填写页面，继续填写数据。这个过程就是重定向。用户可以使用 response 对象的 sendRedirect(URL url)方法实现。

下面的例子有两个页面，一个页面 ch4_12.jsp 输入用户名，另一个页面 ch4_12_show.jsp 负责显示数据，如果没有输入用户名，则重新定向到 ch4_12.jsp 页面。

【例 4.12】重定向示例(其代码参见 ch4_12.jsp 和 ch4_12_show.jsp)。

ch4_12.jsp 文件代码如下：

```
<%@ page contentType="text/html;charset=GB2312" %>
<html>
<head><title>【例 4.12】页面重定向示例</title></head>
<body>
<form method="post" action="ch4_12_show.jsp" name="form1" >
  <br>请输入用户名：
  <input type="text" name="userName" size="10" >
  <input type="submit" name="submit" value="提交">
</form>
</body>
</html>
```

ch4_12_show.jsp 文件代码如下：

```
<%@ page contentType="text/html;charset=GB2312" %>
<html>
<head><title>【例 4.12】页面重定向示例</title></head>
<body>
<%
   String str=request.getParameter("userName");
   if(str==null){
      str="";
   }
   if(str.length()<=0)
      response.sendRedirect("ch4_12.jsp");
   else
      out.print("欢迎"+str+"您来到测试页面！");
%>
```

```
</body>
</html>
```

4.4.4 改变状态码

HTTP 响应消息的第一行称为响应状态行，响应行有 3 位数的一个响应状态码，参见 4.1.1 节内容。用户可以通过 response 对象的 setStatus(int n)方法来改变响应状态行的状态码。浏览器在得到响应后，根据状态码做出相应的处理。

下面的例子修改了响应的状态码。

【例 4.13】修改状态码示例(其代码参见 ch4_13.jsp)。

ch4_13.jsp 文件代码如下：

```
<%@ page contentType="text/html;charset=GB2312" %>
<html>
<head><title>【例 4.13】修改状态码</title></head>
<body>
<%
    response.setStatus(408);
     out.print("请求超时，页面不显示");
%>
</body>
</html>
```

浏览该网页时，浏览器会显示“该网页太忙，无法显示该页面”。

4.5 session 对象

HTTP 是无状态协议，也就是完成一个请求和响应后连接就关闭，服务器并不保存客户原来的连接信息。所以用户浏览器再次访问服务器时，服务器并不知道客户以前什么时间访问过什么。为了保持浏览器与服务器的连接状态，Tomcat 使用内建的 session 对象记录有关的连接信息。session 对象是 HttpSession 接口类的实例，由服务器负责创建和销毁。

4.5.1 对象的 id 与生命周期

当用户浏览器第一次访问服务器 Web 服务目录中的某个页面时，Tomcat 服务器自动创建一个 session 对象，并分配给 session 对象一个 String 类型的 ID 号，同时将这个 id 发送到客户端 cookie 中保存起来，这样就建立了一个服务器和客户浏览器之间的一个会话。如果服务器为客户创建的 session 对象没有销毁，客户再访问同一个“Web 服务”目录中的其他页面时，服务器会根据客户端保存的 id 查找属于该客户的 session 对象，不会为其创建新的 session 对象，这样 session 对象和客户就建立了一一对应关系，好像服务器和客户保持了持续的连接状态。

需要读者注意：同一个客户在同一个 Web 服务目录中的 session 对象是相同的，在不同的 Web 服务目录中的 session 对象是不相同的。有关 Web 服务目录的概念参见第 2 章 Web 应用基础。如果在页面中使用 page 指令设置了 session 属性等于 false，则服务器不为客户创建 session 对象。

session 对象的生命周期是指从客户 session 对象创建到销毁的整个过程，又称为会话周期，它取决于三个因素：客户是否关闭浏览器；session 对象是否调用 invalidate()方法使得 session 无效；session 对象是否达到了最长“发呆状态”时间。三个因素有一个发生，客户的 session 对象就会消失。所谓的最长发呆时间是指服务器允许客户对某个 Web 服务目录发出两次请求之间的时间间隔。客户发呆时间总是从它最后一次访问计时。

session 对象使用下列方法获得 id 及生存时间的信息。

- public String getId()：获取 session 的 ID。
- public long getCreateTime()：获得 session 创建时间，单位毫秒。数值为自 1900 年 7 月 1 日午夜起至创建时相隔的毫秒数。
- public long getLastAccessTime()：获得最后一次被操作的时间，单位毫秒。数值同创建时间。
- public long getMaxInactiveInterval()：获得最长发呆时间，单位秒。
- public long setMaxInactiveInterval()：设置最长发呆时间，单位秒。
- public boolean isNew()：判断是否新建 session。
- public invalidate()：使 session 无效。

下面的例子中 ch4_14_1.jsp 和 ch4_14_2.jsp 页面都在 ch4 服务目录中，从一个页面跳转到另一页面，session 对象的 id 是相同的。关闭浏览器或等待 15 秒后，再重新打开或刷新该例中的网页，比较 session 对象的 id 是否相同。

【例 4.14】session 对象 id 示例(其代码参见 ch4_11_1.jsp 和 ch4_11_2.jsp)，页面效果如图 4.13 所示。

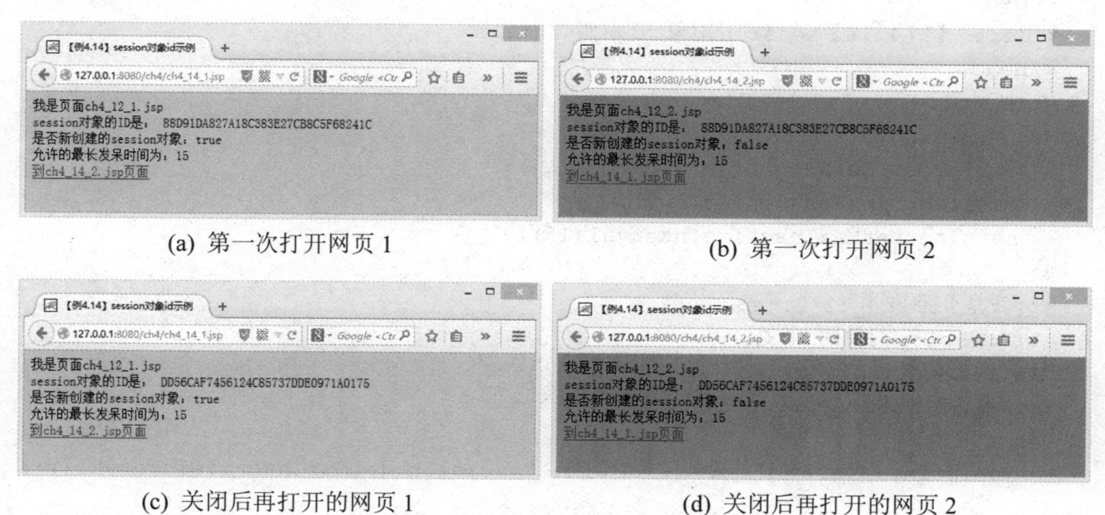

(a) 第一次打开网页 1　　(b) 第一次打开网页 2

(c) 关闭后再打开的网页 1　　(d) 关闭后再打开的网页 2

图 4.13　session 对象的 id 与生命周期

ch4_14_1.jsp 文件代码:

```
<%@ page contentType="text/html; charset=gb2312"  %>
<html>
<head>
<title>【例 4.14】session 对象 id 示例</title>
</head>
<body bgcolor="#55FFFF">
<%
   String strSessionID=session.getId();
   boolean boo=session.isNew();
   session.setMaxInactiveInterval(15);
%>
<p>我是页面 ch4_12_1.jsp
<br>session 对象的 ID 是:
<%= strSessionID %>
<br>是否新创建的 session 对象: <%= boo%>
<%
   out.print("<br>允许的最长发呆时间为: "+session.getMaxInactiveInterval());
%>
<br><a href="ch4_14_2.jsp" >到 ch4_14_2.jsp 页面</a>

</body>
</html>
```

ch4_14_2.jsp 文件代码:

```
<%@ page contentType="text/html; charset=gb2312"  %>
<html>
<head>
<title>【例 4.14】session 对象 id 示例</title>
</head>
<body bgcolor="#00BFFF">
<%
   String strSessionID=session.getId();
   boolean boo=session.isNew();
   session.setMaxInactiveInterval(15);
%>
<p>我是页面 ch4_12_2.jsp
<br>session 对象的 ID 是:
<%= strSessionID %>
<br>是否新创建的 session 对象: <%= boo%>
<%
   out.print("<br>允许的最长发呆时间为: "+session.getMaxInactiveInterval());
%>
<br><a href="ch4_14_1.jsp" >到 ch4_14_1.jsp 页面</a>
</body>
</html>
```

4.5.2　对象存储数据

Tomcat 服务器为每个不同的客户创建不同的 session 对象，用来保存用户会话期间需要保存的数据信息。session 对象保存在 Web 服务器中，可以记录用户的登录名称，或者需要保存的信息。用户可以通过 session 对象的下列方法存储和读取数据信息。

- public void setAttribute(String key,Object obj)：将 obj 对象添加到 session 对象中，并指定参数 key 为该对象的索引关键字。
- public Object getAttribute(String key)：获取索引关键字是 key 的对象，得到对象后应强制转化为原来的类型。
- public Enumeration getAttributeNames()：获得关键字的一个枚举对象，使用 nextElements 方法遍历所有的关键字。
- public void removeAttribute(String name)：移除关键字是 name 的对象。

setAttribute(String key ,Object obj)和 getAttribute(String name)方法是 session 对象非常重要的两个方法。保存用户登录成功状态和用户名是 session 对象的一个典型应用，登录后用户访问其他页面无须再验证身份，或重新填写用户信息等。下面的例子记录了会话的用户名。ch4_15.jsp 页面输入用户名称和密码，判断登录成功与否。登录成功将登录用户名保存到 session 对象中，跳转到 ch4_15_show.jsp 页面显示登录信息。

【例 4.15】保存登录用户名示例(其代码参见 ch4_15.jsp 和 ch4_15_show.jsp)，页面效果如图 4.14 所示。

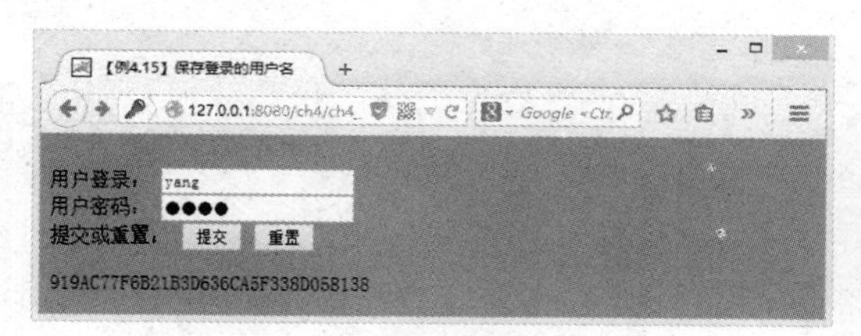

(a) 登录页面

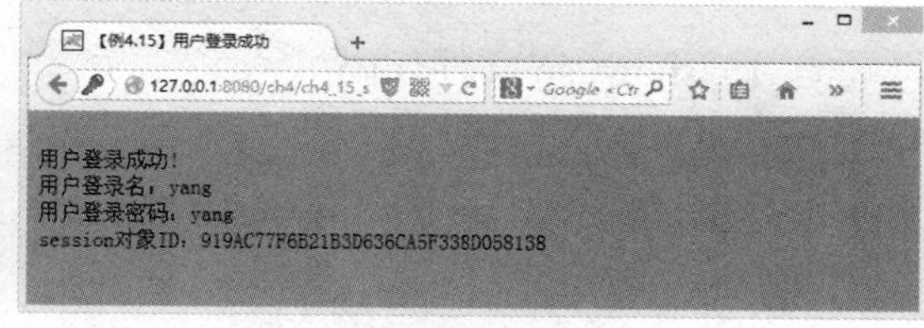

(b) 显示页面

图 4.14　保存用户登录名

ch4_15.jsp 文件代码：

```
<%@ page contentType="text/html;charset=GB2312"%>
<%@ include file="chString.txt" %>
<html>
<head>
<title>【例 4.15】保存登录的用户名</title>
</head>
<body bgcolor="#00BFFF">
<%
   String strUserName=request.getParameter("userName");
   String strPassword=request.getParameter("password");
   String strSessionID=session.getId();
   if (strUserName==null){
```

```
        strUserName="";
    }
    if(strPassword==null){
        strPassword="";
    }
    //登录成功后保存用户名，跳转到某页面
    if(strUserName!=""&&strPassword!=""){
      session.setAttribute("username",chString(strUserName));
        session.setAttribute("password",chString(strPassword));
        response.sendRedirect("ch4_15_show.jsp");
    }
%>
<form method="post" action="ch4_15.jsp" name="form1" >
    <br>用户登录：
    <input type="text" name="userName" size="20">
    <br>用户密码：
    <input type="password" name="password" size="20">
    <br>提交或重置：
    <input type="submit" name="submit"  value="提交" >
    <input type="reset" name="reset" value="重置" >
</form>
<%= strSessionID %>
</body>
</html>
```

ch4_15_show.jsp 文件代码：

```
<%@ page contentType="text/html;charset=GB2312"%>
<%@ include file="chString.txt" %>
<html>
<head>
<title>【例 4.15】用户登录成功</title>
</head>
<body bgcolor="#00BFFF">
<br>用户登录成功！<br>
用户登录名：<%= (String)session.getAttribute("username")%>
<br>用户登录密码：<%= (String)session.getAttribute("password")%>
<br>session 对象 ID：<%= session.getId()%>
</body>
</html>
```

注意：此例用户可以看出，session 对象可以在同一个会话的不同网页中共享变量。利用这些可以记录用户的登录状态和所填写的过信息等，不必让用户重复输入。

4.5.3 对象与 URL 重写

前面介绍了 session 对象的 id 是通过 cookie 保存到客户端浏览器的，这样才保证了服务器 session 对象与客户的一一对应关系。那么什么是 cookie？cookie 是 Web 服务器发送至客户端浏览器的小段文本信息，在日后访问该服务器时浏览器不做任何改变地向服务器返回这些信息。如果用户浏览器禁止使用 cookie(用户可以自行设置)，客户每次访问相同的网页 session 也是互不相同的。那么禁止 cookie 的情况下，如何保证 session 对象与客户的一一对应关系？这就涉及了 URL 重写。

所谓 URL 重写，就是当客户从一个页面连接到同一个 Web 服务目录的另一个页面时，通过向这个新的 URL 添加参数，把 session 对象的 id 作为 URL 参数传带过去，这样就可以保证在该网站的每个页面中的 session 对象完全相同。用户可以直接在浏览器的地址栏写入 URL 时带上 session 对象的 id 和需要传递的参数，例如：

```
http://127.0.0.1/ch4_16.jsp;jsessionid=05CDDE89CC9B56045AD63A7FF6B7CFAE
```

传递参数：

```
http://127.0.0.1/ch4_16.jsp? &username=admin&password=admin
```

在程序片中调用 response 对象调用 encodeURL()或 encodeRedirectURL()方法实现 URL 重写。例如：

```
String strurl=response.encodeRedirectURL("ch4_16.jsp");
<%= strurl%>
```

【例 4.16】URL 重定向示例(其代码参见 ch4_16.jsp)，页面效果如图 4.15 所示。

ch4_16.jsp 文件代码：

```
<%@ page contentType="text/html;charset=GB2312"%>
<html>
<head>
<title>【例 4.16】用户登录成功：页面 1</title>
</head>
<body bgcolor="#00BFFF">
<br>url 重定向示例<br>
<%
  String str=response.encodeURL("ch4_16_show.jsp");
%>
<br>session 对象 ID：<%= session.getId()%>
<br><a href=<%= str%> >到 ch4_16_show.jsp</a>
</body>
</html>
```

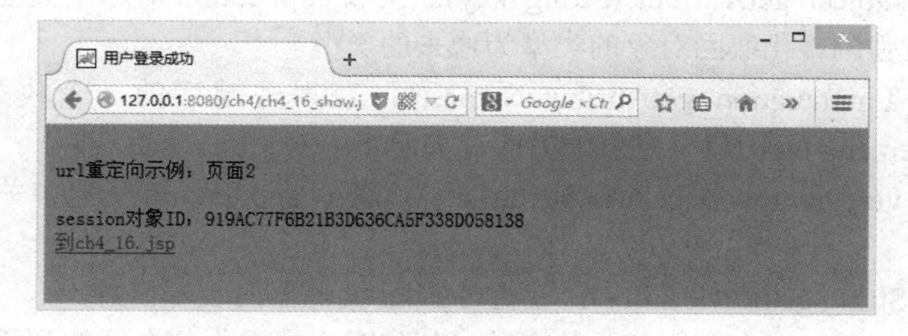

图 4.15　URL 重定向页面显示效果

ch4_16_show.jsp 文件代码：

```
<%@ page contentType="text/html;charset=GB2312"%>
<html>
<head>
<title>【例 4.16】用户登录成功</title>
</head>
<body bgcolor="#00BFFF">
<br>url 重定向示例：页面 2<br>
<%
  String str=response.encodeRedirectURL("ch4_16.jsp");
%>
<br>session 对象 ID：<%= session.getId()%>
<br><a href=<%= str%> >到 ch4_16.jsp</a>
</body>
</html>
```

4.6 application 对象

对于一个 Web 服务目录，JSP 为每个客户创建各自的 session 对象，它们互不相同且数据不共享，当同一个客户在同一个 Web 服务目录中的页面之间切换时，对应的都是一个 session。与 session 对象不同，application 提供了对 ServletContext 对象的访问，它用于多个用户或多个程序之间共享数据。对于一个 Web 服务目录而言，每个用户都共用一个 application 对象，共享 application 对象中存储的数据。不同 Web 服务目录中的 application 对象互不相同。

4.6.1 常用方法

application 对象的生命周期为 Web 服务目录服务启动到关闭，作用范围为所有访问 Web 服务目录的客户。它提供了一些方法，用于对所有用户共享数据的存储和访问。下面是其常用方法介绍。

- public void setAttribute(String key,Object obj)：将参数 obj 对象添加到 application 对象中，并指定 key 为被添加对象的索引关键词。如果添加的对象关键词已经存在，则替换原来添加的对象。
- public Object getAttribute(String key)：获取 application 对象中关键字为 key 的对象。应强制转化取回对象的类型为原来的类型。
- public Enumeration getAttributeNames()：获得关键字的一个枚举对象，使用 nextElement()遍历枚举对象中的所有关键字。
- public void removeAttribute(String key)：删除 application 对象中关键字是 key 的对象。
- public String getServletInfo()：获得 Servlet 编译器的版本信息。

application 对象对所有用户都是共享的，任何对它的操作都会影响到所有的用户，所以在操作它的时候，一般都做同步处理。也就是定义操作方法时，在方法名称前加“synchronized”修饰词，例如：

```
synchronized void sendmessage(){
    operate applilcation …code
}
```

因为有些服务器不支持直接使用 application 对象，所以要用 getServletContext()方法得到它，下面是示例代码：

```
ServletContext application;
application=getServletContext();
```

4.6.2　计数器

第 3 章实现了一个简单的计数器，但是它并不能限制客户通过刷新增加计数器的计数，所以，这不能算一个真正意义上的计数器。所谓的计数器是 Web 服务目录被客户访问的次数，也就是网站被访问次数，客户一个会话周期内的访问被认为是一次访问，客户每次刷新页面不能被计数。

下面用 application 对象和 session 对象来实现一个真正意义上的计数器。实现思路是当一个用户访问 Web 服务目录时，首先判断它是否已经计数，如果未计则计数，否则不计。计数保存在 application 对象中，写计数器变量的函数被同步化。

【例 4.17】计数器示例(其代码参见 ch4_17.jsp)，页面效果如图 4.16 所示。

ch4_17.jsp 文件代码：

```
<%@ page contentType="text/html;charset=GB2312"%>
<html>
<head>
<title>【例 4.17】计数器示例</title>
</head>
<body>
<%!
  synchronized void countPeople(){
      ServletContext application =getServletContext();
       Integer number=(Integer)application.getAttribute("Count");
       if(number==null){
           number=new Integer(1);
           application.setAttribute("Count",number);
       }
       else{
           number=new Integer(number.intValue()+1);
           application.setAttribute("Count",number);
       }
  }
%>
<%
```

```
    Integer personnum=(Integer)application.getAttribute("Count");
    if(session.isNew() | personnum==null){
        countPeople();
    }
    Integer yournumber=(Integer)application.getAttribute("Count");
%>
<p>欢迎访问本站，您是第
<%= yournumber%>个访问用户。
</body>
</html>
```

图 4.16　ch3_17.jsp 页面效果

4.7　上 机 实 训

实训目的

- 掌握内建对象 out 向页面输出各种格式数据的方法。
- 掌握内建对象 response 实现页面跳转及自动跳转的方法。
- 掌握 request 对象获得表单数据等方法。
- 了解表单验证的技术。
- 掌握 session 对象存储和获取存储数据的方法。
- 掌握 application 对象存储和获取存储信息的方法。

实训内容

实训 1　参照第 11 章，完成网上报名系统登录页面设计。

要求：

(1) 登录页面名称为 login.html、登录验证页面 login.jsp 程序设计。

(2) 用户名和密码事先存储在数组中，从表单中获得用户输入的用户名和密码。

(3) 根据验证结果调转页面，成功到 index.jsp 页面，失败到 error.jsp 页面。

实训 2　参照第 11 章，完成考生报名系统报名页面的设计。

要求：

(1) 验证考生报名页面 register.jsp 用户输入数据的合法性。

(2) 获得考生报名信息并显示在 registershow.jsp 页面。

实训总结

通过本章的上机实训，学员应该能够了解 JSP 内建对象的有效作用范围和使用特点；了解表客户端表单验证；掌握 request、response 和 session 对象的使用方法，掌握 request

对象获得表单数据的方法，掌握 response 对象重定向、改变刷新时间方法；掌握使用 session 对象存储和获取数据的方法，用户会话跟踪方法 cookie 和 URL 重写等。

4.8 本 章 习 题

思考题

(1) 如何处理表单提交的汉字？
(2) 一个用户在不同的 Web 服务目录的 session 相同吗？
(3) 内置对象的 4 个作用范围？什么情况下 session 会关闭？
(4) response.sendRedirect(URL)方法的作用？
(5) 是不是所有 Web 服务目录共用一个 application？
(6) 怎样使用 request、session 和 application 对象进行参数存取？
(7) HTTP 请求消息、响应消息与 request 和 response 对象之间的关系？

拓展实践题

使用 JavaScript 对网上报名系统报名页面和用户登录页面的数据输入进行验证。

第 5 章　使用 JavaBean

学习目的与要求：

使用 JavaBean 可以使页面显示和业务逻辑处理有效地分离，并使 Bean 成为有效重复利用的组件，进一步实现代码重用，方便应用系统维护。本章主要学习 JavaBean 的概念及特性，编写和使用 JavaBean，HTML 表单与 Bean 的交互，JavaBean 的典型应用等内容。通过本章的学习，学员要理解 JavaBean 的概念和特点，掌握编写、编译、调试、布置 Bean 的方法；掌握表单与 Bean 的交互方法；程序片中使用 JavaBean 的方法；掌握计数器、购物车 Bean 的编写技术。

5.1　JavaBean 的基本概念

前面已经介绍，JSP+JavaBean 是 JSP 开发的一种典型模式，Sun 的 JavaBean 技术为 JSP 开发提供了极大的方便，使得页面与数据处理实现了真正地分离，本章在 E:\programJsp 文件夹中建立 ch5 目录，并将其指定为虚拟服务目录 ch5，所有的例题均保存在 ch5 目录中。下面就来介绍 JavaBean 的基本概念。

5.1.1　什么是 JavaBean

爱好电脑 DIY 的用户对组装电脑不会陌生。组装一台电脑需要选购多个组件，例如 CPU、主板、内存、硬盘、显卡、电源以及输入输出设备等，用户并不用关心 CPU、硬盘等的内部电路是如何设计的，只要了解它的接口和功能，将这些组件安装到能够支持它高效运转的主板上即可，这就是组件技术在电脑组装上的典型应用。

JavaBean 是 Java 程序设计中的一种组件技术。Sun 公司把 JavaBean 定义为一个可重复使用的软件组件，类似于电脑 CPU、硬盘等组件。从程序员编程的角度看，实际上 JavaBean 组件就是 Java 开发中的一个类，通过封装属性和方法成为具有某种功能和接口的类，简称 Bean。由于 JavaBean 是使用 Java 语言开发的组件，所以具有 Java 应用程序的特点，如与平台无关性，可以在任何安装了 Java 平台的环境中运行；可以实现代码的重复使用等优点。

前面各章的 JSP 页面都是由 HTML 标记和 Java 程序片组成的，如果大量的程序片和 HTML 标记掺杂在一起，会给程序的编写、修改和维护带来很多困难，也不利于团队分工合作开发，页面美工和程序设计人员不能有效地分工协作。在 Web 服务器端使用 JavaBean，将原来页面中程序片完成的功能封装到 JavaBean 中，使其实现业务逻辑封装和数据处理功能，很好地实现了业务逻辑层与视图层的分离。JSP 提供了<jsp:useBean>、<jsp:getProperty>、<jsp:setProperty>等一系列支持 JavaBean 的 JSP 动作指令，使得 JSP 使用 Bean 非常容易和高效。

5.1.2 JavaBean 的规范

JavaBean 分为可视化组件和非可视化组件。在早期，JavaBean 最长使用的领域是可视化领域，如菜单、文本框、按钮等可视化 GUI。随着 B/S 模式软件的流行，非可视化的 JavaBean 越来越显示出优势，被广泛地使用在 JSP 编程中。用户主要设计 Bean 的属性和方法，不必设计组件的外观，数据的显示由 JSP 页面来实现。下面就 Sun 对 JavaBean 的一些规定做简单的介绍。

1. JavaBean 的构造方法

JavaBean 就是符合一定条件的 Java 类，该类必须声明为 public 类，可供其他类实例化。类中如果有构造方法，必须声明为 public 类型且无参数。例如：

```
package  mybean;
import java.io.*;
public class myBeanOne {
    …
    public myBeanOne()
    {
    }
    …
}
```

2. JavaBean 的属性与方法

JavaBean 的属性必须声明为 private，方法必须声明为 public 访问类型。

JavaBean 中用一组 set 方法设置 Bean 的私有属性值，get 方法获得 Bean 的私有属性值。set 和 get 方法名称与属性名称之间必须对应，也就是说：如果属性名称为 xxx，那么 set 和 get 方法的名称必须为 setXxx()和 getXxx()，set 和 get 为前缀与属性名字首字母大写组成 set 和 get 方法名称。对于 boolean 类型的属性，允许使用 is 代替方法名称中的 set 和 get 前缀，但是并不推荐使用，创建 Bean 必须带有包名。使用这些规则编写 JavaBean，能够方便 JSP 知道 JavaBean 属性和方法。

例如，myBeanOne 定义格式如下：

```
package  mybean;
import java.io.*;
public class myBeanOne {
    String name;
    boolean  flag;
    public myBeanOne()
    {   name="";
        flag=true;
    }
    public setName(String s){
        name=s;
```

```
    }
    public String getName(){
        return name;
    }
    public setFlag(boolean flag){
        this.flag=flag;
    }
    public boolean getFlag(){
        return flag;
    }
}
```

5.2 创建与使用 JavaBean

创建 JavaBean 的过程和编写 Java 类的过程基本相似，可以在任何 Java 的编程环境下完成编写、编译和发布的操作，例如 MyEclipse 开发环境等。下面介绍创建与使用 JavaBean 的方法。

5.2.1 创建 JavaBean

创建 JavaBean 要经过编写代码、编译源文件、配置 JavaBean 这样一个过程，这个过程可以借助 Java 开发工具，也可以使用记事本等简单文本编辑器编写代码，在 JDK 1.8.0 下编译并在 Tomcat 7.0 环境下测试。下面以一个求矩形面积的 JavaBean 为例，介绍 JavaBean 的代码编写、编译和配置过程。

例如，创建求矩形面积的 JavaBean。JavaBean 的名字为 MyRectangle。以下是创建 MyRectangle 的源文件。

MyRectangle.java 文件中的代码如下：

```
package mybean.maths;
import java.io.*;
public class  MyRectangle
{
    double length;
    double width;
    double area;
    public MyRectangle(){
        //构造函数
        length=0;
        width=0;
    }
    public void setLength(double length){
        //设置长
        this.length=length;
    }
```

```
    public double getLength(){
        //得到长
        return length;
    }
    public void setWidth(double width){
        //设置宽
        this.width=width;
    }
    public double getWidth(){
        //得到宽
        return width;
    }
    public double getArea(){
        //求矩形面积
        double faceArea=length*width;
        return faceArea;
    }
    /*public static void main(String[] args)
    {
        MyRectangle myrect=new MyRectangle();
        System.out.println(myrect.getArea());
    }*/
}
```

为了能够正确地编译 Bean，可以先建立一个与包对应的目录结构，例如 D:\bean 目录下创建 mybean\maths 目录，在 DOS 窗口切换到 d:\bean 目录中，使用 javac 编译 MyRectangle.java 文件，命令格式为 javac mybean\maths\MyRectang.java，调试命令为 java mybean\maths\MyRectang，编译调试通过后得到字节码文件 MyRectangle.class。

注意：编写 Bean 可以在该类中先加入 main 函数，调试通过再将 main 函数再注释掉，然后再将字节码文件部署到 Tomcat 7.0 环境中，就可以直接在 JSP 中调用了，这样可以有效地降低 Bean 的调试难度。

5.2.2 布置 JavaBean

要想在 JSP 中使用 JavaBean，JSP 引擎必须创建一个 JavaBean 对象，然后在 JSP 页面中才能调用这个创建的 JavaBean。为了让 Tomcat 7.0 找到 JavaBean 的字节码，字节码文件必须保存在特定的目录中，这就是 JavaBean 的布置。

布置 JavaBean 首先要在当前 Web 服务目录中建立：WEB-INF\classes 子目录，用户要注意目录名称的大小写，然后再根据 JavaBean 的包名创建对应的子目录，并将 JavaBean 的字节码文件复制到包对应的子目录中即可。

例如 MyRectangle 的包名为 mybean.maths，则创建的目录结构如图 5.1 所示，然后将 MyRectangle.class 文件复制到 ch5\WEB_INF\classes\mybean\maths 子目录中。

在 Web 服务目录中新建立了 WEB-INF\classes 目录，要重新启动 Tomcat 服务器启用该目录，这样 JSP 页面才能使用 JavaBean 对象。

> 注意：修改了 Bean 的字节码后，要将新的字节码复制到对应的 WEB-INF\classes 目录中，重新启动 Tomcat 服务器才能生效。

```
programJsp
  .metadata
  ch1
  ch2
  ch3
  ch4
  ch5
    WEB-INF
      classes
        mybean
          maths
```

图 5.1　布置 Bean 的目录结构

5.2.3　在 JSP 中使用 JavaBean

在 JSP 中提供了<jsp:useBean>、<jsp:setProperty>和<jsp:getProperty>动作标记来实现对 JavaBean 的操作，也可以在程序片中操作 Bean。使用 Bean 首先要在 JSP 页面中使用 import 指令将 Bean 引入，例如引入 MyRectangle，格式如下：

```
<%@ page import="mybean.maths.*" %>
```

1. <jsp:useBean>动作标记

要想在 JSP 页面中使用 Bean，必须首先使用<jsp:useBean>动作标记在页面中定义一个 JavaBean 的实例，这个被定义实例有一定的生存范围及一个唯一的 id，JSP 页面通过 id 来识别 Bean，也可以在程序片中通过 id.method 形式来调用 Bean 中的 set 和 get 方法。下面是 useBean 动作标记的使用格式：

```
<jsp:useBean id="bean名字" scope="page|request|session|application"
class="packageName.calssName"/>
```

例如：

```
<jsp:useBean id="myrectangle" scope="request"
   class="mybean.maths.MyRectangle"/>
```

id 是定义 Bean 对象的唯一标识。当含有<jsp:useBean>动作标记的页面在服务器上执行时，JSP 引擎首先尝试在 pageContent 对象内查找具有相同 id 和 scope 的 JavaBean 实例，如果有就将这个实例分配客户，如果没有就根据 Bean 的字节码文件，创建一个名字为 id、作用范围为 scope 的新实例添加到 pageContent 对象中，并将这个 Bean 分配给客户。

scope 属性代表了 JavaBean 的作用范围，它可以是 page、session、request 和 application 四个作用范围中的一种。

- 取值为 page 时：Bean 的作用范围是 page。也就是，同一个客户访问的每个页面的 Bean 对象是互不相同的，它们占用不同的存储空间，当客户离开这个页面时，JSP 引擎取消为客户该页面分配的 Bean，释放他所占的内存空间。不同客户访问同一页面，Bean 之间互不干扰。
- 取值为 session 时：Bean 的作用范围是 session。也就是，同一个客户在同一个 Web 服务目录中不同的 JSP 页面中，只要 Bean 的 id 相同，作用范围为 session，该客户得到的 Bean 是同一个。也就是客户在某个页面修改 Bean 的属性，在其他页面，该 Bean 的属性会发生同样的变化；不同客户之间的 Bean 是互不相同的。

- 取值为 request 时：Bean 的作用范围是 request。也就是同一个客户在 request 作用范围内请求与响应页面之间共用一个 id 相同的 Bean，请求响应完成该 Bean 即被释放；不同客户的 Bean 互不相同。
- 取值为 application 时：Bean 的作用范围是 application。也就是 Web 服务目录下所有用户共享一个 Bean，只要所有客户、该服务目录下所有页面使用同一个 id 的 Bean，它们的 Bean 是同一个，任何客户对 Bean 属性的修改都会影响到其他用户。

当在页面中得到了 Bean 的实例后，就可以在程序片中使用 id.method 格式调用 Bean 的方法，修改或得到 Bean 的属性，也可以使用<jsp:setProperty>和<jsp:getProperty>动作标记设置或获得 Bean 的属性值，这样可以减少页面中的程序片。

> 注意：在<jsp:useBean>标记中，其中 class 属性中指定的类必须带有包名；作用域为 session 的 Bean，用户要保证客户端支持 cookie。

2. <jsp:getProperty>动作标记

使用<jsp:getProperty>动作标记可以在 JSP 页面中得到 Bean 实例的属性值，并将其转换为 String 类型的字符串，发送到客户端。使用<jsp:getProperty>动作标记，必须先声明 Bean 的实例，并保证 Bean 中有对应的 getXxx()方法。使用格式如下：

```
<jsp:getProperty name="bean的id" property="bean的属性名" />
```

或

```
<jsp:getPropety name="bean的id" propety="bena的属性名">
</jsp:getProperty>
```

该指令相当于<%= bean.getXxx()%>表达式的作用。

下面的例子 MyRectangle 类创建 Bean，id 为"myrectangle"、scope 为 page。使用了 getProperty 动作指令和表达式两种方式获得 Bean 的属性值。

【例 5.1】使用 Bean 计算矩形面积，页面代码参见 ch5_1.jsp。页面效果如图 5.2 所示。

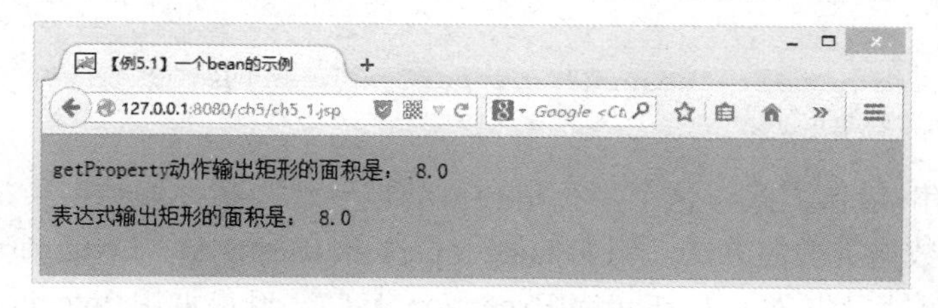

图 5.2 ch5_1.jsp 的页面效果

MyRectangle.java 文件见 5.1.1 节。

ch5_1.jsp 文件中的代码如下：

```
<%@ page contentType="text/html;charset=GB2312" %>
<%@ page import="mybean.maths.MyRectangle" %>
<html>
```

```
<head>
<title>【例 5.1】一个 bean 的示例</title>
</head>
<body bgcolor="cyan" >
<jsp:useBean id="myrectangle" class="mybean.maths.MyRectangle"
    scope="page" />
<%  //设置矩形长
  myrectangle.setLength(2);
   //设置矩形宽
  myrectangle.setWidth(4);
%>
<p>getProperty 动作输出矩形的面积是:
<jsp:getProperty name="myrectangle" property="area"/>
<p>表达式输出矩形的面积是:
<%= myrectangle.getArea() %>
</body>
</html>
```

注意：在 MyRectangle.java 中，area 属性属于关联属性，它依赖于 length 和 width 属性，所以 area 属性只提供了 getArea() 方法。上例在程序片中使用了 myrectangle.setLength(2)和 myrectangle.setWidth(4)设置 myrectangle 的长和宽属性。

3. <jsp:setProperty>动作标记

使用<jsp:setProperty>动作标记可以在 JSP 页面中设置 Bean 的属性，但必须保证 Bean 有对应的 setXxx 方法。setProperty 动作标记设置 Bean 的属性有两种方式。

(1) 第一种，使用表达式或字符串为 Bean 的属性赋值。

使用格式如下：

```
<jsp:setProperty name="beanid"  property="bean 的属性"
    value="<%= expression %>" />
```

此方式下表达式的数据类型必须和 bean 的属性类型相一致。使用字符串为 bean 的属性赋值，格式如下：

```
<jsp:setProperty name="beanid"  property="bean 的属性"
    value="字符串" />
```

value="字符串"赋值方式，这个字符串会自动被转化为属性的数据类型。自动转化会调用 Java 语言的数据类型的方法，例如 Integer.parseInt(String s)、Long.parseLong(String s)等。这些方法都可能发生 NumberFormatException 异常，例如调用 Integer.parseInt("123a")方法时会发生异常。

下面的例子编写一个描述考生信息的 Bean，在页面中使用 setProperty 动作标记的第一种方式为 Bean 属性赋值，将 Student.java 编译后的字节码文件复制到 E:\programJsp\ch5 目录下的 WEB-INF\classes\mybean\maths 子目录中。

【例 5.2】setProperty 动作指令使用的表达式赋值示例(相关代码参见 student.java 和 ch5_2.jsp)，页面效果如图 5.3 所示。

图 5.3　ch5_2.jsp 的页面效果

Student.java 文件中的代码如下：

```
package mybean.maths;
import java.io.*;
public class Student
{
    String name=null;
    long number;
    String sex=null;
    double height,weight;
    public Student(){
        name="yang";
        number=123;
        sex="k";
        height=170;
        weight=80;
    }
    public String getName(){
        return name;
    }
    public void setName(String s){
        name=s;
    }
    public long getNumber(){
        return number;
    }
    public void setNumber(long num){
        number=num;
    }
    public String getSex(){
        return sex;
    }
    public void setSex(String s){
        sex=s;
    }
    public double getHeight(){
        return height;
    }
```

```
    public void setHeight(double height){
        this.height=height;
    }
    public double getWeight(){
        return weight;
    }
    public void setWeight(double weight){
        this.weight=weight;
    }
    /*public static void main(String[] args)
    {
        Student stu=new Student();
        System.out.println(stu.getName());
    }*/
}
```

ch5_2.jsp 文件中的代码如下：

```
<%@ page contentType="text/html;charset=GB2312" %>
<%@ page import="mybean.maths.Student" %>
<html>
<head>
<title>【例 5.2】 setProperty 动作指令示例</title>
</head>
<body bgcolor="cyan" >
<jsp:useBean id="stu" class="mybean.maths.Student" scope="page" />
<%
   long number=12345;
%>
<jsp:setProperty name="stu" property="name" value="杨得力"/>
<jsp:setProperty name="stu" property="sex" value="男"/>
<jsp:setProperty name="stu" property="number" value="<%= number%>"/>
<jsp:setProperty name="stu" property="height" value="<%= 170%>"/>
<jsp:setProperty name="stu" property="weight" value="80"/>
<p>getProperty 动作指令输出学生信息:
<br>学生姓名:
<jsp:getProperty name="stu" property="name"/>
<br>学生学号:
<jsp:getProperty name="stu" property="number"/>
<br>学生性别:
<jsp:getProperty name="stu" property="sex"/>
<br>学生身高:
<jsp:getProperty name="stu" property="height"/>
<br>学生体重:
<jsp:getProperty name="stu" property="weight"/>
</body>
</html>
```

(2) 第二种，使用表单的参数值为 Bean 的属性赋值。

在 Web 程序设计中，客户端通常是使用表单与应用程序交互的，HTML 表单中提交的信息存放在 request 对象中，并且以名称-值对的形式进行传递。使用 setProperty 动作标记，可以实现 HTML 表单元素与 JavaBean 属性的映射，从而实现 Bean 属性与表单输入参数之间的交互。使用这种方法为 Bean 的属性赋值格式如下：

```
<jsp:setProperty name="bean的名字" property ="*" />
```

或：

```
<jsp:setPropety name="bean的名字" property="bean属性名" param="表单参数名"/>
```

property="*"格式要求 Bean 的属性名字必须和表单参数名称一一对应，JSP 引擎会自动将表单参数的字符串值转换为 Bean 对应的属性值。property="Bean 属性名" param="表单参数名"格式，明确指定 Bean 的某个属性设定为表单的某个参数值。关于汉字的处理问题可以在 Bean 的 setXxx(String s)方法中，为 xxx 属性重新编码，也可以将 contentType 属性中的 charset=GB2312 写成 Charset=GB2312。

下面的例子演示了用户登录 LoginBean.java 的使用，ch5_3.jsp 提供用户登录信息输入界面，ch5_3_show.jsp 页面通过调用 LoginBean.java 的 check 方法检查是否合法用户。把 LoginBean.java 编译生成的字节码文件复制到 ch5\WEB-INF\classes\mybean\maths 目录中。

【例 5.3】登录 Bean 示例(相关代码参见 LoginBean.java、ch5_3.jsp 和 ch5_3_show.jsp)，页面效果如图 5.4 和图 5.5 所示。

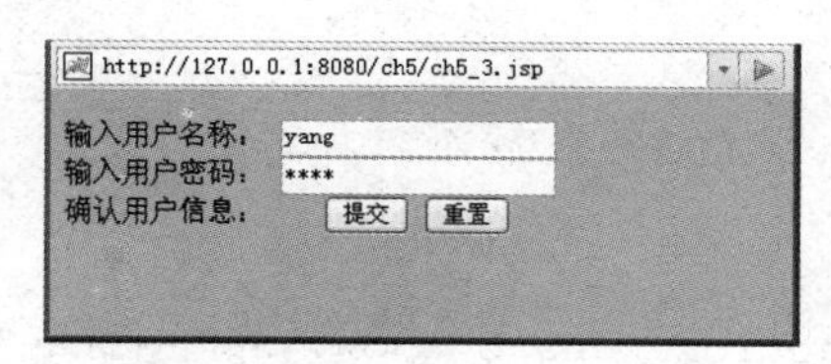

图 5.4　ch5_3.jsp 页面效果

LoginBean.java 文件中的代码如下：

```
package mybean.maths;
import java.io.*;
public class LoginBean
{
    private String loginName=null;
    private String password=null;
    public void LoginBean(){
    }
    public void setLoginName(String s){
        loginName=s;
    }
    public String getLoginName(){
        return loginName;
    }
```

```
    public void setPassword(String s){
        password=s;
    }
    public String getPassword(){
        return password;
    }
    public boolean check(){
        if("yang".equals(loginName)&& "yang".equals(password))
            return true;
        else
            return false;
    }
}
```

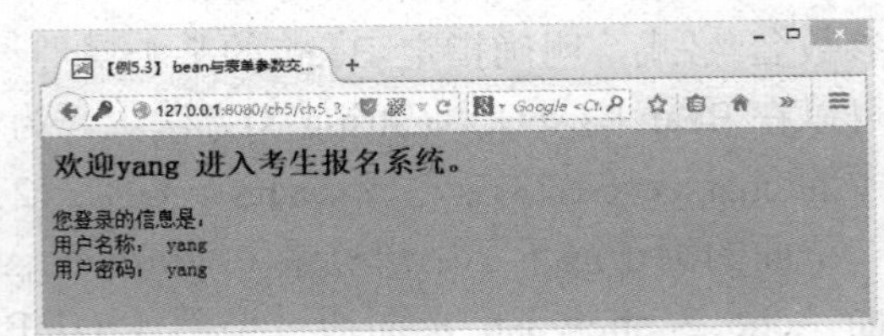

图 5.5　登录成功和失败的页面

ch5_3.jsp 文件中的代码如下：

```
<%@ page contentType="text/html;charset=GB2312" %>
<html>
<head>
<title>【例 5.3】与表单参数值进行交互</title>
</head>
<body bgcolor="cyan" >
<form method="post" action="ch5_3_show.jsp" name="form1" >
  <p>输入用户名称：
   <input type="text" name="loginName" size="20"/>
  <br>输入用户密码：
   <input type="password" name="password" size="20"/>
  <br>确认用户信息：   
   <input type="submit" name="submit" value="提交" size="6" />
   <input type="reset"  name="reset" value="重置" size="6" />
</form>
</body>
</html>
```

ch5_3_show.jsp 文件中的代码如下：

```
<%@ page contentType="text/html;charset=GB2312" %>
<%@ page import="mybean.maths.LoginBean" %>
<html>
<head>
<title>【例 5.3】 bean 与表单参数交互示例</title>
```

```
</head>
<body bgcolor="cyan" >
<jsp:useBean id="stu" class="mybean.maths.LoginBean" scope="page" />
<jsp:setProperty name="stu"  property="*" />
<%
   if (stu.check()){
%>
        <h2>欢迎<jsp:getProperty name="stu" property="loginName"/>
        进入考生报名系统。
        </h2>
<% }
   else{
%>
       <h2>登录失败,单击<a href="javascript:history.back(-1);">
                这里重新登录</a></h2>
<% }
%>
<p>您登录的信息是：
<br>用户名称：
<jsp:getProperty name="stu" property="loginName"/>
<br>用户密码：
<jsp:getProperty name="stu" property="password"/>
</body>
</html>
```

注意： *表单提交后，<jsp:setProperty>动作指令才会执行。*

5.3　JavaBean 的辅助类

上一节介绍了简单 Bean 的创建、布置和使用。用户在实际 Web 应用开发中，编写 Bean 除了要使用 import 语句引入 Java 的标准类，可能还需要自己编写的其他类。用户自己编写的被 Bean 引用的类称为 Bean 的辅助类。此时用户即可以将这些类和创建 Bean 的类编写在一个 Java 源文件中，也可以单独编写然后引入。不管采用哪种形式，都要将源文件编译后产生的全部字节码文件复制到 Web 服务目录 classes 的相应的目录中(见 5.2.2 节)。

在下面的例子中，使用 Bean 保存用户的多条留言，在写 Bean 的类文件 messList.java 时，需要一个封装单条留言信息的辅助类 message 和一个字符串转换辅助类 chString，这两个类分别用于处理留言中的汉字和封装留言信息。客户通过 ch5_4.jsp 页面向 ch5_4_submit.jsp 页面提交留言信息，单击“提交留言”，ch5_4_submit.jsp 页面获得信息后保存到 Bean 中，留言 Bean 的范围为 application。单击“查看留言”按钮，ch5_4_show.jsp 页面显示保存在 Bean 中的留言。

在本例中使用了泛型类 ArrayList，下面的语句声明和构造一个保存 message 类对象的数组列表 messList：

```
ArrayList<message> messList=new ArrayList<message>();
```

使用 add()方法可以将元素添加到数组列表中：

```
message mtemp=new message();
messList.add(mtemp);
```

可以使用下面的语句访问数组列表中的每一个元素：

```
for(message me:messList){
    do something with me;
}
```

【例 5.4】 Bean 的辅助类示例(相关代码参见 messList.java.ch5_4.jsp、ch5_4_submit.jsp 和 ch5_4_show.jsp)，程序运行效果如图 5.6 所示。

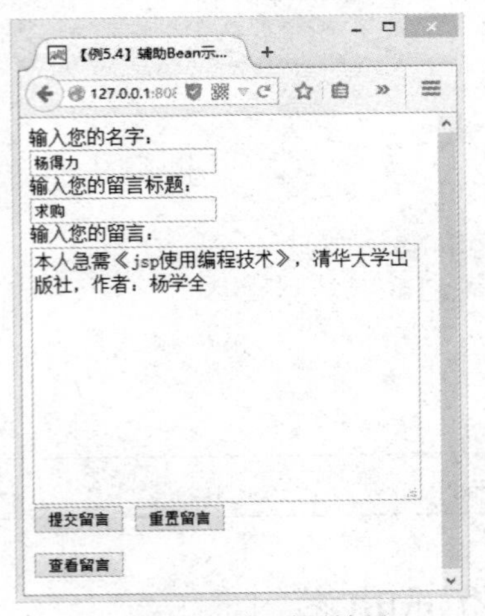

(a) 提交留言页面效果

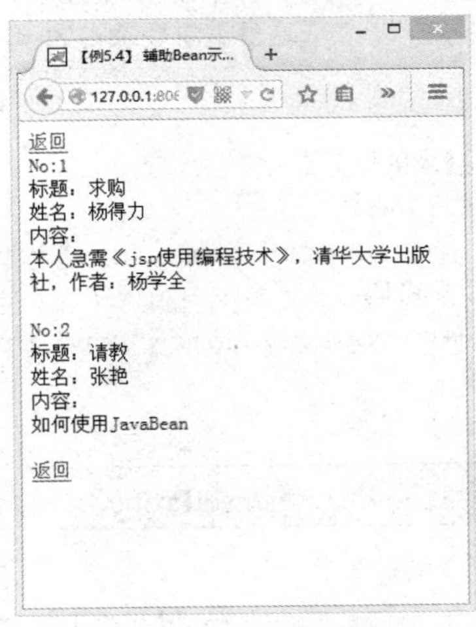

(b) 显示留言页面效果

图 5.6　Bean 的辅助类示例

messList.java 文件中的代码如下：

```
package mybean.maths;
import java.util.ArrayList;
class chString{
    public String handleString(String s){
        String str=s;
        try{
            byte b[]=str.getBytes("ISO-8859-1");
            str=new String(b);
        }
        catch(Exception ee){}
        return str;
    }
}
class message{
```

```
    String title=null;
    String author=null;
    String mess=null;
    long num;
    chString chstr=new chString();
    public String getTitle() {
        return title;
    }
    public void setTitle(String title) {
        this.title = chstr.handleString(title);
    }
    public String getAuthor() {
        return author;
    }
    public void setAuthor(String author) {
        this.author=chstr.handleString(author);
    }
    public String getMess() {
        return mess;
    }
    public void setMess(String mess) {
        this.mess =chstr.handleString(mess);
    }
    public long getNum() {
        return num;
    }
    public void setNum(long num) {
        this.num = num;
    }
}
public class messList {
    ArrayList<message> messL=null;
    long number;
    public messList(){
    messL=new ArrayList<message>();
    number=0;
    }
    synchronized public void add(String author,String title,
            String mess){
            if(!("".equals(author)||"".equals(title)
                ||"".equals(mess))){
        number++;
        message mtemp=new message();
        mtemp.setAuthor(author);
        mtemp.setMess(mess);
        mtemp.setTitle(title);
        mtemp.setNum(number);
```

```
            messL.add(mtemp);
            }
        }
        public StringBuffer getMessSet(){
            StringBuffer strWatch=new StringBuffer();
            for(message me:messL){
                strWatch.append("<br>");
                strWatch.append("No:"+me.getNum());
                strWatch.append("<br>标题：");
                strWatch.append(me.getTitle());
                strWatch.append("<br>姓名：");
                strWatch.append(me.getAuthor());
                strWatch.append("<br>内容：<br>");
                strWatch.append(me.getMess()+"<br>");
            }
            return strWatch;
        }
}
```

ch5_4.jsp 文件中的代码如下：

```
<%@ page contentType="text/html;charset=GB2312" %>
<html>
<head><title>【例 5.4】辅助 Bean 示例留言板</title>
</head>
<body>
<form action="ch5_4_submit.jsp" method=post name="form1">
    输入您的名字：<br>
    <input type="text" name="author" />
    <br>输入您的留言标题：<br>
    <input type="text" name="title" />
    <br>输入您的留言：<br>
    <textarea  name="mess" rows="10" cols="36"
        warp="physical"></textarea><br>
    <input type="submit" value="提交留言" name="submit"/>
    <input type="reset" value="重置留言" name="reset" />
</form>
<form action="ch5_4_show.jsp" method="post" name="form2">
    <input type="submit" value="查看留言" name="look">
</form>
</body>
</html>
```

ch5_4_submit.jsp 文件中的代码如下：

```
<%@ page contentType="text/html;charset=GB2312" %>
<%@ page import="mybean.maths.*" %>
<html>
<head><title>【例 5.4】 辅助 Bean 示例留言板 submit.jsp</title>
```

```
</head>
<body>
<jsp:useBean id="peoplemess" class="mybean.maths.messList"
      scope="application"/>
<%
  String author=request.getParameter("author"),
        title=request.getParameter("title"),
         mess=request.getParameter("mess");
  if(author==null)
     author="佚名";
  if(title==null)
     title="无标题";
  if(mess==null)
     mess="无内容";
  peoplemess.add(author,title,mess);
  out.print("您的信息已提交！");
%>
<a href="ch5_4.jsp" >返回</a>
</body>
</html>
```

ch5_4_show.jsp 文件中的代码如下：

```
<%@ page contentType="text/html;charset=GB2312" %>
<%@ page import="mybean.maths.*" %>
<html>
<head><title>【例 5.4】 辅助 Bean 示例留言板 show.jsp</title>
</head>
<body>
<jsp:useBean id="peoplemess" class="mybean.maths.messList"
    scope="application"/>
<a href="ch5_4.jsp" >返回</a>
<jsp:getProperty name="peoplemess" property="messSet"/>
<br>
<a href="ch5_4.jsp">返回</a>
</body>
</html>
```

注意：messList.java 编译后会生成 messList.class、message.class 和 chString.class 三个字节码文件，要将它们一起复制到 ch5\WEB-INF\classes\mybena\maths 目录中。

5.4 JSP 与 JavaBean 模式实例

5.4.1 计数器 Bean

在 4.6.2 节使用 application 对象和 session 对象实现了一个计数器，它可以限制客户通过不断刷新页面或再次访问该目录的其他页面来增加计数器的计数。本节使用 Bean 来实现一个同样功能的计数器。为了演示一个页面使用多个 Bean 的技术，本例使用两个 Bean 实现计数。负责计数的 Bean 是 application 范围的 CounterBean，负责检查是否应该计数的 Bean 是 CountPeople。CountPeople 首先判断该会话是否已经计数，如果没有计数，则调用 CounterBean 的实例 Counter()方法计数，否则不计数。计数保存在 application 范围内的 Bean 中。

【例 5.5】在 ch5 服务目录下有 ch5_5.jsp 和 ch5_5_two.jsp 两个计数页面，首次请求这两个页面之中的任何一个，都可以使计数器计数。如果从其中一个页面转到另一个页面不会计数。页面显示效果如图 5.7 所示。

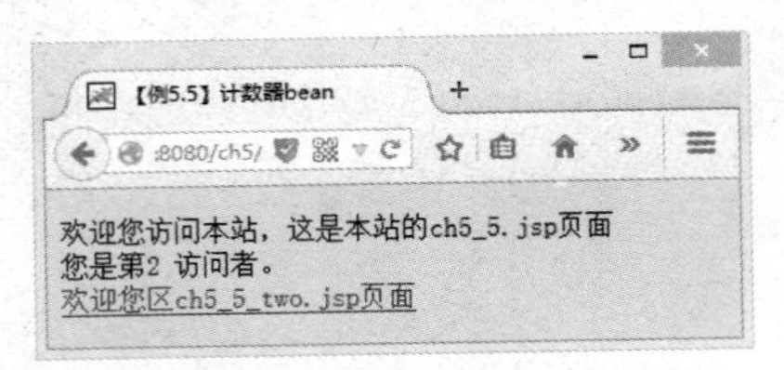

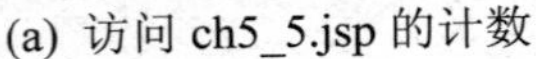

(a) 访问 ch5_5.jsp 的计数

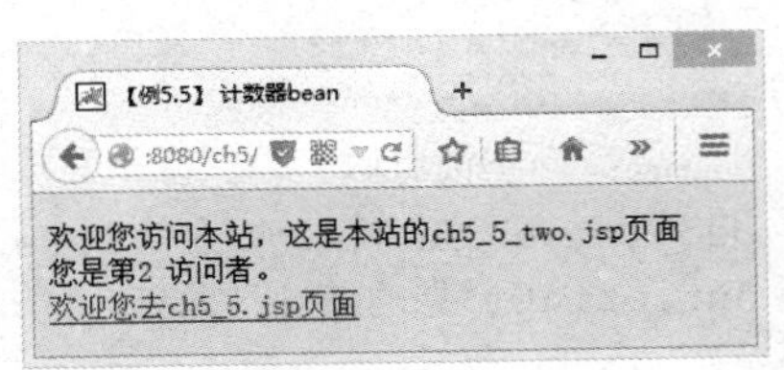

(b) 访问 ch5_5_two.jsp 的计数

图 5.7 “计数器 Bean”的页面效果

1. 创建 CounterBean 和 CountPeople

CounterBean.java 文件中的代码如下：

```
package mybean.maths;
public class CounterBean {
    long counter=0;
    public long getCounter() {
        return counter;
    }
    synchronized public void  Counter() {
       this.counter ++;
    }
}
```

CountPeople.java 文件中的代码如下：

```
package mybean.maths;
import mybean.maths.CounterBean;
public class CountPeople {
    boolean isCome=false;
    public long getCount(CounterBean countB){
        if (isCome==false){
          countB.Counter();
          isCome=true;
        }
        return countB.getCounter();
    }
}
```

2. 编写使用两个 Bean 的页面

ch5_5.jsp 文件中的代码如下：

```
<%@ page contentType="text/html;charset=GB2312" %>
<%@ page import="mybean.maths.CounterBean" %>
<%@ page import="mybean.maths.CountPeople" %>
<jsp:useBean id="countA" class="mybean.maths.CounterBean"
        scope="application" />
<jsp:useBean id="countS" class="mybean.maths.CountPeople"
        scope="session"/>
<html>
<head><title>【例 5.5】计数器 bean</title>
</head>
<body bgcolor="yellow"> <font size="3">
<%  long num;
    num=countS.getCount(countA);
%>
    <p>欢迎您访问本站，这是本站的 ch5_5.jsp 页面
    <br>您是第<%= num%>
     访问者。
     <br><a href="ch5_5_two.jsp">欢迎您去 ch5_5_two.jsp 页面</a>
</font>
</body>
</html>
```

ch5_5_two.jsp 文件中的代码如下：

```
<%@ page contentType="text/html;charset=GB2312" %>
<%@ page import="mybean.maths.CounterBean" %>
<%@ page import="mybean.maths.CountPeople" %>
<jsp:useBean id="countA" class="mybean.maths.CounterBean"
scope="application" />
<jsp:useBean id="countS" class="mybean.maths.CountPeople"
scope="session"/>
<html>
<head><title>【例 5.5】 计数器 bean</title>
</head>
<body bgcolor="yellow"> <font size="3">
<%  long num;
    num=countS.getCount(countA);
%>
    <p>欢迎您访问本站，这是本站的 ch5_5_two.jsp 页面
    <br>您是第<%= num%>
     访问者。
     <br><a href="ch5_5.jsp"> 欢迎您去 ch5_5.jsp 页面</a>
</font>
</body>
</html>
```

5.4.2 购物车 Bean

在电子商务网站中，用户选择商品首先要放入购物车中，对于购物车中的商品，用户可以根据需要进行商品数量的更改、商品的删除等操作，购物车就像一台虚拟的超市购物小推车。电子商务应用程序中的购物车，其本质是一个 Scope 为 session 的 JavaBean，用于保存用户会话期间选定商品的关键字和数量信息，并提供访问信息的接口，允许用户增加、删除和修改关键字及对应数量。购物车是电子商务 Web 应用程序开发的关键技术之一。下面以一个网上书店为例，讲述使用 Hashtable 类实现购物车 Bean 的程序编写方法。

Hashtable，即散列表，是 java.util 包中的一个类，用于存储“关键字/值”对，一般使用 String 对象作为关键字，其他对象作为关键字的关联的值。使用 Hashtable 类必须先导入 java.util 包。下面是它的一些常用的方法。

(1) Hashtable<T>()：声明和构造一个散列表对象，语句如下：

```
Hashtable<String, Integer> myGoods=new Hashtable<String, Integer>();
```

(2) put(key,value)：把关键字和值对插入散列表中，语句如下：

```
String goods_id="123";
int iTemp=10;
myGoods.put(goods_id,new Integer(iTemp));
```

(3) containsKey(key)：判断关键散列表中是否有某关键字，有返回 true，否则返回 false，语句如下：

```
String goods_id="123";
if (myGoods.containsKey(goods_id)){//是否有"123",有则返回true
    do something myGoods;
}
```

(4) get(key)：获得关键字相关联的对象，有则返回关键字关联值，否则返回 null。语句如下：

```
int iTemp=((Integer)myGoods.get(goods_id)).intValue();
```

(5) remove(key)：删除指定 key 及 key 的值，如果删除成功返回被删除关键字关联的值，删除失败返回 null。语句如下：

```
String goods_id="123";
if(myGoods.remove(goods_id)==null){}
```

(6) keys()：返回散列表中关键字的枚举，返回值为 Enumeration 类型。语句如下：

```
Hashtable<String, Integer> myGoods=new Hashtable<String, Integer>();
Enumeration<String> enumkeys=myGoods.keys();
```

(7) size()：返回散列表中的项数，返回值是 int 类型数据。语句如下：

```
Hashtable<String, Integer> myGoods=new Hashtable<String, Integer>();
int count=myGoods.size();
```

Hashtable 使用枚举接口(Enumeration)来遍历元素序列。主要使用 hasMoreElements 和 nextElement 这两个方法。hasMoreElements()如果还有更多的元素可以查看，则返回 true，否则返回 false；nextElement()方法返回要查看的下一个元素，如果 hasMoreElements()返回 false 不能调用 nextElement()方法。例如遍历散列表中所有元素序列的语句如下：

```
Hashtable<String, Integer> list= Hashtable<String, Integer>();
Enumeration<String> enumkeys=list.keys();
while(enumkeys.hasMoreElements()){
    String goods=(String)enumkeys.nextElement();
    do something with goods;
}
```

【例 5.6】购物车 JavaBean 及其应用。页面效果如图 5.8、图 5.9 和图 5.10 所示。

首先建立一个商品类 Goods，该类用于封装单个商品的序号(id)、商品名称(name)和商品价格(price)信息。然后建立一个名字为 GoodList 的 Bean，该 Bean 的属性为数组列表 ArrayList<Goods>，封装了多个商品对象。GoodList 的作用范围为 application，用于保存多个商品的信息。

Goods.java 文件中的代码如下：

```
package mybean.maths;
public class Goods{
    String goods_id=null;
    String goods_name=null;
    double goods_price;
    public String getGoods_id() {
        return goods_id;
    }
    public void setGoods_id(String goods_id) {
        this.goods_id = goods_id;
    }
    public String getGoods_name() {
        return goods_name;
    }
    public void setGoods_name(String goods_name) {
        this.goods_name = goods_name;
    }
    public double getGoods_price() {
        return goods_price;
    }
    public void setGoods_price(double goods_price) {
        this.goods_price = goods_price;
    }
}
```

GoodsList.java 文件中的代码如下：

```
package mybean.maths;
import java.util.ArrayList;
```

```
public class GoodsList {
    //声明 bean 的属性，类型为商品对象的数组列表
    ArrayList<Goods> goodsL=new ArrayList<Goods>();
    public ArrayList<Goods> getGoodL() {
        return goodsL;
    }
    public void setGoodL(ArrayList<Goods> goodsL) {
        this.goodsL = goodsL;
    }
    public GoodsList(){//构造函数，建立 10 本书的商品对象数组列表
        for(int i=0;i<10;i++){
        Goods goodsTemp=new Goods();
        goodsTemp.setGoods_id(String.valueOf(i+1));
        goodsTemp.setGoods_name("Java 语言编程技术"+String.valueOf(i+1));
        goodsTemp.setGoods_price(68+i);
            goodsL.add(goodsTemp);
        }
    }
}
```

shoppingCart.java 文件中的代码如下：

```
package mybean.maths;
import java.util.*;
import java.io.*;
public class shoppingCart implements Serializable{
    Hashtable<String, Integer> myGoods=new Hashtable<String, Integer>();
    public shoppingCart(){}//构造函数
    //只存关键字和数量，其他价格等信息从 GoodList 中得到
    public void addGoods(String goods_id,int goods_count ){
        //商品放入购物车，商品存在则累加，否则新建
        if(myGoods.containsKey(goods_id)){
            int iTemp=((Integer)myGoods.get(goods_id)).intValue();
            iTemp=iTemp+goods_count;
            myGoods.put(goods_id,new Integer(iTemp));
        }
        else{
            myGoods.put(goods_id, new Integer(goods_count));
        }
    }
    public boolean minusGoods(String goods_id,int goods_count){
        //商品从购物车中拿出
        if(myGoods.containsKey(goods_id)){
            //存在该商品，则减数量
            int iTemp=((Integer)myGoods.get(goods_id)).intValue();
            iTemp=iTemp-goods_count;
```

```
            if(iTemp<=0){
                deleteGoods(goods_id);
            }
            else{
                myGoods.put(goods_id, new Integer(iTemp));
            }
            return true;
        }
        else{//不存在该商品
            return false;
        }
    }
    public boolean deleteGoods(String goods_id){
        //删除购物车中的一件商品
        if(myGoods.remove(goods_id)==null){
            return false;
        }
        else{
            return true;
        }
    }
    public Hashtable<String,Integer> listMyGoods(){
        //得到购物车中所有的商品
      return myGoods;
    }
}
```

编译基础类 Goods 和 GoodsList、shoppingCart 两个 Bean，并布置字节码文件。接下来编写 JSP 页面。该例的购物流程为查看商品(代码参见 ch5_6_book.jsp)→购买商品(代码参见 ch5_6_buy.jsp)→添加到购物车(代码参见 ch5_6_add.jsp)或者删除(代码参见 ch5_6_delete.jsp)或者减少(代码参见 ch5_6_minus.jsp)→继续购买(代码参见 ch5_6_book.jsp)。

ch5_6_book.jsp 文件中的代码如下：

```
<%@ page language="java" contentType="text/html;charset=GB2312"%>
<%@ page import="java.util.*" %>
<%@ page import="mybean.maths.GoodsList" %>
<%@ page import="mybean.maths.Goods" %>
<jsp:useBean id="glApp" class="mybean.maths.GoodsList"
    scope="application"/>
<html>
  <head>
  </head>
  <body>
  我要<a href="ch5_6_buy.jsp" >查看购物车</a><br>
```

```
<%
    ArrayList<Goods> goodsList=glApp.getGoodL();
     out.print("<table border>");
     out.print("<tr><td colspan=4 align=center>");
     out.print("网上书店图书列表</td></tr>");
     out.print("<tr><td width=40>"+"序号"+"</td>");
     out.print("<td width=300>"+"书名"+"</td>");
     out.print("<td width=80>"+"价格"+"</td>");
     out.print("<td width=80>"+"购买吗? "+"</td>");
     out.print("</tr>");
    for(Goods me:goodsList){
     out.print("<tr>");
        out.print("<td>" +me.getGoods_id()+"</td>");
        out.print("<td>"+me.getGoods_name()+"</td>");
        out.print("<td>"+me.getGoods_price()+"</td>");
        out.print("<td><a href='ch5_6_buy.jsp?goods_id="+
             me.getGoods_id()+"'>我要购买</a></td>");
        out.print("</tr>");
     }
    out.print("<table>");
 %>
 </body>
</html>
```

该页面没有用到购物车 shoppingCart，程序片从范围为 application(应用程序)的 GoodsList 中得到商品的所有信息，并以表格的形式显示出来，如图 5.8 所示。页面左上角有一个“查看购物车”的链接，单击它链接到查看购物车 ch5_6_buy.jsp 页面。每个商品的后面是放入购物车的超级链接“我要购买”，单击它链接到放入购物车 ch5_6_buy.jsp 页面，查看购物车和放入购物车是一个页面，购物车页面效果如图 5.9 所示。 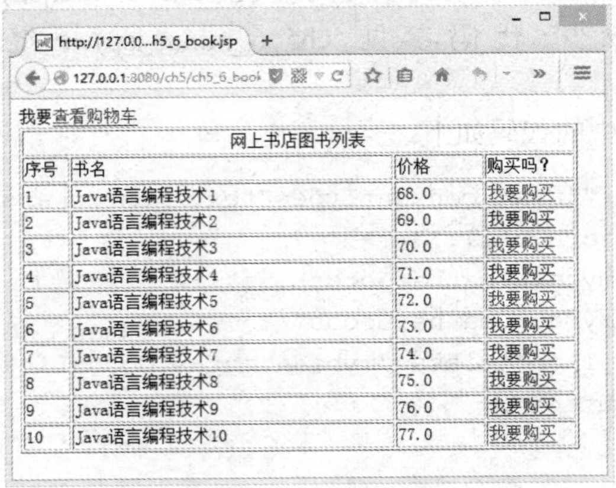

我要查看购物车

网上书店图书列表			
序号	书名	价格	购买吗？
1	Java语言编程技术1	68.0	我要购买
2	Java语言编程技术2	69.0	我要购买
3	Java语言编程技术3	70.0	我要购买
4	Java语言编程技术4	71.0	我要购买
5	Java语言编程技术5	72.0	我要购买
6	Java语言编程技术6	73.0	我要购买
7	Java语言编程技术7	74.0	我要购买
8	Java语言编程技术8	75.0	我要购买
9	Java语言编程技术9	76.0	我要购买
10	Java语言编程技术10	77.0	我要购买

图 5.8　ch5_6_book.jsp 的页面效果

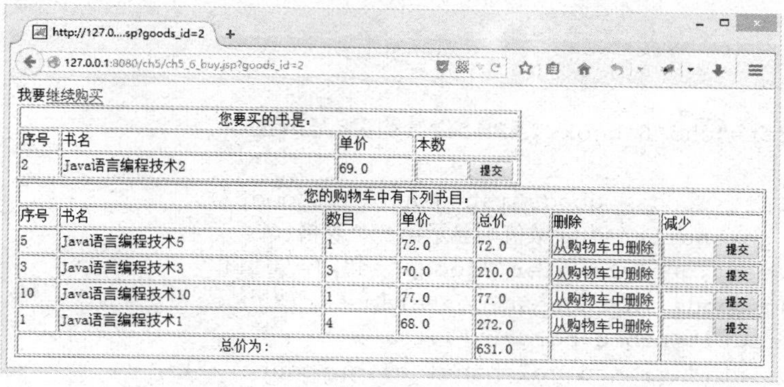

http://127.0....sp?goods_id=2

127.0.0.1:8080/ch5/ch5_6_buy.jsp?goods_id=2

我要继续购买

您要买的书是：			
序号	书名	单价	本数
2	Java语言编程技术2	69.0	提交

您的购物车中有下列书目：						
序号	书名	数目	单价	总价	删除	减少
5	Java语言编程技术5	1	72.0	72.0	从购物车中删除	提交
3	Java语言编程技术3	3	70.0	210.0	从购物车中删除	提交
10	Java语言编程技术10	1	77.0	77.0	从购物车中删除	提交
1	Java语言编程技术1	4	68.0	272.0	从购物车中删除	提交
总价为：				631.0		

图 5.9　ch5_6_buy.jsp 的页面效果

ch5_6_buy.jsp 文件中的代码如下：

```
<%@ page language="java" contentType="text/html;charset=GB2312"%>
<%@ page import="java.util.*"%>
<%@ page import="mybean.maths.GoodsList" %>
<%@ page import="mybean.maths.Goods" %>
<%@ page import="mybean.maths.shoppingCart" %>
<jsp:useBean id="glApp" class="mybean.maths.GoodsList"
    scope="application"/>
<jsp:useBean id="spcart" class="mybean.maths.shoppingCart"
    scope="session"/>
<%!
   //汉字处理函数
   public String handleString(String s){
     String str=s;
     try{
        byte b[]=str.getBytes("ISO-8859-1");
        str=new String(b);
     }
     catch(Exception ee){}
       return str;
   }
   //查找特定关键字的商品对象
   public Goods searchGood(ArrayList<Goods> goodsList,
       String goods_id){
     Goods gds=null;
        for(Goods me:goodsList){
         if(goods_id.equals(me.getGoods_id())){
             gds=me;
             break;
        }
        }
        return gds;
   }
%>
<html>
```

```
<head>
</head>
<body>
我要<a href="ch5_6_book.jsp" >继续购买</a><br>
<%
   ArrayList<Goods> goodsList=glApp.getGoodL();
   Goods gds=null;//有参数值则显示第一个表格
   if(request.getParameter("goods_id")!=null){
    String goods_id=request.getParameter("goods_id");
    out.print("<table border>");
    out.print("<tr><td colspan=4 align=center>");
    out.print("您要买的书是：</td></tr>");
    out.print("<tr>");
        out.print("<td width=40>序号</td>");
        out.print("<td width=300>书名</td>");
        out.print("<td width=80>单价</td>");
        out.print("<td width=100>本数</td>");
    out.print("</tr>");
    gds=searchGood(goodsList,goods_id);
    out.print("<tr>");
      out.print("<td>"+goods_id+"</td>");
      out.print("<td>"+gds.getGoods_name()+"</td>");
      out.print("<td>"+gds.getGoods_price()+"</td>");
      out.print("<form action=ch5_6_add.jsp method=post>");
      out.print("<td>"+"<input type=input name=book_count size=6>");
     out.print("<input type=hidden name=goods_id value="+goods_id+">");
      out.print("<input type=submit name=submit value=提交
></td></form>");
      out.print("</tr>");
      out.print("</table>");
    }
   double all_price=0;
   out.print("<table border>");
   out.print("<tr><td colspan=7 align=center>");
   out.print("您的购物车中有下列书目：</td></tr>");
   out.print("<tr>");
       out.print("<td width=40>序号</td>");
       out.print("<td width=300>书名</td>");
       out.print("<td width=80>数目</td>");
       out.print("<td width=80>单价</td>");
       out.print("<td width=80>总价</td>");
       out.print("<td width=100>删除</td>");
       out.print("<td width=100>减少</td>");
   out.print("</tr>");
   //遍历购物车中所有商品的 key
   Hashtable<String, Integer> list=spcart.listMyGoods();
   Enumeration<String> enumkeys=list.keys();
   while(enumkeys.hasMoreElements()){
    String goods=(String)enumkeys.nextElement();
```

```
        goods=handleString(goods);
        out.print("<tr>");
            out.print("<td>"+goods+"</td>");
            gds=searchGood(goodsList,goods);//从商品数组列表找到商品
            out.print("<td>"+gds.getGoods_name()+"</td>");
            out.print("<td>"+list.get(goods)+"</td>");
            out.print("<td>"+gds.getGoods_price()+"</td>");
            out.print("<td>");
            out.print(gds.getGoods_price()*list.get(goods)+"</td>");
            out.print("<td> <a href='ch5_6_delete.jsp?goods_id=");
            out.print(goods+"'>从购物车中删除</a></td>");
            out.print("<form action=ch5_6_minus.jsp method=post>");
            out.print("<td>"+"<input type=input name=book_count size=6>");
            out.print("<input type=hidden name=goods_id value="+goods+">");
            out.print("<input type=submit value=提交></td></form>");
            out.print("</tr>");
            all_price=all_price+list.get(goods)*gds.getGoods_price();
        }
        out.print("<tr>");
        out.print("<td colspan=4 align=center>"+"总价为:"+"</td>");
        out.print("<td>"+all_price+"</td><td> </td><td> </td>");
        out.print("</tr>");
        out.print("</table>");
    %>
    </body>
</html>
```

购物车 shoppingCart 的 scope 为 session(会话)，保存了用户会话期间选择商品的信息。页面中声明了两个表格，一个表格显示要添加到购物车的信息，要求客户输入购买个数，然后单击“提交”链接到 ch5_6_add.jsp 页面，将信息放入购物车中。下边的表格显示购物车中所有商品的信息，提供删除和减少操作的链接，单击“删除”链接到删除页面(ch5_6_delete.jsp)，单击“减少”链接到减少页面(ch5_6_minus.jsp)。查看购物车时，页面根据 goods_id 参数值决定是否显示第一个表格。

ch5_6_add.jsp 向购物车中添加商品，该页面接受两个参数，一个是商品 goods_id，另一个是商品数量 book_count。页面效果如图 5.10(a)所示。

ch5_6_add.jsp 文件中的代码如下：

```
<%@ page language="java" contentType="text/html;charset=GB2312"%>
<%@ page import="java.util.*"%>
<%@ page import="mybean.maths.shoppingCart" %>
<jsp:useBean id="spcart" class="mybean.maths.shoppingCart"
    scope="session"/>
<html>
<head>
</head>
<body>
我要<a href="ch5_6_book.jsp" >继续购买</a><br>
<%
```

```
    String goods_id=request.getParameter("goods_id");
    try{
        int goods_count=
          Integer.parseInt(request.getParameter("book_count"));
        spcart.addGoods(goods_id,goods_count);
        out.print("成功放入购物车!");
    }
    catch(Exception e){
        out.print("输入的数字不正确! ");
    }
%>
<br>我要<a href="ch5_6_buy.jsp">查看购物车</a>
</body>
</html>
```

ch5_6_minus.jsp 减少购物车中的商品，该页面接受两个参数，书的 goods_id 和减少的数量参数 book_count，页面效果如图 5.10(b)所示。

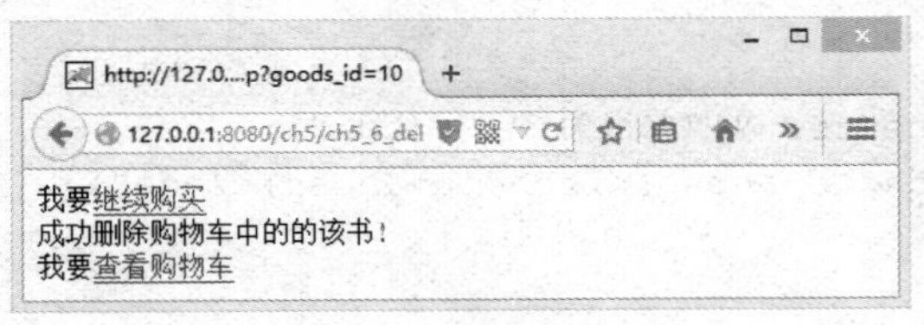

(a) ch5_6_minus.jsp 的页面效果

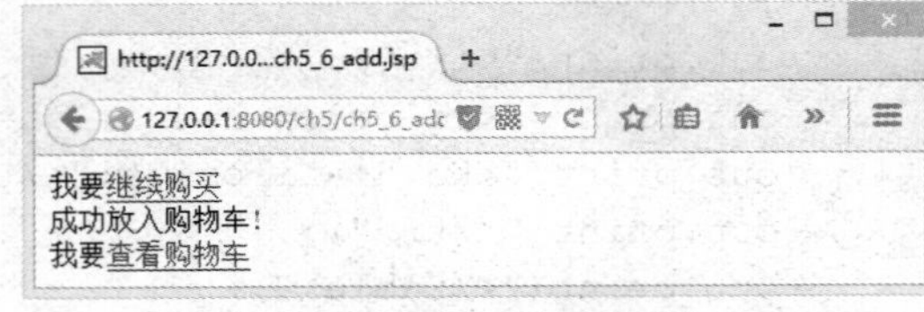

(b) ch5_6_add.jsp 的页面效果

图 5.10 添加、删除购物车中的书目

ch5_6_minus.jsp 文件中的代码如下：

```
<%@ page language="java" contentType="text/html;charset=GB2312"%>
<%@ page import="java.util.*"%>
<%@ page import="mybean.maths.shoppingCart" %>
<jsp:useBean id="spcart" class="mybean.maths.shoppingCart"
    scope="session"/>
<html>
<head>
</head>
<body>
我要<a href="ch5_6_book.jsp" >继续购买</a><br>
<%
  String goods_id=request.getParameter("goods_id");
  try{
      int goods_count=
        Integer.parseInt(request.getParameter("book_count"));
      spcart.minusGoods(goods_id,goods_count);
      out.print("成功从购物车中取出!");
  }
  catch(Exception e){
      out.print("输入的数字不正确! ");
  }
```

```
%>
<br>我要<a href="ch5_6_buy.jsp">查看购物车</a>
</body>
</html>
```

ch5_6_delete.jsp 删除购物车中该商品，接受书的 goods_id 参数，根据这个参数将商品删除。其代码如下：

```
<%@ page language="java" contentType="text/html;charset=GB2312"%>
<%@ page import="java.util.*"%>
<%@ page import="mybean.maths.shoppingCart" %>
<jsp:useBean id="spcart" class="mybean.maths.shoppingCart"
    scope="session"/>
<html>
<head>
</head>
<body>
我要<a href="ch5_6_book.jsp" >继续购买</a><br>
<%
  String goods_id=request.getParameter("goods_id");
  try{
      if(spcart.deleteGoods(goods_id)){
          out.print("成功删除购物车中的该书!");
      }
      else{
          out.print("删除失败! ");
      }
  }
  catch(Exception e){
      out.print("输入参数不正确! ");
  }
%>
<br>我要<a href="ch5_6_buy.jsp">查看购物车</a>
</body>
</html>
```

注意：该例在实际开发中商品信息会存入数据库中，通过数据库访问程序得到商品对象的数组列表，其页面和 Bean 基本相同。关于数据库访问的内容见第 7 章。

5.5 上机实训

实训目的

- 理解 JavaBean 的概念和特性。
- 掌握编写 JavaBean 的方法和注意事项。
- 掌握程序片中使用 Bean 的方法。
- 掌握表单与 JavaBean 的参数交互方法。

- 掌握 JavaBean 的辅助类的编写方法与注意事项。
- 掌握计数器和购物车编写技术。

实训内容

实训 1 编写一个计算圆周长和面积的 Bean，使用一个 JSP 页面调用该 Bean。

要求：

(1) 用表单输入圆的半径。

(2) 使用 JSP 动作标记获得并处理表单参数，得到圆的面积和周长。

实训 2 参照第 11 章，完成网上报名系统登录 JavaBean 的设计。

要求：

(1) 登录页面名称为 login.html、登录验证页面 login.jsp 程序设计。

(2) JavaBean 存储用户名和密码，提供用户密码验证方法。从表单中获得用户输入的用户名和密码，调用 Bean 的方法进行验证。

(3) 根据验证结果跳转页面，若成功转到 index.jsp 页面，若失败转到 error.jsp 页面。

实训 3 将 5.4.2 购物车实例中的汉字处理、遍历数组列表中元素的方法封装到 Bean 中，重新实现购物车示例。

要求：

(1) 改变页面的风格。

(2) 将原来程序片中的某些 out 输出，重写为 HTML 输出。

实训总结

通过本章的上机实训，学员应该能够理解 JavaBean 的概念和特点；掌握 JavaBean 的编写、编译、布置和使用方法；掌握 Bean 与表单参数的交互方法；掌握计数器、购物车等常用 JavaBean 的编写方法。

5.6 本章习题

思考题

(1) JavaBean 和一般意义上的 Java 类有何区别？

(2) 如何实现一个 Bean 的属性与表单参数的关联？

(3) 如何在页面的程序片中使用 Bean？

(4) 试述 request、session 和 application 有效范围的 Bean 生命周期。

(5) 怎样编写、编译、调试和布置 Bean？

拓展实践题

完成 5.4.2 购物车实例中的数据输入部分的有效性检查代码。

第6章 文件访问

学习目的与要求：

在 Web 程序开发中，经常需要对文件进行操作，例如将文件上传到服务器、从服务器下载文件、分页显示文件内容等。本章主要学习 Java 输入/输出流的概念，File 类对文件属性和目录的管理技术，字节流类、字符流类和文件的随机访问等内容。通过本章的学习，学员要理解输入/输出流的概念，掌握使用字节流、字符流类读写文件的 JSP 编程方法；掌握随机读写文件类的使用方法，掌握常用的上传文件、下载文件以及文件内容分页显示的编程技术。

JSP 通过 Java 的输入/输出流来实现文件操作。本章采用 JSP+Bean 的模式来讲解文件操作，JSP 负责数据的表示，Bean 负责文件的读写操作，Bean 的包名为 mybean.file。学习本章前，先建立一个虚拟服务目录 ch6，并在 ch6 中建立 WEB-INF\classes\mybean\file 目录结构，所有 JSP 文件存入 ch6 目录，Bean 的字节码文件复制到相应目录。

6.1 输入/输出流概述

6.1.1 流的概念

输入/输出意味着程序可能有多种不同类型的输入/输出设备，如：磁盘文件，网络套接字，外部设备等。在 Java 中，用 java.io 包中提供的“流”类来完成各种 I/O 操作。

流是一个很形象的概念，当程序需要读取(输入)数据的时候，就会开启一个通向数据源的通道，这个数据源可以是文件，内存，或是网络连接等。类似地，当程序需要写入(输出)数据的时候，就会开启一个通向目的地的通道。流就好像是在计算机的输入/输出之间建立的一条通道，数据好像在这其中“流”动一样。

采用流的机制可以使数据有序地进行输入/输出，即每一个数据都必须等前面的数据读入或写出之后才能被读写。每次读写操作处理的都是序列中剩余的未处理数据中的第一个，而不能够随意选择输入/输出的位置。

另外，使用流的机制可以增强程序的安全性。当数据流建立好以后，如果程序是数据流的源，不用管数据的目的地是哪里(可能是显示器、打印机或远程网络客户端)，把接收方看成一个“黑匣子”，只要提供数据就可以了。类似地，如果程序是数据流的目的地，同样不用关心数据流的源是哪里，只要在数据流中提取自己所需要的数据就可以了。

Java 提供的流类，从功能上看，它可以分为输入流/输出流(即 I/O 流)两种。从所操作的数据单位来看，流类可分为字节流和字符流两类。

6.1.2 输入流与输出流

在进行数据输入时，可以使用流通道把源中的数据送到应用程序。这里把读取数据的地方称为源，这条通道称为输入流，代表从外设流入计算机的数据序列。即利用输入流将源中的数据送到应用程序，如图 6.1 所示。

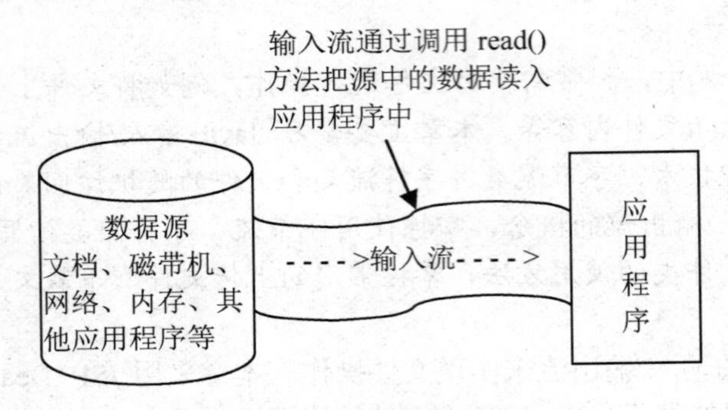

图 6.1 输入流示意图

在进行数据输出时，可以使用流通道把应用程序中的数据送往目的地。这里的目的地指需要写入数据的地方，此时的通道称为输出流，代表从计算机流向外设的数据序列。即利用输出流将应用程序中的数据送往目的地，如图 6.2 所示。

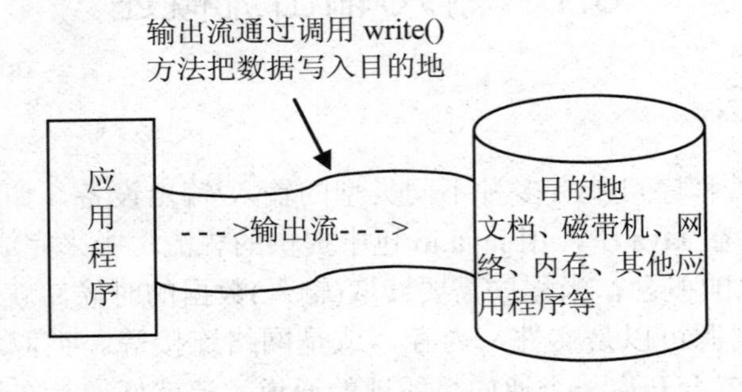

图 6.2 输出流示意图

6.1.3 字节流与字符流

Java 提供的流类(在 java.io 包)，从功能上看，它可以分为输入流/输出流(即 I/O 流)两种。从所操作的数据单位来看，流类可分为字节流和字符流两类。所有输入流类都是抽象类 InputStream(字节输入流)或抽象类 Reader(字符输入流)的子类。所有输出流都是 OutputStream(字节输出流)或抽象类 Writer(字符输出流)的子类。

如果程序需要输入的是 8 位的 byte 数据，则需要利用字节流来处理。字节流一般是用来读入或写出二进制数据。如：影像，声音。另外，像压缩文件等“非纯文本”的数据，也可以用此种流类来进行处理。如果程序需要处理的是“纯文本”类的信息，建议使用字符流来进行处理。原因是它们能够支持 Unicode 标准字符集的各种字符。

6.2 File 类

File 类直接处理文件和文件系统，它并不涉及文件的读写操作，而是获取或设置文件本身的属性信息，例如文件所在的目录、文件的长度、文件的读写权限、文件的日期时间等。目录也被当成是一个 File 对象，File 可以对目录进行管理操作。

6.2.1 File 类的重要属性与方法

1. 属性

常用属性如下。

- public static final String pathSeparator：系统相关的路径分隔字符串。
- public static final char pathSeparatorChar：系统相关的路径分隔符，用于分隔多个路径。
- public static final String separator：系统相关的文件分隔字符串。
- public static final char separatorChar：系统相关的文件分隔符。

在 Windows 系统中，pathSeparator 和 pathSeparatorChar 分别是“;”和‘;’；而 separator 和 separatorChar 分别是“\”和‘\’。

2. 构造方法

- public File(String pathname)：用参数 pathname 创建一个文件对象。
- public File(String parent，String child)：用父目录 parent 及子路径 child 创建一个文件对象。
- public File(File dirObj，String child)：用对象 dirObj 指定目录及子路径创建一个文件对象。

注意：创建一个 File 对象，并不会在某个物理路径下创建一个文件或目录，而只是在内存中生成一个 File 类的实例对象，创建目录或文件需要执行相应的方法。

3. File 类的常用方法

File 类定义了很多获取文件对象的标准属性的方法。主要方法如下。

- public boolean canRead()：判断应用程序能否读取给定路径下的文件对象。
- public boolean canWrite()：判断应用程序能否向给定路径下的文件对象写入数据。
- public boolean exists()：判断给定路径下的文件是否存在。
- public boolean isDirectory()：判断 File 对象对应的路径是否为目录。
- public boolean isFile()：判断 File 对象对应的路径是否为文件。
- public boolean isHidden()：判断 File 对象对应的文件属性是否为隐藏。
- public long lastModified()：返回 File 对象对应的文件最后修改的时间。
- public long length()：获取 File 对象对应的文件的长度。

- public boolean createNewFile()：创建一个新文件。如果创建成功返回 true，否则返回 false(该文件已经存在)。
- public String[] list()：获取 File 对象对应的目录下的文件名及目录名。用字符串的形式返回该目录下的文件和子目录名称。
- public String[] list(FilenameFilter obj)：用字符串的形式返回该目录下指定类型的所有文件和子目录名称。
- public boolean mkdir()：创建 File 对象对应的路径，如果创建成功返回 true，否则返回 false(该目录已经存在)。
- public File[] listFiles()：获取 File 对象对应的目录下的文件名及目录名。用 File 对象的形式返回该目录下的文件和子目录。
- public File[] listFiles(FilenameFilter obj)：该方法用 File 对象的形式返回对象对应目录指定类型的所有文件。
- public boolean renameTo(File dest)：重命名 File 对象对应的文件。
- public boolean setReadOnly()：设置 File 对象对应的文件为只读。
- public String toString()：返回 File 对象对应路径的字符串。

6.2.2 查询文件属性

前面已经介绍了 File 类常用的属性和方法，下面通过一个例子演示查询文件属性的方法。本例查询文件属性的操作代码写在了 JSP 页面的程序片中。

【例 6.1】File 类查询文件属性方法示例。页面效果如图 6.3 所示。

图 6.3 查询文件属性的页面效果

ch6_1.jsp 文件代码如下：

```
<%@ page contentType="text/html;charset=GB2312" %>
<%@ page import="java.util.*" %>
<%@ page import="java.io.*" %>
<html>
<head>
<title>【例 6.1】File 类查询文件属性方法示例</title>
</head>
<body>
<%
   File dir=new File("E:\\programJsp\\ch6");
   File f1=new File(dir,"ch6_1_file.txt");
```

```
    try
    {
      dir.mkdir(); //创建目录
      f1.createNewFile(); //创建文件
      out.println("<br>目录: "+dir);
      out.println("<br>文件: "+f1+"的属性:");
      out.println("<br>名字:"+f1.getName());
      out.println("<br>路径:"+f1.getPath());
      out.println("<br>只读:"+f1.canRead());
      Date date=new Date(f1.lastModified());
      out.println("<br>最后修改日期:"+date.toString());
    }
    catch(Exception e)
    {
      System.out.println(e.toString());
    }
%>
</body>
</html>
```

说明：以上程序运行后，首先在 E:\programJsp\ch6 目录下创建一个文本文件 ch6_1_file.txt。然后查询文件的各种属性。程序运行结果如图 6.3 所示。

6.2.3　目录管理

File 对象提供了管理目录的一些方法，例如创建目录 mkdir()、列出目录中的文件 list()或 listFiles()、删除目录的 delete()方法等，这为用户在程序中操作目录带来了方便。

list()和 listFiles()方法不仅可以列出 File 对象对应目录的所有文件及子目录，还可以列出指定类型的文件，列出指定类型文件的调用格式为：

```
String str[]=list(FilenameFilter obj);
```

或

```
File file1[]=listFiles(FilenameFilter obj);
```

FilenameFilter 是一个接口，该接口有一个方法：

```
public boolean accept(File dir,String name);
```

当向 File 对象的 list(FilenameFilter obj)方法传递了一个 FilenameFilter 接口对象时，list 用接口对象的 accept()方法判断文件是否符合要求，符合要求则列出。例如实现接口 FilenameFilter 的 Filename 类的定义语句如下：

```
class Filename implements FilenameFilter{
    String str=null;
    public void Filename(String s){
        str="."+s;
```

```
    }
    public boolean accept(File dir,String name){
        return name.endsWith(str);
    }
}
```

下面是目录管理的一个示例，该例与例 6.1 不同，使用 JavaBean 来实现新建、删除、列出文件等目录操作，JSP 页面负责信息输入和显示。

【例 6.2】 列出目录中的所有文件和子目录(相关代码参见 manageFile.java 和 ch6_2.jsp)，页面效果如图 6.4 所示。

manageFile.java 文件代码如下：

```
package mybean.file;
import java.io.*;
public class manageFile {
    String mdDirName=null;
    String ltDirName=null;
    String dlDirName=null;
    String parentPath=null;
    StringBuffer allFilesDirs=null;
    public String getParentPath() {
        return parentPath;
    }
    public void setParentPath(String parentPath) {
        this.parentPath = parentPath;
    }
    public manageFile(){
        allFilesDirs=new StringBuffer();
    }
    public String getMdDirName() {
        return mdDirName;
    }
    public void setMdDirName(String mdDirName) {
        this.mdDirName = mdDirName;
        if(this.mdDirName!=null){
            File dir=new File(parentPath,
                    this.mdDirName);
            dir.mkdir();
        }
    }
    public String getLtDirName() {
        return ltDirName;
    }
    public void setLtDirName(String ltDirName) {
        this.ltDirName = ltDirName;
    }
    public String getDlDirName() {
```

```
        return dlDirName;
    }
    public void setDlDirName(String dlDirName) {
        this.dlDirName = dlDirName;
        if(this.dlDirName!=null){
            File dir=new File(parentPath,
                    this.dlDirName);
            dir.delete();
        }
    }
    public StringBuffer getAllFilesDirs() {
        if(ltDirName!=null){
            File dir=new File(parentPath,
                    ltDirName);
            File ltF[]=dir.listFiles();
            for(int ii=0;ii<ltF.length;ii++){
                if(ltF[ii].isDirectory()){
                    allFilesDirs.append("<br>子目录:"+
                            ltF[ii].getName());
                }
            }
            for(int ii=0;ii<ltF.length;ii++){
                if(ltF[ii].isFile()){
                    allFilesDirs.append("<br>文件:"+
                            ltF[ii].getName());
                }
            }
        }
        return allFilesDirs;
    }
}
```

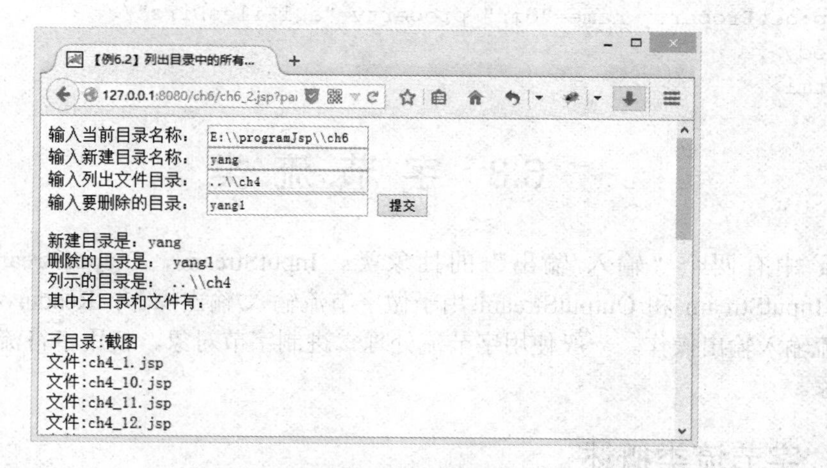

图 6.4　列出文件和子目录

ch6_2.jsp 文件代码如下：

```
<%@ page contentType="text/html;charset=GB2312" %>
<%@ page import="mybean.file.manageFile" %>
<jsp:useBean id="dir" class="mybean.file.manageFile" scope="page" />
<html>
<head>
<title>【例 6.2】列出目录中的所有文件和子目录</title>
</head>
<body>
<jsp:setProperty name="dir" property="parentPath" param="parentPath"/>
<jsp:setProperty name="dir" property="mdDirName" param="mdDirName"/>
<jsp:setProperty name="dir" property="ltDirName" param="ltDirName"/>
<jsp:setProperty name="dir" property="dlDirName" param="dlDirName"/>
<form action="" name="form1" >
  输入当前目录名称：
  <input type="text" name="parentPath" value="E:\programJsp\ch6"/>
  <br>输入新建目录名称：
  <input type="text" name="mdDirName" value="yang"/>
  <br>输入列出文件目录：
  <input type="text" name="ltDirName" value="..\ch4" />
  <br>输入要删除的目录：
  <input type="text" name="dlDirName" value="yang1"/>
  <input type="submit" name="submit" value="提交"/>
</form>
新建目录是：<jsp:getProperty name="dir" property="mdDirName"/>
<br>删除的目录是：
<jsp:getProperty name="dir" property="dlDirName"/>
<br>列示的目录是：
<jsp:getProperty name="dir" property="ltDirName"/>
<br>其中子目录和文件有：<br>
<jsp:getProperty name="dir" property="allFilesDirs"/>
</body>
</html>
```

6.3 字节流类

Java 中有四个“输入/输出”的抽象类，InputStream、OutputStream、Reader 和 Writer。InputStream 和 OutputStream 用于做字节流输入/输出操作，Reader 和 Writer 用于做字符流输入输出操作。一般使用字节流处理二进制字节对象，使用字符流处理字符或字符串对象。

6.3.1 字节流类概述

字节流类 InputStream 和 OutputStream 被声明为抽象类，所以，真正用来作数据输入和输出处理的是它的各个具体子类。这些子类可以针对不同的输入进行处理。常用的字节流类如表 6.1 所示。

表 6.1　字节流类

流　　类	描　　述
InputStream	描述流输入的抽象类
OutputStream	描述流输出的抽象类
ByteArrayInputStream	从字节数组读取的输入流
ByteArrayOutputStream	向字节数组写入的输出流
FileInputStream	读取文件的输入流
FileOutputStream	写文件的输出流
DataInputStream	包含读取 Java 标准数据类型方法的输入流
DataOutputStream	包含编写 Java 标准数据类型方法的输出流
BufferedInputStream	缓冲输入流
BufferedOutputStream	缓冲输出流
PrintStream	包含 print()和 println()的输出流

InputStream 是一个定义了字节输入的抽象类。它提供了其所有子类共用的一些方法来统一基本的读操作。但作为抽象类，它不能直接生成对象，只能通过其子类来生成程序中需要的对象。

InputStream 类中的方法如下。

- public abstract int read()：从输入流中读取字节数据。如果读取的下一个字节数据成功，则返回 0～255 之间的一个整型值；如果遇到流结尾，则返回-1。此方法是 abstract 的，必须在子类中实现。
- public int read(byte[] b)：从输入流中读取 b.length 个字节数据，并存放在参数代表的数组 b 中。返回成功读取的字节数。如果遇到流结尾，则返回-1。
- public int read(byte[] b, int off, int len)：从输入流中读取 len 个字节数据，并存放在参数代表的数组 b[off]开始处。返回成功读取的字节数。如果遇到流结尾，则返回-1。
- public int available()：返回当前输入流中可读取的字节数。子类需要覆盖此方法，否则返回值总是 0。
- public void mark(int readlimit)：标记当前输入流的位置，该标记在从流中读取 readlimit 个字节前都保持有效。
- public void reset()：将输入流的位置移到上次以 mark()方法标记的位置。
- public long skip(long n)：跳过 n 个字节。返回实际跳过的字节数。
- public void close()：关闭当前输入流，同时释放相关的系统资源。

注意：以上所有方法在出错条件下都将产生一个 IOException 异常。

OutputStream 是一个定义了字节输出的抽象类。它提供了其所有子类共用的一些方法来统一基本的写操作。但作为抽象类，它同样不能直接生成对象，只能通过其子类来生成程序中需要的对象。

OutputStream 类中的方法如下。

- public abstract void write(int b)：将 int 类型表示的字节数据 b 写入输出流中。
- public void write(byte[] b)：将字节数组 b 中的数据写入输出流中。
- public void write(byte[] b, int off, int len)：将字节数组 b 中以 b[off]为起点的 len 个字节元素的值写入输出流中。
- public void flush()：将缓冲区中的数据全部写入输出流中，即刷新缓冲区。
- public void close()：关闭当前输入流，同时释放相关的系统资源。

注意：以上所有的方法在出错条件下都将抛出 IOException 异常。

字节流的子类可以分为两大类。

(1) 负责不同的数据存储类型的类，如 File、StringBuffer、ByteArray 等。

(2) 对数据进行额外处理的类，如 Buffered、Data 等。

下面，以 File 数据存储类型为例介绍字节流的使用。

6.3.2 以 File 存储类型为例介绍字节流与缓冲流的使用

当需要从文件中读取数据时，将用到 FileInputStream 或 FileOutputStream。FileInputStream 类是从 InputStream 类中派生出来的简单的输入类。该类的常用方法是从 InputStream 类继承而来的。同样，FileOutputStream 类是从 OutputStream 类派生出来的简单的输出类。该类的所有方法都是从 OutputStream 类继承而来的。

1. FileInputStream 类

当编写的程序需要从文件中读取“非文本性质”的信息时，需要用到 FileInputStream 类。FileInputStream 的构造方法如下。

- public FileInputStream(File file)：利用 File 对象创建文件输入流，参数 file 指被打开并被读取数据的文件。
- public FileInputStream(String name)：利用给定的文件名 name 创建文件输入流，参数 name 指被打开并被读取数据的文件的全称路径。

2. BufferedInputStream 类

缓冲流把内存缓冲器连接到输入输出流，允许 Java 程序对多个字节同时操作，这样提高了读写效率。通过 BufferedInputStream 类可以实现缓冲输入流。使用 FileInputStream 类和 BufferedInputStream 类结合操作文件，可以极大地提高文件读入效率。

BufferedInputStream 类的构造函数如下。

- public BufferedInputStream(InputStream InputStreamName)：利用给定的输入流 InputStream 创建缓冲输入流，默认缓冲区大小为 8192 字节。
- public BufferedInputStream(InputStream InputStreamName，int BufferSize)：利用给定的输入流创建缓冲文件输入流，参数 BufferSize 以字节为单位指定缓冲区的大小。

把缓冲区与文件输入流联系的代码如下：

```
FileInputStream f0=new FileInputStream("E:\programJsp\ch6\file1.txt");
或者：
File f1=new File("E:\programJsp\ch6\file1.txt");
FileInputStream finput2=new FileInputStream(f1);
BufferedInputStream buffer1=new BufferedInputStream(finput2);
```

【例 6.3】使用字节流或缓冲流将服务器磁盘驱动器中的某个文件的内容读入(相关文件参见 readFile.java 和 ch6_3.jsp)，并显示在页面上，如图 6.5 所示。

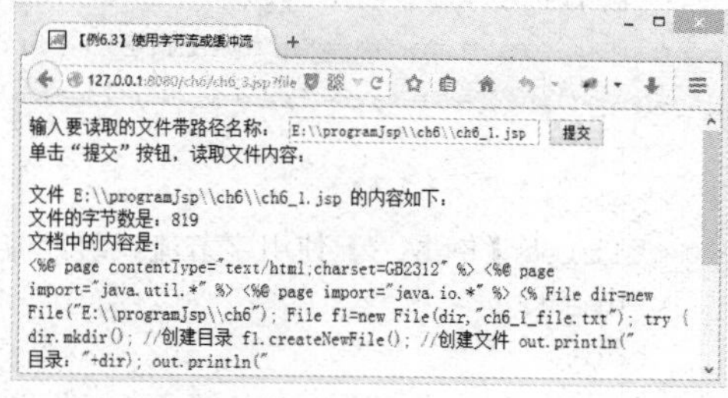

图 6.5　读取文件内容页面效果

readFile.java 文件代码如下：

```
package mybean.file;
import java.io.*;
public class readFile {
    StringBuffer allFilesmess=null;
    int size;
    String fileName=null;
    public String getFileName() {
        return fileName;
    }
    public void setFileName(String
       fileName) {
        this.fileName = fileName;
    }
    public readFile(){
        allFilesmess=new StringBuffer();
    }
    public StringBuffer getAllFilesmess()
      {
        try{
          File file1=new File(fileName);
          FileInputStream finput=new FileInputStream(file1);
          int size;
          size=finput.available();
          BufferedInputStream buffer1=new BufferedInputStream(finput);
          byte b[]=new byte[90];
          int n=0;
          allFilesmess.append("文件的字节数是："+size);
          allFilesmess.append("<br>文档中的内容是：<br>");
          while((n=buffer1.read(b))!=-1){
           allFilesmess.append(new String(b,0,n));
          }
          buffer1.close();
          finput.close();
        }
```

```
        catch(Exception e)
        {
          allFilesmess.append("读文件发生错误！");
        }
        return allFilesmess;
      }
}
```

ch6_3.jsp 文件代码如下：

```
<%@ page contentType="text/html;charset=GB2312" %>
<%@ page import="mybean.file.readFile" %>
<jsp:useBean id="readf" class="mybean.file.readFile" scope="page" />
<html>
<head>
<title>【例 6.3】使用字节流或缓冲流</title>
</head>
<body>
<form action="" name="form1" >
  输入要读取的文件带路径名称：
  <input type="text" name="fileName" size="30"
      value="E:\programJsp\ch6\ch6_1.jsp"/>
  <input type="submit" name="submit" value="提交"/>
  <br>单击"提交"按钮，读取文件内容：
</form>
<jsp:setProperty name="readf" property="fileName" param="fileName"/>
文件
<jsp:getProperty name="readf" property="fileName"/>
的内容如下：<br>
<jsp:getProperty name="readf" property="allFilesmess"/>
</body>
</html>
```

注意：InputStream 的方法在缓冲流类 BufferedInputStream 中仍可以使用。如 read(byte b[])等。

3. FileOutputStream 类

当需要从“非文本性质”数据写入文件中时，将用到 FileOutputStream 类。

FileOutputStream 的构造方法如下。

- public FileOutputStream(String name)：利用 File 对象创建文件输入流。参数 name 指被打开并且需要写入数据的文件名。
- public FileOutputStream(File file)：利用 File 对象创建文件输入流。参数 file 指被打开并且数据写入的文件。
- public FileOutputStream(File file，boolean append)：利用 File 对象创建文件输入流。参数 file 指被打开并且数据写入的文件；参数 append 指追加数据的方式，如果值为 true，则追加到文件末尾，否则添加到文件头。

FileOutputStream 对象的创建不依赖于文件是否存在。如果文件不存在，FileOutputSteam 会在打开输出文件之前创建它。如果 FileOutputStream 试图打开一个只读文件，会引发 IOException 异常。

4. BufferedOutputStream 类

BufferedOutputStream 类是用于输出缓冲的缓冲流类，和 BufferedInputStream 一样所有 OutputSteam 的方法它都可以使用。用输出缓冲流进行写入操作，必须调用 flush()方法，将缓冲区的数据存入文件中。BufferedOutputStream 类的 flush()和 close()方法与 OutputStream 相同，这里不再赘述。

BufferedOutputStream 类的构造函数有两种。

```
BufferedOutputStream(OutputStream outputStreamName);
BufferedOutputStream(OutputStream outputStreamName,int BufferSize);
```

第一种创建的是默认缓冲区大小为 512 字节的缓冲流，第二种可以创建指定缓冲区大小的缓冲流。

【例 6.4】利用字节流将内容写入文件。此例使用了两个 Bean，writeFile 将内容写入文件，readFile 从文件中读入内容。ch6_4.jsp 页面效果如图 6.6 所示。

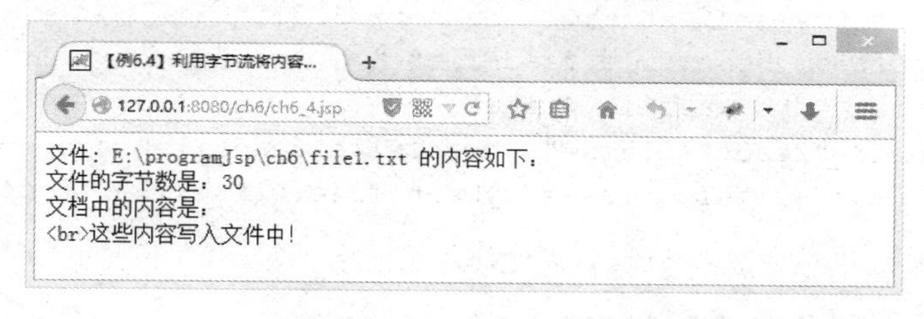

图 6.6　内容写入文件效果

readFile.java 与例 6.3 中的 Bean 相同，这里体现了组件的复用特点，其代码参见例 6.3，这里省略。

writeFile.java 文件代码如下：

```
package mybean.file;
import java.io.*;
public class writeFile {
    String fileName=null;
    String filesMess=null;
    int size;
    public String getFileName() {
        return fileName;
    }
    public void setFileName(String fileName) {
        this.fileName = fileName;
    }
    public String getFilesMess() {
        return filesMess;
    }
```

```
    public void setFilesMess(String filesMess) {
        this.filesMess = filesMess;
        try{
            FileOutputStream outf=new FileOutputStream(fileName);
            BufferedOutputStream bufferout=
                new BufferedOutputStream(outf);
            byte b[]=this.filesMess.getBytes();
            bufferout.write(b);
            bufferout.flush();
            bufferout.close();
            outf.close();
        }
        catch(Exception e){
            System.out.print(e.toString());
        }
    }
}
```

ch6_4.jsp 文件代码如下：

```
<%@ page contentType="text/html;charset=GB2312" %>
<%@ page import="mybean.file.readFile" %>
<%@ page import="mybean.file.writeFile" %>
<jsp:useBean id="readf" class="mybean.file.readFile" scope="page" />
<jsp:useBean id="writef" class="mybean.file.writeFile" scope="page"/>
<html>
<head>
<title>【例 6.4】利用字节流将内容写入文件</title>
</head>
<body>
<jsp:setProperty name="writef" property="fileName"
     value="E:\programJsp\ch6\file1.txt" />
<jsp:setProperty name="writef" property="filesMess"
     value="&lt;br&gt;这些内容写入文件中！"/>
<jsp:setProperty name="readf" property="fileName"
     value="E:\programJsp\ch6\file1.txt"/>
文件：
<jsp:getProperty name="readf" property="fileName"/>
的内容如下：<br>
<jsp:getProperty name="readf" property="allFilesmess"/>
</body>
</html>
```

注意： 程序首先生成了一个文件输出流对象，指向 E:\programJsp\ch6\file1.txt 文件。然后构造了一个输出缓冲流，使用输出缓冲流对象的 write 方法将内容写入文件缓冲区，调用 flush()方法将缓冲区的数据写入文件。最后要关闭缓冲流和输出流对象。如果 file1.txt 文件不存在，则首先创建它，然后再将内容保存到其中。

6.4 字符流类

字节流为输入/输出编程提供了强有力的支持，但它并不能处理 Unicode 编码的数据。字节流会把汉字作为两个字节来处理，如果错位就会出现乱码。而 Unicode 把一个汉字看成是一个字符来编码，Java 提供了字符流来专门处理 Unicode 编码的汉字。

6.4.1 字符流概述

字符流类指 Reader 和 Writer 类。由于 Java 采用 16 位的 Unicode 字符编码，因此需要提供基于字符的输入/输出操作。同样，这两个类被声明为抽象类，只提供一系列用于字符流处理的接口，不能生成这两个类的对象。所以真正用来作数据输入和输出处理的是它的各个具体的子类。这些子类可以针对不同的输入进行处理，如表 6.2 所示。

表 6.2 字符流类

流 类	描 述
Reader	字符流输入的抽象类
Writer	字符流输出的抽象类
FileReader	读取文件的输入流
FileWriter	写文件的输出流
CharArrayReader	从字符数组读取的输入流
CharArrayWriter	向字符数组写入的输出流
StringReader	读取字符串的输入流
StringWriter	写字符串的输出流
BufferedReader	缓冲输入流
BufferedWriter	缓冲输出流
FilterReader	过滤输入流
FilterWriter	过滤输出流
InputStreamReader	将字节输入流转换为字符输入流
OutputStreamReader	将字节输出流转换为字符输出流

Reader 是定义字符输入流的抽象类。它提供了其所有子类共用的一些方法来统一基本的读操作。常用方法如下。

- public int read()：从输入流中读入一个字符，如果读取成功，则返回一个 0～65535 之间的整型表示。如果遇到输入流尾，则返回-1。
- public int read(char[] cbuf)：从输入流中读取一些字符并存放在字符数组 cbuf 中。如果读取成功，返回读取的字符的数目；如果遇到输入流尾，则返回-1。
- public abstract int read(char[] cbuf，int off，int len)：从输入流中读取一些字符并存放在字符数组 cbuf[off]开始处，如果读取成功，返回读取的字符数；如果遇到流结尾，则返回-1。参数 len 代表读取的最大字符数。

- public boolean ready()：判断此输入流是否已就绪，并能被读取。
- mark()、reset()、skip()和 close()方法与 InputStream 类相同，不再讲述。

注意：以上所有方法在出错条件下都将产生一个 IOException 异常。

Writer 类是定义字符输出的抽象类。其常用方法如下。

- public void write(int c)：将一个以整型表示的字符写入输出流。
- public void write(char[] cbuf)：将字符数组 cbuf 中的数据写入输出流。
- public abstract void write(char[] cbuf，int off，int len)：将字符数组中以 cbuf[off]为起点的 len 个字符写入输出流。
- public void write(String str)：将字符串 str 写入输出流。
- public void write(String str，int off，int len)：将字符串中以 off 为起点的 len 个字符写入输出流。
- flush()和 close()方法与 OutputStream 相同，不再讲述。

6.4.2 以 File 存储类型为例介绍字符流和字符缓冲流的使用

1. FileReader 类

FileReader 是 Reader 的子类，可以读取“文本性质”文件的内容。现在以此为例介绍字符流的使用。

FileReader 的构造方法如下。

- public FileReader(File file)：建立文件输入流，其源文件指 file，是一个带路径的完整的文件名。
- public FileReader(String fileName)：建立文件输入流，其源文件指由 fileName 参数命名的文件。

2. BufferedReader 类

为了提高文件的读取效率，Java 中提供了字符输入缓冲流类 BufferedReader，BufferedReader 类与 FileReader 类结合起来使用，提高文件读取程序的效率。

BufferedReader 构造方法如下。

public BufferedReader(FileReader fileReaderName)：利用给定的字符输入流 FileReader 创建字符缓冲输入流。

BufferedReader 除了 Reader 类中的方法外，还提供了 public String readLine()方法。该方法读入一行文本。这里的“一行”指字符串以“\n”或“\r”作结尾。以字符串类型返回读入的字符，但不包含该行的结尾字符。如果已到达流结尾而未读入任何文字，则返回 null。下面是一个用字符输入流和字符输入缓冲流读取文件的实例。

【例 6.5】用字符输入流读取文件。该例编写了一个读文件的 Bean(textReader)，它负责用字符流读入文件内容，ch6_5.jsp 负责文件内容显示，页面显示效果如图 6.7 所示。

textReader.java 文件代码如下：

```
package mybean.file;
import java.io.*;
```

```
public class textReader {
    String fileName=null;
    String allFilesmess=null;
    public textReader(){
        allFilesmess=new String();
    }
    public String getFileName() {
        return fileName;
    }
    public void setFileName(String fileName) {
        this.fileName = fileName;
    }
    public String getAllFilesmess() {
        try{
            StringBuffer temp=new StringBuffer();
            if(fileName!=null){
            String strTemp=null;
            FileReader fr=new FileReader(fileName);
            BufferedReader buffer1=new BufferedReader(fr);
            while((strTemp=buffer1.readLine())!=null){
                byte bb[]=strTemp.getBytes();
                strTemp=new String(bb);
                temp.append("\n"+strTemp);
            }
            allFilesmess="<textarea rows=8 cols=62>"+temp+"</textarea>";
            buffer1.close();
            fr.close();
            }
        }
        catch(IOException ee){
            System.out.print(ee.toString());
        }
        return allFilesmess;
    }
}
```

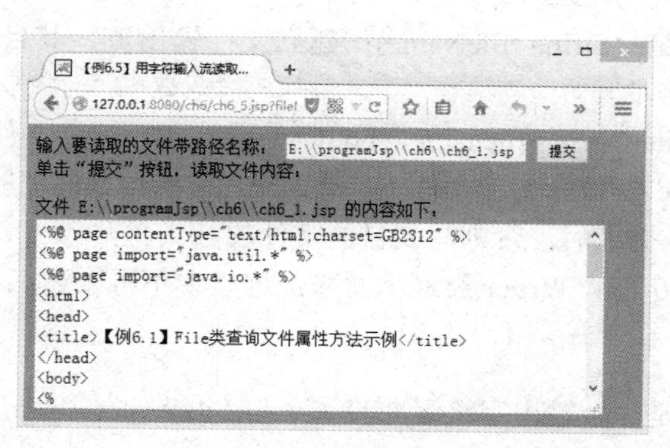

图 6.7　用字符输入流读取文件

ch6_5.jsp 文件代码如下：

```
<%@ page contentType="text/html;charset=GB2312" %>
<%@ page import="mybean.file.textReader" %>
<jsp:useBean id="txtreadf" class="mybean.file.textReader" scope="page"
/>
<html>
<head>
<title>【例 6.5】用字符输入流读取文件</title>
</head>
<body bgcolor="#00ff00">
<form action="" name="form1" method="get" >
  输入要读取的文件带路径名称：
  <input type="text" name="fileName" size="30"
     value="E:\programJsp\ch6\ch6_1.jsp"/>
  <input type="submit" name="submit" value="提交"/>
  <br>单击"提交"按钮，读取文件内容：
</form>
<jsp:setProperty name="txtreadf" property="fileName" param="fileName"/>
文件
<jsp:getProperty name="txtreadf" property="fileName"/>
的内容如下：<br>
<jsp:getProperty name="txtreadf" property="allFilesmess"/>
</body>
</html>
```

3. FileWriter 类

FileWriter 是 Writer 的子类，可以将数据写入“文本性质”的文件中。这个类的所有方法返回的都是 void 值，如果出错则产生 IOException 异常。

FileWriter 的构造方法如下。

- public FileWriter(File file)：建立文件输出流，其目的文件指 file。
- public FileWriter(File file，boolean append)：建立文件输出流，其目的文件指 file。参数 append 指数据追加的位置，如果为 true，续接于原有数据之后；否则写到目的文件的开头。
- public FileWriter(String fileName)：建立文件输出流，其目的文件指由 fileName 参数命名的文件。
- public FileWriter(String filename，boolean append)：建立文件输出流，其目的文件指由 fileName 参数命名的文件。参数 append 指数据追加的位置，如果为 true，续接于原有数据之后；否则写到目的文件的开头。

FileWriter 的常用方法从 Writer 类继承而来，这里便不再叙述。

4. BufferedWriter 类

BufferedWriter 类是一个输出字符缓冲流类，与 FileWriter 类结合使用，可以提高程序读写文件效率。该类的构造方法如下。

```
BufferedWriter(Writer WriterObjectName)
```

创建一个默认大小的缓冲输出流。BufferedWriter 新增加的方法有：public void newLine()：写一个“行结束符”。行结束符由系统的“line.separator”属性决定。例如在 Windows 平台上，行结束符是“\n”。

下面是一个使用字符输出流的例子，该例有两个 Bean，textWriter 负责将内容用字符流写入文件，textReader 封装了字符流读取文件内容的操作，页面 ch6_6.jsp 负责显示写入文件的内容。

【例 6.6】用字符输出流写文件(其代码参见 textReader.java、textWriter.java 和 ch6_6.jsp)，页面效果如图 6.8 所示。

textReader.java 的内容参见例 6.5，这里省略。

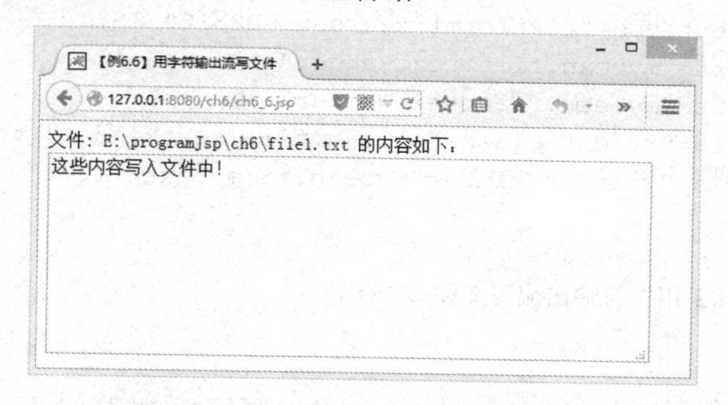

图 6.8　显示写入文件的内容

textWriter.java 文件代码如下：

```
package mybean.file;
import java.io.*;
public class textWriter {
    String fileName=null;
    String allFilesmess=null;
    public textWriter(){
        allFilesmess=new String();
    }
    public String getFileName() {
        return fileName;
    }
    public void setFileName(String fileName) {
        this.fileName = fileName;
    }
    public void setAllFilesmess(String s) {
        allFilesmess=s;
        try{
            if(fileName!=null){
            FileWriter wr=new FileWriter(fileName);
            BufferedWriter bufferw=new BufferedWriter(wr);
            bufferw.write(allFilesmess);
```

```
                bufferw.flush();
                bufferw.close();
                wr.close();
                }
            }
            catch(IOException ee){
                System.out.print(ee.toString());
            }
        }
}
```

ch6_6.jsp 文件代码如下：

```
<%@ page contentType="text/html;charset=GB2312" %>
<%@ page import="mybean.file.textReader" %>
<%@ page import="mybean.file.textWriter" %>
<jsp:useBean id="readf" class="mybean.file.textReader" scope="page" />
<jsp:useBean id="writef" class="mybean.file.textWriter" scope="page"/>
<html>
<head>
<title>【例 6.6】用字符输出流写文件</title>
</head>
<body>
<jsp:setProperty name="writef" property="fileName"
      value="E:\programJsp\ch6\file1.txt" />
<jsp:setProperty name="writef" property="allFilesmess"
      value="这些内容写入文件中！"/>
<jsp:setProperty name="readf" property="fileName"
      value="E:\programJsp\ch6\file1.txt"/>
文件：
<jsp:getProperty name="readf" property="fileName"/>
的内容如下：<br>
<jsp:getProperty name="readf" property="allFilesmess"/>
</body>
</html>
```

注意：如果当前目录下没有 E:\programJsp\ch6\file1.txt 文件，则首先创建该文件，然后将当前文件的内容保存到 E:\programJsp\ch6\file1.txt 文件中。

6.5 随机读写文件

6.5.1 随机存取文件

以前所讲的流类，只能对文件进行顺序读写。如果需要对文件内容进行随机读写时，则需要用到随机存取文件的 RandomAccessFile 类。

RandomAccessFile 类与前面讲的流类不同，它既不是输入流类 InputSteam 的子类，也不是输出流类 OutputStream 的子类，但它同时实现了 DataInput 和 DataOutput 接口，其创建的对象的数据源既可以作为源也可以作为目的地。即当需要对一个文件进行读写操作时，可以创建一个指向该文件的 RandomAccessFile 对象。这样既可以从这个对象中读取该文件的数据，也可以将数据写入该文件。

RandomAccessFile 类的构造方法如下。

- public RandomAccessFile(File file，String mode)：创建一个随机存取文件流。参数 file 指明流的源(也是流的目的地)。参数 mode 指明打开文件的方式。包括 4 个常量参数：r 代表只读、rw 代表可读写、rwd 代表同步更新文件内容、rws 代表同步更新文件内容及属性。
- public RandomAccessFile(String name，String mode)：创建一个随机存取文件流。参数 name 用来确定一个文件名，指明流的源(也是流的目的地)。参数 mode 与上述方法相同。

在 RandomAccessFile 类中，程序可以通过改变文件指针的位置来达到随机访问文件内容的目的。可以通过以下方法对文件指针进行操作。

- public long getFilePointer()：返回当前的文件指针的位置。
- public void seek(long pos)：设置文件指针的位置在 pos 处，参数 pos 指文件指针距离文件开头的字节位置。
- public int skipBytes(int n)：使文件指针向前移动 n 个字节。

由于 RandomAccessFile 类同时实现了 DataInput 和 DataOutput 接口，所以它可以对文件进行读写操作。在该类中，定义了各种数据类型的读写方法，如下所示。

- public long length()：返回文件的长度。
- public int read()：从文件中读取一个字节的数据。
- public int read(byte[] b，int off，int len)：从文件的 off 处读取 len 个字节，将其存储在字节数组 b 中。
- public final boolean readBoolean()：从文件中读取一个布尔值。
- public final byte readByte()：从文件中读取一个字节。
- public final char readChar()：从文件中读取一个 Unicode 字符。
- public final double readDouble()：从文件中读取一个双精度值。
- public final float readFloat()：从文件中读取一个单精度值。
- public final int readInt()：从文件中读取一个整型值。
- public final long readLong()：从文件中读取一个长整型值。
- public final String readLine()：从文件中读取一个文本行。
- public void write(byte[] b)：向文件写入字节数组 b 中 b.length 个数据，写入的位置是当前文件指针所在处。
- public void write(byte[] b，int off，int len)：向文件写入字节数组 b[off]开始处的 len 个数据。
- public final void writeBoolean(boolean v)：向文件写入一个布尔型数据。
- public final void writeChar (int v)：向文件写入一个字符型数据。

- public final void writeInt(int v)：向文件写入一个整型数据。
- public final void writeLong(long v)：向文件写入一个长整型数据。
- public final void writeFloat(float v)：向文件写入一个单精度数据。
- public final void writeDouble(double v)：向文件写入一个双精度数据。
- public final void writeChars(String s)：向文件写入一个字符串。
- public final void writeUTF(String str)：用 UTF 编码方式向文件写入一个字符串。

在对象的创建期间，如果发生错误，比如找不到文件，则产生 FileNotFoundException 异常，如果是输入输出的操作期间发生错误，则产生 IOException 异常。

6.5.2 随机读写文件示例

下面通过一个例子来介绍随机读写文件的方法。

【例 6.7】编写一个 Bean，Bean 的名称为 randfileBean。在 Bean 中的 getAllFilesmess 方法中，首先声明了一个随机读写文件对象 randomFile1 对象，指向 fileName 文件，并得到文件的长度和当前文件指针位置。使用 readline 方法读取文件。文件的中文内容要转换成 ISO-8859-1 编码。程序运行效果如图 6.9 所示。

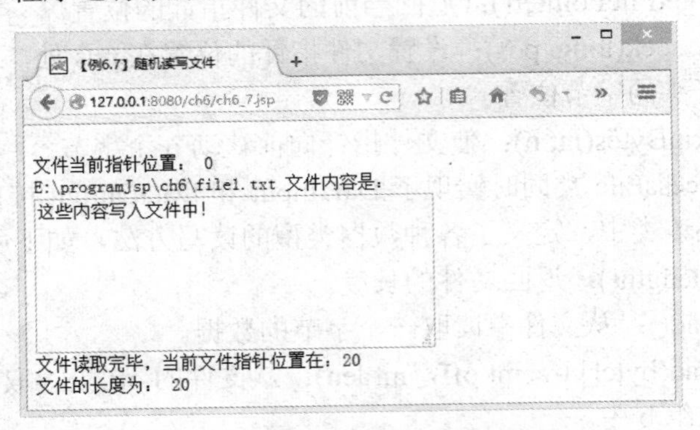

图 6.9 随机访问文件页面效果

randfileBean.java 文件代码如下：

```
package mybean.file;
import java.io.*;
public class randfileBean {
    String fileName=null;
    long currentpoint;
    long fileLength;
    String allFilesmess=null;
    public long getCurrentpoint() {
        return currentpoint;
    }
    public randfileBean(){
        allFilesmess=new String();
    }
    public String getFileName() {
```

```
        return fileName;
    }
    public void setFileName(String fileName) {
        this.fileName = fileName;
    }
    public long getFileLength() {
        return fileLength;
    }
    public String getAllFilesmess() {
        try{
            StringBuffer strbufTemp=new StringBuffer();
            RandomAccessFile randomFile1=new
                        RandomAccessFile(fileName,"rw");
            fileLength=randomFile1.length();
            currentpoint=randomFile1.getFilePointer();
            String strtemp=null;
            byte bb[]=new byte[100];
            while((strtemp=randomFile1.readLine())!=null){
                bb=strtemp.getBytes("ISO-8859-1");
                strtemp=new String(bb);
                strbufTemp.append("\n"+strtemp);
            }
            allFilesmess="文件内容是：<br>"+"<textarea cols=40 rows=6>"+
                strbufTemp+"</textarea>";

           allFilesmess+="<br>文件读取完毕，当前文件指针位置在："+
               randomFile1.getFilePointer()+"<br>";
           randomFile1.close();

        }
        catch(FileNotFoundException exp){
            allFilesmess="文件不存在！";
            System.out.print(exp.toString());
        }
        catch(IOException ee){
            allFilesmess="文件读写错误";
            System.out.print(ee.toString());
        }
        return allFilesmess;
    }
}
```

ch6_7.jsp 文件代码如下：

```
<%@ page contentType="text/html;charset=GB2312" %>
<%@ page import="mybean.file.randfileBean" %>
<jsp:useBean id="randf" class="mybean.file.randfileBean" scope="page" />
<html>
```

```
<head>
<title>【例 6.7】随机读写文件</title>
</head>
<body>
<jsp:setProperty name="randf" property="fileName"
     value="E:\programJsp\ch6\file1.txt" />
<br>文件当前指针位置:
<jsp:getProperty name="randf" property="currentpoint"/>
<br>
<jsp:getProperty name="randf" property="fileName" />
<jsp:getProperty name="randf" property="allFilesmess"/>
文件的长度为:
<jsp:getProperty name="randf" property="fileLength" />
</body>
</html>
```

6.6 文件操作案例

6.6.1 上传文件

文件上传是 Web 应用程序需要具备的功能，本节讲解 HTML 文件类型的表单结合随机读写操作类 RandomAccessFile 来实现文件上传的编程技术。

为了让用户能够选择上传文件，JSP 提供了 File 类型的文件表单，File 类型表单使用格式如下:

```
<form action="JSP 页面或 Servlet" method="post|get"
      ENCTYPE="mutipart/form-data" >
   <input type="file" name="表单元素名称">
</form>
```

uploadFileBean 负责将文件上传到服务器。在接受文件上传信息的页面使用 request.getInputStream()方法得到上传文件输入流，Bean 通过这个输入流将上传表单域信息及文件内容保存到临时文件中。按照 HTTP 的规定，临时文件的前四行和最后的五行是表单域的信息，其中第二行有上传文件的文件名信息，第四行结束位置到倒数第六行结束位置之间的内容是文件本身部分。下面是一个临时文件的头部和最后几行。

```
1 行 -----------------------------22372610224060
2 行 Content-Disposition: form-data; name="test"; filename="SP_A0502.jpg"
3 行 Content-Type: image/jpeg
4 行 ....
□□□□□□□□□□□□□□□□□□□□□□□□□□□□□□
□□□□□□□□□□□□□□□□□□□□□□□□□□□□□□
5 行 -----------------------------22372610224060
4 行 Content-Disposition: form-data; name="submit"
3 行 ....
```

```
2 行　提交
1 行　-----------------------------22372610224060--
```

Bean 使用 RandomAccessFile 类随机访问文件的第二行，取得上传文件名，定位到文件的第四行结束位置，开始获取上传文件的内容部分，直到临时文件的倒数第六行结束位置止，将得到的文件内容部分保存为一个新文件。

【例 6.8】 上传文件示例(相关代码参见 uploadFileBean.java、ch6_8.jsp 和 ch6_8_show.jsp)，页面效果如图 6.10、图 6.11 所示。

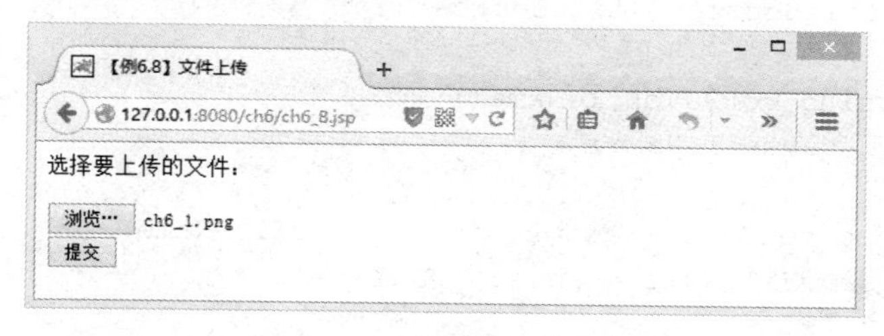

图 6.10　选择上传文件

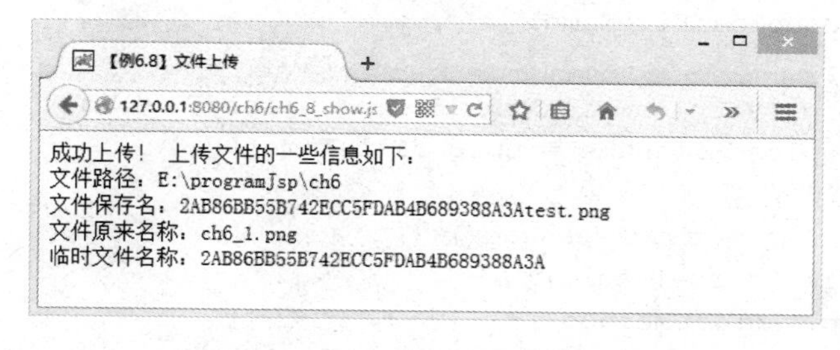

图 6.11　上传文件并显示结果信息

uploadFileBean.java 文件代码如下：

```
package mybean.file;
import java.io.*;
public class uploadFileBean {
    private String driverPath=null;
    private String tempFileName=null;
    private String saveFileName=null;
    private InputStream fileSource=null;
    private String uploadFileName=null;
    private String backMessage=null;
    boolean flag=false;
    public boolean isFlag() {
        return flag;
    }
    public String getTempFileName() {
        return tempFileName;
    }
    public void setTempFileName(String tempFileName) {
        this.tempFileName = tempFileName;
```

```
        }
        public String getDriverPath() {
            return driverPath;
        }
        public void setDriverPath(String driverPath) {
            this.driverPath = driverPath;
        }
        public void setFileSource(InputStream fileSource) {
            this.fileSource = fileSource;
        }
        public String getSaveFileName() {
            return saveFileName;
        }
        public void setSaveFileName(String saveFileName) {
            this.saveFileName = saveFileName;
        }
        public String getUploadFileName() {
            return uploadFileName;
        }
        public void setUploadFileName(String uploadFileName) {
            this.uploadFileName = uploadFileName;
        }
        public String getBackMessage() {
            if (fileSource!=null){
                uploadFileMethod();
            }
            else{
                backMessage="未选择文件！";
            }
            return backMessage;
        }
        public boolean uploadFileMethod(){
            try{//上传文件保存到临时文件中
                File f1=new File(driverPath,tempFileName);
                FileOutputStream fos=new FileOutputStream(f1);
                byte b[]=new byte[10000];
                int n;
                while((n=fileSource.read(b))!=-1){
                    fos.write(b, 0, n);
                }
                fos.close();
                fileSource.close();
                //读取临时文件中第二行的内容
                RandomAccessFile random=new RandomAccessFile(f1,"r");
                int second=1;
                String secondLine=null;
                while(second<=2){
```

```
        secondLine=random.readLine();
        second++;
    }
    //得到上传文件的文件名
    int position=secondLine.lastIndexOf('=');
    uploadFileName=
       secondLine.substring(position+2, secondLine.length()-1);
    //转换编码，识别汉字文件名
    byte cc[]=uploadFileName.getBytes("iso-8859-1");
    uploadFileName=new String(cc);
    //得到上传文件的扩展名
    int extposition=uploadFileName.lastIndexOf('.');
    String extName=uploadFileName.substring(extposition+1,
            uploadFileName.length());
    //获取上传临时文件第四行回车符的位置
    random.seek(0);
    long forthEndPosition=0;
    int forth=1;
    while((n=random.readByte())!=-1&&forth<=4){
        if(n=='\n'){
            forthEndPosition=random.getFilePointer();
            forth++;
        }
    }
    //删除重名的文件
    saveFileName=saveFileName.concat("."+extName);
    File dir=new File(driverPath);
    dir.mkdir();
    File file[]=dir.listFiles();
    for(int k=0;k<file.length;k++){
        if(file[k].getName().equals(saveFileName)){
            file[k].delete();
        }
    }
    //按新文件名保存文件
    File savingFile=new File(driverPath,saveFileName);
    RandomAccessFile random2=
       new RandomAccessFile(savingFile,"rw");
    //在临时文件中获得上传文件结束位置
    random.seek(random.length());
    long endPosition=random.getFilePointer();
    long mark=endPosition;
    int j=1;
    while((mark>=0)&&(j<=6)){
        mark--;
        random.seek(mark);
        n=random.readByte();
```

```
                    if(n=='\n'){
                        endPosition=random.getFilePointer();
                        j++;
                    }
                }
                random.seek(forthEndPosition);
                long startPoint=random.getFilePointer();
                while(startPoint<endPosition-1){
                    n=random.readByte();
                    random2.write(n);
                    startPoint=random.getFilePointer();
                }
                random2.close();
                random.close();
                f1.delete();
                flag=true;
                backMessage="成功上传! ";
            }
            catch(Exception exp)
            {
                System.out.println("uploadFileMethod"+exp.toString());
                backMessage="上传失败";
                flag=false;
            }
            return flag;
        }
}
```

ch6_8.jsp 文件代码如下:

```
<%@ page contentType="text/html;charset=GB2312" %>
<html>
<head>
<title>【例 6.8】文件上传</title>
</head>
<body>
<p>选择要上传的文件: <br>
<form action="ch6_8_show.jsp" method="post"
     ENCTYPE="multipart/form-data">
   <input type="file" name="test" size="45">
   <br><input type="submit" name="submit" value="提交">
</form>
</body>
</html>
```

ch6_8_show.jsp 文件代码如下:

```
<%@ page contentType="text/html;charset=GB2312" %>
<%@ page import="mybean.file.uploadFileBean" %>
```

```
<%@ page import="java.io.*" %>
<jsp:useBean id="upfile" class="mybean.file.uploadFileBean"
     scope="page" />
<html>
<head>
<title>【例 6.8】文件上传</title>
</head>
<body>
<%   String tempFileName=request.getSession().getId();
    String saveFileName=tempFileName+"test";
    InputStream fileSource=request.getInputStream(); %>
<jsp:setProperty name="upfile" property="driverPath"
     value="E:\programJsp\ch6"/>
<jsp:setProperty name="upfile" property="saveFileName"
     value="<%=saveFileName%>"/>
<jsp:setProperty name="upfile" property="fileSource"
     value="<%=fileSource %>"/>
<jsp:setProperty name="upfile" property="tempFileName"
     value="<%=tempFileName %>"/>
<jsp:getProperty name="upfile" property="backMessage" />
<%    out.print("上传文件的一些信息如下：<br>");
     out.print("文件路径：");
     out.println(upfile.getDriverPath()+"<br>");
     out.print("文件保存名：");
     out.print(upfile.getSaveFileName()+"<br>");
     out.print("文件原来名称：");
     out.print(upfile.getUploadFileName()+"<br>");
     out.print("临时文件名称：");
     out.print(upfile.getTempFileName()+"<br>");  %>
</body>
</html>
```

6.6.2　下载文件

很多网站都具有文件下载功能，一般用户单击链接后出现另存为对话框，然后保存下载文件。当某个 JSP 页面用 response.setHeader()方法添加下载的头给客户端浏览器，浏览器就会打开下载对话框。response.setHeader()方法添加下载头格式如下：

```
response.setHeader("content-disposition","attachment;
        filename="下载文件名");
```

response 对象的 setContentType()用来定义服务器发送给客户端的 MIME 类型。Word 文件的 MIME 类型是“application/msword”，Excel 文件的 MIME 类型是 application/msexcel。客户端在与服务器的交互中，根据 MIME 类型来决定进行怎样的处理。如果没有指定 MIME 类型，客户端根据文件扩展名或内容猜测其类型进行处理。

打开下载对话框后，JSP 页面通过 response.getOutputStream()方法得到向客户端输出的输出流，通过输出流向客户端发送数据。

下面的示例，将 E:\programJsp\ch6 目录中的所有 word 文档在页面中列表供用户下载。单击“我要下载”，即打开文件下载页面 ch6_9_down.jsp。ch6_9_down.jsp 页面使用 downFileBean 下载文件。

【例 6.9】下载文件示例(相关代码参见 downFileBean.java、FileNameAccept.java、ch6_9.jsp 和 ch6_9_show.jsp)，页面效果如图 6.12、图 6.13 所示。

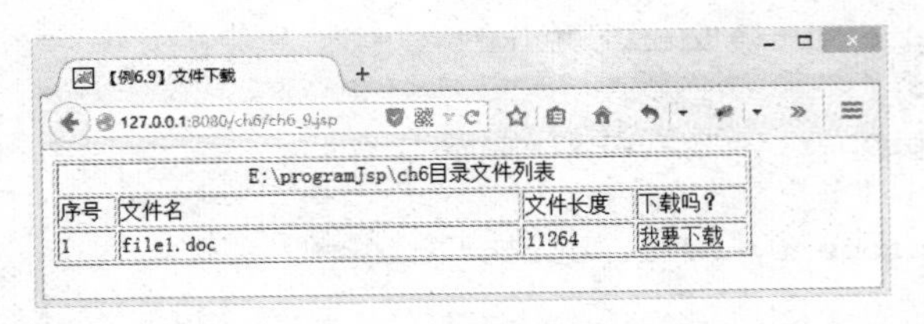

图 6.12　下载文件列表页面效果

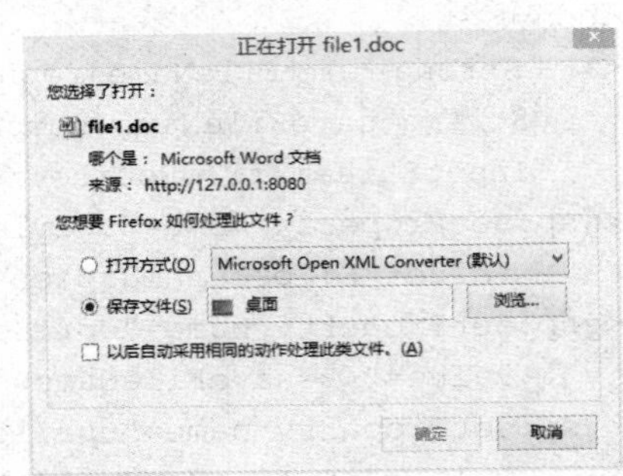

图 6.13　下载对话框

downFileBean.java 文件代码如下：

```
package mybean.file;
import java.io.*;
import javax.servlet.http.*;
public class downFileBean {
    HttpServletResponse response;
    String fileName;
    String dirPath;
    String fileMIME;
    public void setFileMIME(String fileMIME) {
        this.fileMIME = fileMIME;
    }
    public void setResponse(HttpServletResponse response) {
        this.response = response;
    }
    public void setDirPath(String dirPath) {
        this.dirPath = dirPath;
    }
    public void setFileName(String fileName) {
        this.fileName = fileName;
        try{
            //得到向客户端输出的输出流
            OutputStream outStream1=response.getOutputStream();
            //输出文件用的字符数组
           byte bb[]=new byte[600];
           //要下载的文件
           File fileload=new File(dirPath,this.fileName);
           //客户端打开保存文件对话框
           response.setHeader("content-disposition",
```

```
                "attachment;filename="+fileName);
            response.setContentType(fileMIME);
            long filelength=fileload.length();
            String length=String.valueOf(filelength);
            //通知客户端文件长度
            response.setHeader("content_length",length);
            //读取文件并发送给客户端
            FileInputStream in=new FileInputStream(fileload);
            int n=0;
            while((n=in.read(bb))!=-1){
               outStream1.write(bb,0,n);
            }
        }
        catch(Exception exp){
            System.out.print(exp.toString());
        }
    }
}
```

FileNameAccept.java 文件代码如下：

```
package mybean.file;
import java.io.*;
public class FileNameAccept implements FilenameFilter{
    public boolean accept(File dir, String filename) {
       boolean boo=false;
       if(filename.endsWith(".doc")||filename.endsWith(".DOC")){
        boo=true;
       }
       return boo;
    }
}
```

ch6_9.jsp 文件代码如下：

```
<%@ page language="java" contentType="text/html;charset=GB2312"%>
<%@ page import="mybean.file.FileNameAccept" %>
<%@ page import="java.util.*" %>
<%@ page import="java.io.*" %>
<html>
  <head>
      <title>【例 6.9】文件下载</title>
      </head>
  <body>
  <%
     File file1=new File("E:\programJsp\ch6");
     File filelist[]=file1.listFiles(new FileNameAccept());
     int count=0;
      out.print("<table border>");
      out.print("<tr><td colspan=4 align=center>");
      out.print("E:\programJsp\ch6 目录文件列表</td></tr>");
      out.print("<tr><td width=40>"+"序号"+"</td>");
```

```
        out.print("<td width=300>"+"文件名"+"</td>");
        out.print("<td width=80>"+"文件长度"+"</td>");
        out.print("<td width=80>"+"下载吗？"+"</td>");
        out.print("</tr>");
        for(File me:filelist){
            count++;
        out.print("<tr>");
           out.print("<td>" +count+"</td>");
           out.print("<td>"+me.getName()+"</td>");
           out.print("<td>"+me.length()+"</td>");
           out.print("<td><a href='ch6_9_down.jsp?filename="+
                 me.getName()+"'>我要下载</a></td>");
           out.print("</tr>");
        }
        out.print("<table>");
    %>
</body>
</html>
```

ch6_9_down.jsp 文件代码如下：

```
<%@ page contentType="text/html;charset=GB2312" %>
<%@page import="mybean.file.downFileBean"%>
<jsp:useBean id="downf" class="mybean.file.downFileBean"
    scope="page"></jsp:useBean>
<html>
<head>
    <title>【例 6.9】文件下载</title>
    </head>
<body>
<jsp:setProperty name="downf" property="dirPath"
   value="e:\programJsp\ch6"/>
<jsp:setProperty name="downf" property="response"
   value="<%=response%>"></jsp:setProperty>
<jsp:setProperty name="downf" property="fileMIME"
   value="application/msword"/>
<jsp:setProperty name="downf" property="fileName"
   value='<%=request.getParameter("filename") %>'/>
</body>
</html>
```

6.6.3 文件内容分页显示

当客户浏览一个较大的文件时，比如浏览网络小说，如果将全部内容显示到浏览器中，会使页面拉得很长，浏览起来很不方便。本节介绍一个分页显示文件内容的例子，它使用了一个分页 pageFileBean 和一个页面 ch6_10.jsp。Bean 负责文件内容的分页读取，ch6_10.jsp 页面负责文件内容的显示，并提供“下一页”和“上一页”链接，每次调用 ch6_10.jsp 页面都传递参数 page(当前页数)。

【例 6.10】文件内容分页显示示例(相关代码参见 pageFileBean.java 和 ch6_10.jsp)，

页面效果如图 6.14 所示。

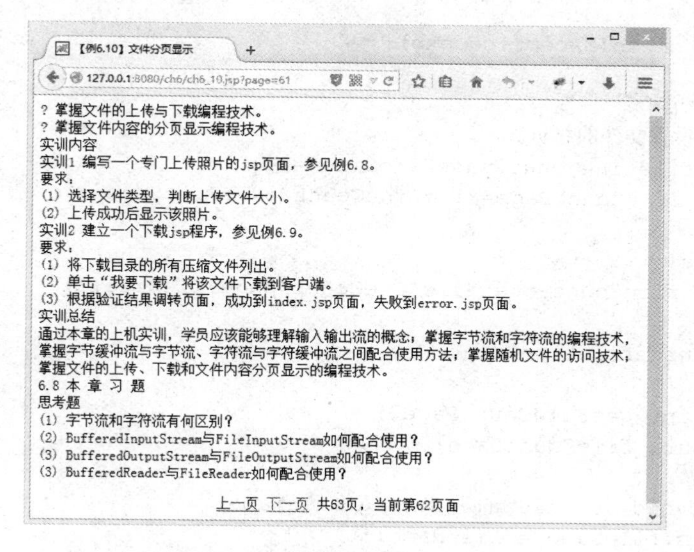

图 6.14　分页读取文件内容

pageFileBean.java 文件代码如下：

```
package mybean.file;
import java.io.*;
public class pageFileBean {
    String fileName=null;
    String dirPath=null;
    int pageFileSize=20;
    int fileLineCount=0;
    int fileCountPage=0;
    int pages=0;
    public void setPageFileSize(int pageFileSize) {
        this.pageFileSize = pageFileSize;
    }
    public int getPages() {
        return pages;
    }
    public void setPages(int pages) {
        this.pages = pages;
    }
    public int getFileLineCount() {//获得文件的行数
        File f=new File(dirPath,fileName);
        if(f.exists()){
            try{
                FileReader fileReader1=new FileReader(f);
                BufferedReader buffer1=new BufferedReader(fileReader1);
                while((buffer1.readLine())!=null){
                 fileLineCount++;
                }
                buffer1.close();
```

```
                fileReader1.close();
            }
            catch(Exception exp){

            }
        }//设置文件的总页数
        if(fileLineCount%pageFileSize==0){
            fileCountPage=fileLineCount/pageFileSize;
        }
        else{
            fileCountPage=fileLineCount/pageFileSize+1;
        }
        return fileLineCount;
    }
    public int getFileCountPage() {
        return fileCountPage;
    }
    public void setFileName(String fileName) {
        this.fileName = fileName;
    }
    public void setDirPath(String dirPath) {
        this.dirPath = dirPath;
    }
    public String getFileContents(){//读取文件的当前内容
        StringBuffer strbuffer=new StringBuffer();
        String strTemp=null;
        File f=new File(dirPath,fileName);
        if(f.exists()){
            try{
                FileReader fileReader1=new FileReader(f);
                BufferedReader buffer1=new BufferedReader(fileReader1);
                for(int i=0;i<pages*pageFileSize;i++){
                    buffer1.readLine();
                }
                int i=0;
                while(i<pageFileSize&&
                      (strTemp=buffer1.readLine())!=null){
                    strbuffer.append(strTemp+"<br>");
                    i++;
                }
            }
            catch(Exception exp){
                strbuffer.append("读文件发生错误！");
            }
        }
        else{
            strbuffer.append("文件中没有内容！");
        }
        return strbuffer.toString();
    }
}
```

ch6_10.jsp 文件代码如下：

```
<%@ page language="java" contentType="text/html;charset=GB2312"%>
<%@ page import="mybean.file.pageFileBean" %>
<%@ page import="java.util.*" %>
<jsp:useBean id="pagefile" class="mybean.file.pageFileBean"
   scope="page"/>
<html>
<head>
<title>【例 6.10】文件分页显示</title>
</head>
 <body>
 <%
   String strpage=request.getParameter("page");
   if (strpage==null){
    strpage="0";
   }
   int diPage=Integer.parseInt(strpage);
 %>
 <jsp:setProperty name="pagefile" property="dirPath"
     value="e:\programJsp\ch6"/>
 <jsp:setProperty name="pagefile" property="fileName"
     value="text.txt"/>
 <jsp:setProperty name="pagefile" property="pages"
     value="<%=diPage %>"/>
 <%
    int fileLineCount=pagefile.getFileLineCount();
    int getFileCountPage=pagefile.getFileCountPage();
 %>
 <jsp:getProperty name="pagefile" property="fileContents"/>
 <p align="center">
 <%
   if(diPage>0)
   {%>
    <a href="ch6_10.jsp?page=<%=diPage-1 %>">上一页</a>
 <%}%>
 <%if(diPage<getFileCountPage-1)
   {%>
    <a href="ch6_10.jsp?page=<%=diPage+1%>">下一页</a>
 <%}%>
   共<%=pagefile.getFileCountPage()%>页，当前第<%=diPage+1 %>页面
 </body>
 </html>
```

6.7 上机实训

实训目的

- 理解 Java 输入/输出流的概念。

- 掌握字节输入/输出流与字节缓冲输入/输出流的配合使用。
- 掌握随机文件访问技术。
- 掌握文件的上传与下载编程技术。
- 掌握文件内容的分页显示编程技术。

实训内容

实训 1 编写一个专门上传照片的 JSP 页面，参见例 6.8。

要求：

(1) 选择文件类型，判断上传文件大小。

(2) 上传成功后显示该照片。

实训 2 建立一个下载 JSP 程序，参见例 6.9。

要求：

(1) 将下载目录的所有压缩文件列出。

(2) 单击“我要下载”将该文件下载到客户端。

(3) 根据验证结果调转页面，成功到 index.jsp 页面，失败到 error.jsp 页面。

实训总结

通过本章的上机实训，学员应该能够理解输入/输出流的概念；掌握字节流和字符流的编程技术，掌握字节缓冲流与字节流、字符流与字符缓冲流之间配合使用方法；掌握随机文件的访问技术；掌握文件的上传、下载和文件内容分页显示的编程技术。

6.8 本 章 习 题

思考题

(1) 字节流和字符流有何区别？

(2) BufferedInputStream 与 FileInputStream 如何配合使用？

(3) BufferedOutputStream 与 FileOutputStream 如何配合使用？

(4) BufferedReader 与 FileReader 如何配合使用？

(5) BufferedWriter 与 FileWriter 如何配合使用？

(6) RandomAccessFile 访问文件有何特点？

(7) 如何处理文件读写操作中发生的异常？

拓展实践题

完善例 6.10，列出某目录所有文本文件，单击“阅读”链接，打开该文件实现分页阅读。

第 7 章　JSP 中使用数据库

学习目的与要求：

在 Web 程序开发中，几乎离不开数据库操作。数据库在数据的查询、修改、保存与安全方面扮演着重要的角色。本章学习 JDBC 应用程序接口，利用 JDBC 访问数据库，常用的数据库增、删、改、查编程技术等内容。通过本章的学习，学员要理解 JDBC 应用程序编程接口，掌握使用纯 Java 驱动程序操作数据库的编程方法，掌握对数据的增加、删除、修改和查询编程技术，以及查询记录的分页面显示和数据库连接池编程技术。

本章使用的 Web 服务目录为 ch7(创建服务目录见第 1 章 1.3 节)，为了使用 Bean，在当前 Web 服务目录 ch7 下建立 WEB-INF\classes 目录，然后根据包名在 classes 目录建立相应的子目录，本章使用的包为 mybean\database。本书使用的是 MYSQL5.0 数据库，对其他数据库的连接和使用也做了一些简单介绍。

7.1　JDBC 概述

7.1.1　什么是 JDBC

JDBC 是 Java 数据库连接(Java Data Base Connectivity)技术的简称，指 Java 同许多数据库之间连接的一种标准。它是由 Sun 定义的技术规范，并由 Sun 及其 Java 合作伙伴开发的与平台无关的标准数据库访问接口。JDBC 为数据库应用开发人员、数据库前台工具开发人员提供了一种标准的 JavaAPI，使开发人员可以用纯 Java 语言编写完整的数据库应用程序。

通过使用 JDBC，开发人员可以很方便地将 SQL 语句传送给几乎任何一种数据库。用 JDBC 编写的程序能够自动地将 SQL 语句传送给相应的数据库管理系统。不仅如此，使用 Java 编写的应用程序可以在任何支持 Java 的平台上运行，不必在不同的平台上编写不同的应用程序。Java 和 JDBC 的结合可以让开发人员在开发数据库应用时真正实现一次编写，处处运行(Write Once，Run Everywhere)。

简单地说，JDBC 能够完成下列三件事。

(1) 与一个数据库建立连接(Connection)。

(2) 向数据库发送 SQL 语句(Statement)。

(3) 处理数据库返回的结果(Resultset)。

JDBC API 是能体现 SQL 最基本抽象概念的、最直接的 Java 接口，它建构在 ODBC 的基础之上。熟悉 ODBC 的编程人员学习 JDBC 非常容易，因为 JDBC 保持了 ODBC 的基本设计特征。它们最大的区别是 JDBC 是基于 Java 的风格和特点，并强化了 Java 的风格和特点。JDK1.1 以后，SQL 类包(也就是 JDBC API)已成为 Java 语言的标准部件，在程序中要引入 java.sql.*。

7.1.2 JDBC 的构成

JDBC 主要由两部分组成：一部分是访问数据库的高层接口，即通常所说的 JDBC API；另一部分是由数据库厂商提供的使 Java 程序能够与数据库连接通信的驱动程序，即 JDBC Database Driver(JDBC 数据库驱动程序)。如图 7.1 表示了 Java 应用程序、JDBC 及数据库之间的关系。

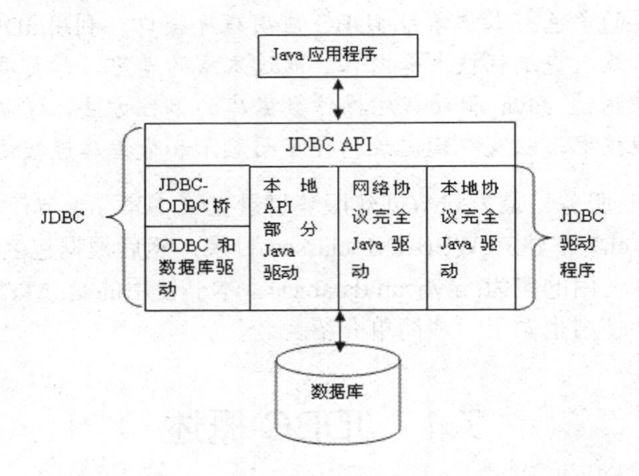

图 7.1　Java 应用程序、JDBC 及数据库的关系

1. JDBC API

JDBC 定义了表示数据库连接、SQL 句柄、预编译的 SQL 句柄、执行存储过程的 SQL 句柄、记录集、记录集元数据和数据库元数据的 Java 接口。这些接口提供了标准的数据库访问功能。这些 JDBC API 是高层的 API，它独立于数据库。开发人员可以不必写一个程序访问 Oracle，写另一个程序访问 DB2，再写一个程序访问 Microsoft 的 SQL Server。而只需写一个程序就够了。

2. JDBC 数据库驱动程序

JDBC 数据库驱动程序必须针对特定的数据库系统实现 JDBC API 中定义的方法。在使用 JDBC API 编写访问数据库的 Java 程序时，调用的方法实际上是由 JDBC 数据库驱动程序实现的。

目前，JDBC 驱动程序可以分为以下四类。

(1) JDBC-ODBC 桥(JDBC-ODBC Bridge)加 ODBC 驱动程序。

JDBC-ODBC 桥驱动程序为 Java 应用程序提供了一种把 JDBC 调用映射为 ODBC 调用的方法。这种类型的驱动使 Java 应用可以访问所有支持 ODBC 的 DBMS。它适合于企业内部网、用 Java 编写的三层结构的应用程序服务器。应用系统使用该类驱动时，必须将 ODBC 驱动程序加载到使用该驱动程序的每个客户机上。而且，使用这种驱动程序，Java 程序要调用底层 ODBC 驱动管理器，ODBC 驱动以及数据库客户商的本地代码，这会造成 Java 应用程序的平台无关性、安全性降低，可移植性差等局限。

(2) 部分用 Java 来编写的本地 API 驱动程序。

这种类型的驱动程序把客户机 API 上的 JDBC 调用转换为对特定的数据库如 Oracle、Sybase、DB2 等或其他 DBMS 的调用。这种类型的驱动程序也需要调用本地代码，因此，它也失去了 JDBC 平台无关性的好处。

(3) JDBC 网络纯 Java 驱动程序。

这种类型的驱动程序将 JDBC 调用转换为与 DBMS 无关的网络协议，之后这种协议又被某个特定的数据库服务器转换为一种 DBMS 协议。这种网络服务器中间件能够将它的纯 Java 客户机连接到多种不同的数据库上。所用的具体协议取决于提供者。通常，这是最为灵活的 JDBC 驱动程序。

(4) 本地协议纯 Java 驱动程序。

这种类型的驱动程序将 JDBC 调用直接转换为 DBMS 所使用的网络协议。这将允许从客户机上直接调用 DBMS 服务器。由于许多这样的协议都是专用的，因此数据库产品提供者自己将是这种驱动程序的主要来源。

7.2 JDBC 应用程序接口简介

JDBC 应用程序接口(基于 JDBC 1.0)是实现 JDBC 标准支持数据库操作的类与方法的集合。它包括 java.sql 和 javax.sql 两个包。其中，java.sql 包含 JDBC 2.0 核心 API 及 JDBC 3.0 增加的部分；javax.sql 包含 JDBC 2.0 和 JDBC 3.0 标准的扩展 API。因为 JDBC 标准从一开始就设计良好，所以 JDBC 1.0 到目前为止都没有改变，后续 JDBC 标准都是在 JDBC 1.0 的基础上进行的扩展。这里，主要介绍 JDBC 的常用类与接口。

7.2.1 JDBC 的驱动程序管理器——DriverManager 类

DriverManager 类是 JDBC 的管理层，作用于用户和驱动程序之间。它提供管理一组 JDBC 驱动程序所需要的基本服务，可以跟踪可用的驱动程序，并在数据库和相应的驱动程序之间建立连接。

对于简单的应用程序，主要使用该类的 getConnection 和 forName 方法。

(1) public static Connection getConnection(String url，String user，String password)。

该方法建立与数据库的连接。其中参数 url 表示数据库资源的地址，user 代表建立数据库连接所需要的用户名，password 是建立数据库连接所需要的密码。如果用户名和密码不正确，数据库连接将不能够建立。

在这里，需要讲述 url 参数的命名方法：url 由 3 部分组成，各部分间用冒号分隔，如下所示：

```
jdbc:<子协议>:<子名称>
```

其中：

- jdbc——协议名称。在 JDBC URL 中的协议总是 jdbc。

- <子协议>——驱动程序名或数据库连接机制的名称。如 odbc，它是用于指定 ODBC 风格的数据库资源名称的 URL 而保留的。
- <子名称>——一种标识数据库的方法。子名称可以依不同的子协议而变化。它还可以有子名称的子名称(含有驱动程序编程人员所选的任何内部语法)。使用子名称的目的是为定位数据库提供足够的信息。

在使用该方法以前，必须加载 JDBC 数据库驱动程序，并在 DriverManger 类中注册，以便 DriverManger 类可以管理该驱动程序，并用其和数据库服务器建立数据库连接。加载数据库的驱动程序需要 Class 类中的以下方法。

(2) public static Class forName(String className)。

显式加载驱动程序类 className，className 为驱动程序的类名。

例如：

```
Class.forName("com.mysql.jdbc.Driver");
```

注意： 需要把 JDBC 驱动程序的 class 文件的路径添加到 Java 编译器的环境变量 classpath 中，否则，在编译时可能会出现错误。至于驱动程序的名字字符串是由驱动程序的编写者提供的，并非由该驱动程序的使用者指定。

7.2.2 JDBC 与数据库的连接——Connection 接口

Connection 接口是 java.sql 包中最重要的接口。Connection 对象代表与数据库的连接。一个应用程序可与单个数据库有一个或多个连接，或者可与许多数据库有连接。

数据库连接一旦建立，就可用来向它所涉及的数据库发送 SQL 语句。Connection 的作用即是创建向数据库发送 SQL 语句的三个接口。这三个接口分别如下。

(1) Statement 接口：用于发送简单的 SQL 语句，如 select、insert 或 update 语句。

(2) PreparedStatement 接口：它继承自 Statement 接口，主要用于发送预编译的带有一个或者多个输入参数的 SQL 语句。PreparedStatement 接口拥有一组方法，可以将输入参数的值传送到数据库中。

(3) CallableStatement 接口：用于执行数据库的 SQL 存储过程(一组可通过名称来调用的 SQL 语句，就像函数的调用)。

Connection 接口提供了多个同名但是参数不同的方法创建以上的三个接口，主要如下。

- public Statement createStatement();
- public Statement createStatement(int resultSetType,int resultSetConcurrency);
- public PreparedStatement prepareStatement(String sql);
- public PreparedStatement prepareStatement(String sql,int resultSetType, int resultSetConcurrency);
- public CallableStatement prepareCall(String sql);
- public CallableStatement prepareCall(String sql,int resultSetType, int resultSetConcurrency)。

在以上的方法中，参数 resultSetType 指返回记录集的类型，它可以是前后滚动的记录集，或者是只能向前滚动的记录集；resultSetConcurrency 指发送的 SQL 语句的权限，是属于不修改数据库资料的 select 语句，还是允许更新的 update，insert；参数 sql 指需要执行的 SQL 语句。

- boolean isClosed()：用于判断 Connection 对象是否已经被关闭。
- void commit()：用于提交 SQL 语句，确认从上一次提交以来所进行的修改。
- void close()：断开连接，释放资源。
- void rollback()：取消 SQL 语句，取消当前事务中进行的修改。
- DatabaseMetaData getMetaData()：获取一个 DatabaseMetaData 对象。

例如：

```
Class.forName("com.mysql.jdbc.Driver");
Connection conn=DriverManager.getConnection("jdbc:mysql://localhost:端口
号/数据库名称",
"用户名","密码");
conn.close();
```

7.2.3 执行 SQL 语句——Statement 接口

Statement 接口对象用于将普通的 SQL 语句发送到数据库中。实际上有三种 Statement 接口对象：Statement、PreparedStatement(从 Statement 接口继承而来)和 CallableStatement (从 PreparedStatement 接口继承而来)。它们都可在给定的数据库连接上发送/执行 SQL 语句：Statement 接口对象用于执行不带参数的简单的 SQL 语句；PreparedStatement 接口对象用来执行带或不带 IN 参数的预编译 SQL 语句；CallableStatement 接口对象用于执行对数据库存储过程的调用。

Statement 接口对象创建完毕后，主要使用以下方法。

- public ResultSet executeQuery(String sql)：用于执行产生单个结果集的语句，如 select。
- public int executeUpdate(String sql)：用于执行 insert、update 或 delete 语句以及 SQL DLL(数据定义)语句。当执行 insert、update 或 delete 语句时将返回一个整数，用于表示受影响的行数，返回结果为 0 时表示操作失败；当执行 SQL DLL 语句，如 create table 时，由于它不操作行，返回值将总为 0。
- public boolean execute(String sql)：用于执行返回多个结果集、多个更新计数或二者结合的 SQL 语句。
- public int[] executeBatch()：用于执行几个 SQL 语句。Statement 接口使用 addBatch(String sql)方法将几个 SQL 语句添加到一个语句块中，然后一同提交给数据库服务器，同时执行。当执行多个 SQL insert 语句时效率将很高，因为一次提交比多次提交要节省很多系统资源。
- void close()：释放 Satement 对象的数据库和 JDBC 资源。

PreparedStatement 接口对象主要使用以下方法。

- public ResultSet executeQuery(String sql)：使用 Select 命名对数据库进行查询。
- public int executeUpdate(String sql)：用于执行 insert、update 或 delete 语句对数据库进行新增、删除和修改操作。
- void setXXX(int parameterIndex, XXX x)：设定 XXX 数据类型值给

PreparedStatement 类对象的 IN 参数。

例如：

```
PrepareStatement pstmt=conn.prepareStatement(
"update examdata set sex=? where id=?");
pstmt.setString(1,"男");
pstmt.setString(2,"106301196702021211");
```

CallableStatement 接口主要用于执行存储过程，其主要方法同 Statement。存储过程有两种参数，IN 参数和 OUT 参数。CallableStatement 接口继承了 PerparedStatement 接口的 setXXX 方法对 IN 参数赋值。在 CallableStatement 接口中使用 OUT 参数，要做两件事情，一是对 OUT 参数进行类型注册，二是获取 OUT 参数的值。

CallableStatement 提供了多种方法进行类型注册，下面是常用的两种。

- registerOutParameter(String parameterStringName, int sqlType);
- registerOutParameter(String parameterStringName, int sqlType，int scale)。

例如：

```
registerOutParameter("id", java.sql.Types.VARCHAR);
```

调用 Connection 中的方法 prepareCall 可以创建一个 CallableStatement 对象。

例如：

```
CallableStatement cstmt=conn.prepareCall("{call test(?,?)}");
cstmt.registerOutParameter("id",Types.VARCHAR);
cstmt.execute();
String Strid=getString("id");
cstmt.close();
conn.close();
```

7.2.4 数据结果集——ResultSet 接口

ResultSet 接口用于获取执行 SQL 语句(或数据库存储过程)返回的结果。它的实例对象是符合 SQL 语句条件的所有行。

常用的方法如下。

(1) public boolean next()：用于数据库游标移动到结果集的下一行，使之成为当前行。如果当前行为最后一行，返回值为 false；否则，返回 true。

(2) 在定位到结果集中的某行后，就可以读取数据。对于不同数据类型，要使用不同的读取方法 getXXX()。JDBC 提供了两种形式。

- public XXX getXXX(String columnName)：XXX 代表任意的数据类型，参数 columnName 代表列名。
- public XXX getXXX(int columnIndex)：XXX 代表任意的数据类型，参数 columnIndex 代表列号。

(3) ResultSetMetaData getMetaData()：获取结果集的列编号、类型和属性。

前面介绍了 JDBC API 的常用接口，这些接口的一般使用过程为：DriverManager 类提供的 getConnection(参数)方法可以得到接口 Connection，而接口 Connection 的

createStatement()方法、prepareStatement(参数)方法、prepareCall(参数)方法可以得到执行数据查询语句的Statement接口、prepareStatement接口和CallableStatement接口，最后使用ResultSet接口接收查询结果。使用JDBC API一般是先声明各种接口的对象，再依次赋值，例如：

```
private Statement stmt = null;
private ResultSet rs = null;
private PreparedStatement prpSql=null;
Class.forName("com.mysql.jdbc.Driver");
Connection conn=
DriverManager.getConnection("jdbc:mysql://localhost:端口号/数据库名称",
"用户名","密码");
stmt = conn.createStatement();
rs = stmt.executeQuery("select * from database");
while(rs.next())
{   ...
rs.getString(1);
}
rs.close();
stmt.close();
conn.close();
```

7.2.5 数据库元数据——DatabaseMetaData和ResultSetMetaData

DatabaseMetaData接口可以从数据库管理系统中获得数据库的信息，例如数据库所有的表名、存储过程名等信息。这个类提供了许多方法来取得这些信息，常用的方法如下。

- String getDatabaseProductName()：获得数据库的名称。
- String getDatabaseProductVersion()：获得数据库的版本代号。
- String getTypeInfo()：获取数据库中可能得到的所有数据类型的描述。
- boolean supportsStoredProcedures()：检查数据库是否支持存储过程。
- String getUserName()：获取数据库当前用户的名称。
- ResultSet getCooumns(String databasename, String dboname, String tablename, String columnname)：获得表字段信息，以列的方式存储在一个ResultSet对象中。

ResultSetMetaData接口用来获取数据库表的结构。通过它提供的一些常用方法，可以获得ResultSet对象中的类型和属性信息的对象。常用方法如下。

- int getColumnCount()：返回此resultSet对象中的列数。
- String getColumnName()：获取指定的列名。
- int getColumnType(int column)：检索指定列的SQL类型。
- String getTableName(int column)：获取指定列的名称。
- int getColumnDisplaySize(int column)：以字符为单位，指示指定列的最大标准宽度。
- boolean isAutoIncrement(int column)：指示列是否进行自动编号。
- int isNullable(int column)：指示列中的值是否可以为空。

7.3 利用 JDBC 访问数据库

所有的利用 JDBC 访问数据库的程序都具有以下的流程。

(1) 加载 JDBC 驱动程序。

(2) 建立和数据库的连接。

(3) 执行 SQL 语句。

(4) 存放处理结果。

(5) 与数据库断开连接。

7.3.1 通过 JDBC-ODBC 桥连接来访问数据库

JDBC 与数据库连接最常见的一种方法是采用 JDBC-ODBC 桥连接。由于 ODBC 被广泛地使用，因此使用这种方式，可以使 JDBC 有能力访问几乎所有类型的数据库。

JDBC-ODBC 桥提供了对 JDBC 2.0 的有限支持，不支持较新版本的 JDBC 规范。JDBC-ODBC 桥被认为是一个过渡，因此是不建议使用的产品，仅包含在 JDK 中，JRE 中并未内置。建议开发者使用数据库商提供的 JDBC 驱动，或使用一个商业 JDBC 驱动，来代替 JDBC-ODBC 桥。读者需注意，在本教材所使用的 JDK 8 中，已经不再包含 JDBC-ODBC 桥。本文使用更低版本的 JDK 1.7 为例，介绍使用 JDBC-ODBC 连接数据库。下面以连接 Access 数据库为例，介绍 JDBC-ODBC 连接数据库。

现在，已建立一个简单的 Access 数据库：booklib，其中的一个表名为 book，如图 7.2 所示。数据库文件保存在 E:\programJSP\ch7 目录中。

字段名称	数据类型	说明
Bid	文本	图书编号
Name	文本	书名
Author	文本	作者
Publish	文本	出版社
Price	数字	单价

图 7.2 book 表结构

1. 配置数据源

首先要配置 ODBC 数据源。选择数据源(ODBC)(Windows 7 路径为“控制面板→管理工具→ODBC 数据源”)如图 7.3 所示。双击 ODBC 数据源图标，出现 ODBC 数据源管理器，选择“用户 DSN”选项卡，单击“添加”按钮，如图 7.4 所示的界面。

为新增的数据源选择驱动程序，Access 数据库选择 Driver do Microsoft Access Driver(*.mdb)，单击“完成”按钮，如图 7.5 所示的界面。在“数据源名”文本框输入数据源的名称，这个名称可以和数据库的名称不同，但是该名称是以后 Java 程序识别数据库的依据，如图 7.6 所示的界面。

图 7.3　管理工具

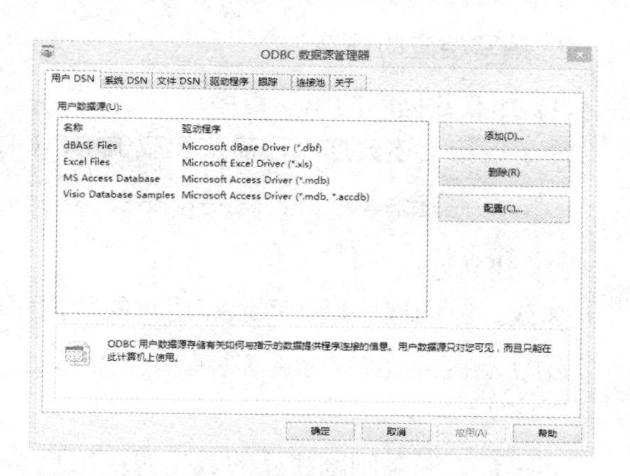

图 7.4　用户数据源对话框

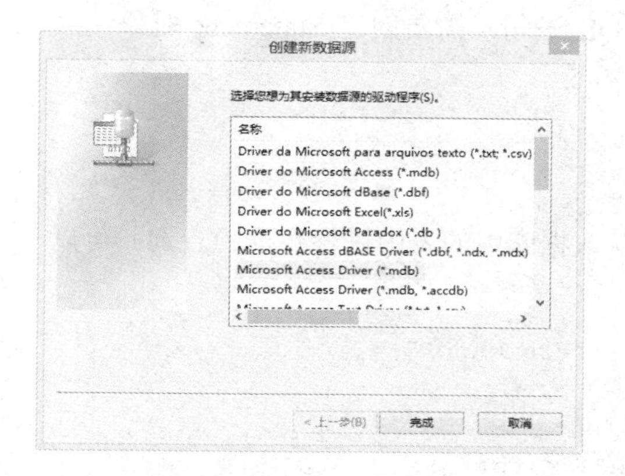

图 7.5　添加新的数据源

图 7.6　输入数据源名称

在“说明”文本框输入相应数据库的简介，可以不填。单击“选择(S)...”按钮可以选择已有数据库，如图 7.7 所示；单击“创建(C)...”按钮可以新建数据库。如果和数据库进行连接时需要用到用户名和密码，需要单击“高级(A)...”按钮，如图 7.8 所示。在登录名称和密码栏处分别输入相应的用户名和密码，单击“确定”按钮。返回选择数据库界面，单击“确定”按钮即完成创建数据源的工作。

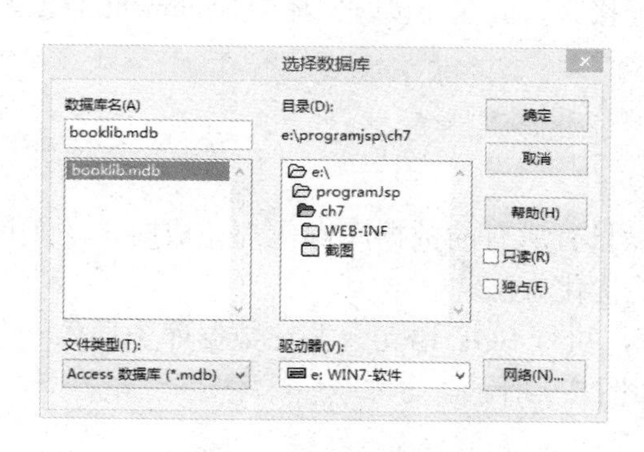

图 7.7　选择数据库

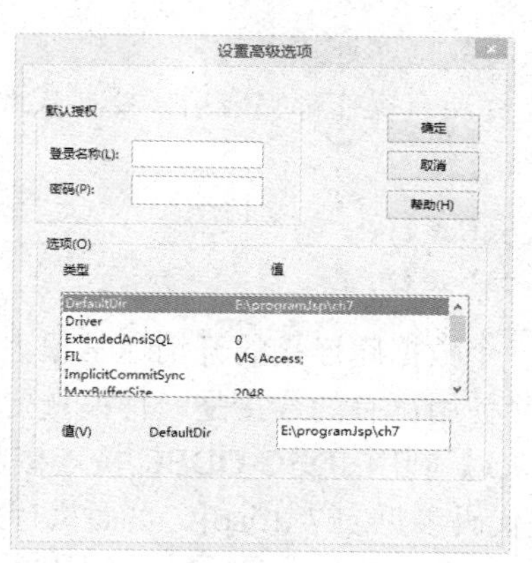

图 7.8　设置登录名和密码

2. 编写相应的程序

(1) 加载数据库的驱动程序。

为了连接上述数据库，需要加载 JDBC-ODBC 桥驱动程序，可以使用的语句如下：

```
Class.forName("sun.jdbc.odbc.JdbcOdbcDriver");
```

(2) 建立连接。

加载好驱动程序后，可以和用户的数据库进行连接。此时使用命令：

```
Connection conn=DriverManager.getConnection("jdbc:odbc:数据源","用户名","
密码");
```

其中数据源是刚刚建立的数据源；用户名和密码如果没有的话，可设置为空串。

(3) 向数据库发送 SQL 语句。

为了使用 Statement 接口对象向数据库发送 SQL 语句，需要创建 Statement 对象。可以使用命令：

```
Statement stmt=conn.createStatement();
```

(4) 处理查询结果。

创建 Statement 对象后，即可利用 SQL 语句从数据库中的表中进行数据的查询。可以使用的语句是：

```
ResultSet rs=stmt.executeQuery("select * from book");
while(rs.next()) //通过 next()方法使游标移到下一行
{  //通过 getString("列名")方法获取 Name 列的值
  out.println("<td>"+rs.getString("Name"+"</td>"));
  //通过 getFloat(列号)获取数据类型为 float 列号为 5 的值
  out.println("<td>"+rs.getFloat(5)+"</td>");
}
```

若想对数据库中表的数据进行更新时，可以使用命令：

```
stmt.executeUpdate("insert into book("TP001","Java2 核心技术","程峰","机械
工业出版社",75)");
```

(5) 关闭数据库连接。

在数据库所有操作都完成后，要显式地关闭连接。一般在连接时先释放 Statement 对象。如：......

```
stmt.close();
conn.close();
```

因为一个数据库连接的开销通常很大，所以只有所有的数据库操作都完成时，才关闭连接。重复使用已有的连接是一种很重要的性能优化方法。

【例 7.1】利用 JDBC-ODBC 桥连接数据库，执行 SQL 语句，并在数据库中插入一条记录(相关代码参见 ch7_1.jsp)，页面执行效果如图 7.9 所示。

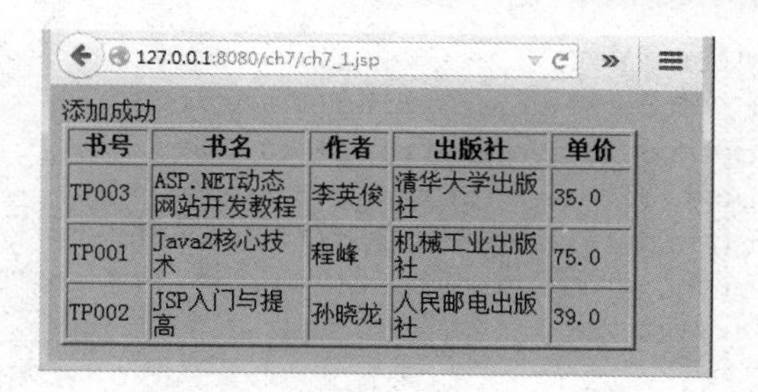

书号	书名	作者	出版社	单价
TP003	ASP.NET动态网站开发教程	李英俊	清华大学出版社	35.0
TP001	Java2核心技术	程峰	机械工业出版社	75.0
TP002	JSP入门与提高	孙晓龙	人民邮电出版社	39.0

图 7.9　ch7_1.jsp 页面效果

ch7_1.jsp 文件内容如下：

```
<%@ page contentType="text/html;charset=GB2312" %>
<%@ page import="java.sql.*" %>
<html>
<head>
<title>【例 7.1】利用 JDBC-ODBC 桥连接数据库</title>
</head>
<body bgcolor=cyan>
<% Connection conn;
   Statement stmt=null;
   ResultSet rs;
   int n;
   try{
      Class.forName("sun.jdbc.odbc.JdbcOdbcDriver");
   }
   catch(ClassNotFoundException e){
      out.print(e);
   }
   try{ //和数据库建立连接
      conn=DriverManager.getConnection("jdbc:odbc:book","","");
      stmt=conn.createStatement(); //创建 Statement 对象，用于执行 SQL 语句
      n=stmt.executeUpdate(
     "insert into book values('TP003',
                'ASP.NET 动态网站开发教程','李英俊','清华大学出版社',35)");
      //向表中插入一条记录
      if(n>0)   out.println("添加成功");
      else      out.println("添加失败");
      rs=stmt.executeQuery("select * from book");
      out.print("<table border=2>");
      out.print("<tr>");
        out.print("<th width=50>"+"书号");
        out.print("<th width=100>"+"书名");
        out.print("<th width=50>"+"作者");
        out.print("<th width=100>"+"出版社");
        out.print("<th width=50>"+"单价");
      out.print("</tr>");
      while(rs.next()){
```

```
        out.print("<tr>");
          out.print("<td>"+rs.getString(1)+"</td>");
          out.print("<td>"+rs.getString(2)+"</td>");
          out.print("<td>"+rs.getString(3)+"</td>");
          out.print("<td>"+rs.getString("publish")+"</td>");
          out.print("<td>"+rs.getFloat("price")+"</td>");
        out.print("</tr>");
        }
        out.print("</table>");
        stmt.close();  //关闭数据库
        conn.close();
    }
    catch(Exception e){
        out.println(e.toString());
    }
%></body>
</html>
```

程序运行前，数据库 book 表中的记录内容如图 7.10 所示。程序运行后，book 表的内容如图 7.11 所示。

book：表

Bid	Name	Author	Publish	Price
TP001	Java2核心技术	程峰	机械工业出版社	75
TP002	JSP入门与提高	孙晓龙	人民邮电出版社	39
				0

记录: 1 共有记录数: 2

图 7.10　程序运行前 book 表中的内容

book：表

Bid	Name	Author	Publish	Price
TP001	Java2核心技术	程峰	机械工业出版社	75
TP002	JSP入门与提高	孙晓龙	人民邮电出版社	39
TP003	ASP.NET动态网站开发教程	李英俊	清华大学出版社	35

记录: 3 共有记录数: 3

图 7.11　程序运行后 book 表中的内容

7.3.2　利用本地协议纯 Java 驱动程序连接数据库

MySQL 数据库是一个优秀的免费网络数据库，应用非常广泛，功能强大，不依赖于平台是一个开源项目，本节以访问 MySQL 数据库为例介绍。

1. 安装 MySQL 数据库

在 MySQL 官网 http://dev.mysql.com/downloads/，可进行 MySQL 各种版本的下载，本书使用的是 MySQL 的社区服务版，为免费版本，供学习交流使用，安装文件为 mysql-installer-community-5.6.17.0.msi，该版本可以安装在 Windows 平台。

双击安装包，进行默认安装即可，安装过程中部分组件需联网下载，请确保安装过程已经连接互联网。同时需注意记住安装过程中所输入的 MySQL 的 root 用户密码。

2. 启动 MySQL

如果 MySQL 安装过程中没有改变 MySQL 服务的名称，用户可在“控制面板→管理工具→服务”中找名称为 MySQL5.6 的服务，便可启动和关闭 MySQL 服务。

3. 创建数据库

在 MySQL5.6 可通过 MySQL Workbench 6.1 CE 控制面板和命令提示符两种方式对

MySQL 数据库进行管理。本教材使用命令行的方式进行演示。

启动 MySQL 服务后，可以在提示符下输入 sql 语句来创建数据库和表，sql 语句结束必须使用“;”符号。在编辑 sql 语句的过程中可以使用/c 终止当前 sql 语句的编辑。

在 MySQL 监视器提示符下，首先输入用户密码，方可进行数据库操作。输入密码后成功后，输入创建图书管理数据库 booklib 的 SQL 语句“create database booklib;”并执行，如图 7.12 所示。

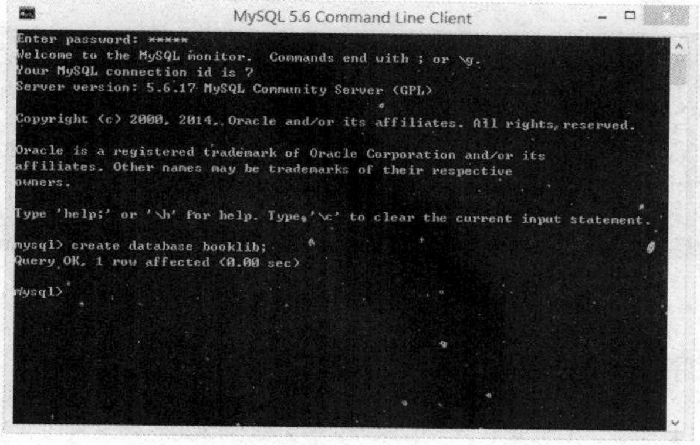

```
MySQL 5.6 Command Line Client
Enter password: *****
Welcome to the MySQL monitor.  Commands end with ; or \g.
Your MySQL connection id is 7
Server version: 5.6.17 MySQL Community Server (GPL)

Copyright (c) 2000, 2014, Oracle and/or its affiliates. All rights reserved.

Oracle is a registered trademark of Oracle Corporation and/or its
affiliates. Other names may be trademarks of their respective
owners.

Type 'help;' or '\h' for help. Type '\c' to clear the current input statement.

mysql> create database booklib;
Query OK, 1 row affected (0.00 sec)

mysql>
```

图 7.12　创建数据库 booklib 的 MS-DOS 窗口

4. 创建数据库中的表

要创建某个建数据库中的表，首先要打开该数据库，使之成为当前数据库，然后再使用 SQL 语句来创建表。打开 booklib 数据库的命令为：

```
use  booklib
```

在 booklib 数据库中创建 book 表的 SQL 语句为：

```
CREATE TABLE  book (
   bid  char(20) NOT NULL,
   name char(20) default NULL,
   author char(10) default NULL,
   publish varchar(50) default NULL,
   price float default NULL,
   PRIMARY KEY  (bid)
) ENGINE=InnoDB  DEFAULT  CHARSET=utf8;
```

建表操作如图 7.13 所示。

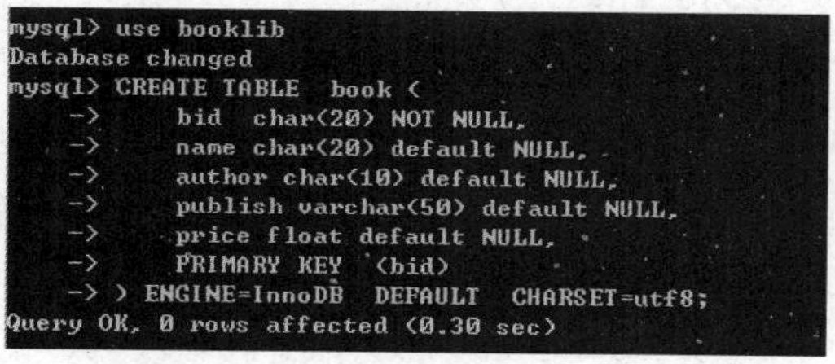

```
mysql> use booklib
Database changed
mysql> CREATE TABLE  book (
    ->      bid  char(20) NOT NULL,
    ->      name char(20) default NULL,
    ->      author char(10) default NULL,
    ->      publish varchar(50) default NULL,
    ->      price float default NULL,
    ->      PRIMARY KEY  (bid)
    -> ) ENGINE=InnoDB  DEFAULT  CHARSET=utf8;
Query OK, 0 rows affected (0.30 sec)
```

图 7.13　创建 book 表的 MS-DOS 窗口

在数据库 booklib 中建立 book 表之后，就可以使用 SQL 语句对表中数据进行增、删、改、查操作。

向 book 表中添加两条记录的操作如图 7.14 所示。查询操作如图 7.15 所示。

```
mysql> insert into book(bid,name,author,publish,price)
    -> values('tp1002','computer','han','apress',25.0);
Query OK, 1 row affected (0.06 sec)
```

图 7.14　在 book 表中插入数据的 MS-DOS 窗口

```
mysql> select * from book;
+--------+----------+--------+---------+-------+
| bid    | name     | author | publish | price |
+--------+----------+--------+---------+-------+
| tp1002 | computer | han    | apress  |    25 |
+--------+----------+--------+---------+-------+
1 row in set (0.00 sec)
```

图 7.15　查询数据的 MS-DOS 窗口

5. 编写相应的程序——连接 MySQL 数据库

采用本地协议纯 Java 驱动程序，首先要从 MySQL 的官方网站下载驱动程序，本教材使用的是 mysql-connector-java-5.1.30-bin.jar。然后将其放置到本机的 jre 安装目录的 lib 文件夹下(本教材为：C:\Program Files\Java\jre8\lib\ext)，并在 classpath 中指定其所在路径，或者将其复制到 Tomcat 7.0 安装目录的 lib 子目录中(C:\Program Files\Apache Software Foundation\Tomcat 7.0\lib)。准备工作做好后，参照 7.3.1 节的程序编写步骤编写相应的数据库访问程序。使用纯 Java 驱动程序访问数据库，只是加载的驱动程序和连接 url 稍有不同。MySQL 服务器占用的端口默认为 3306。加载驱动程序的代码为：

```
try{ Class.forName("com.mysql.jdbc.Driver");
   }
```

创建连接的代码如下:

```
try{ //和数据库建立连接
       conn=DriverManager.getConnection(
"jdbc:mysql://localhost:3306/booklib","root","");
        ...
       conn.close();
   }
catch(Exception e){
       out.println(e.toString());
   }
```

【例 7.2】利用纯 Java 驱动程序连接 MySQL 数据库，查询 book 表中的所有记录。假设：MySQL 数据库的名称为 booklib，表同 Access 数据库中的表结构相同，表名称为 book，相关代码参见 ch7_2.jsp，页面效果如图 7.16 所示。

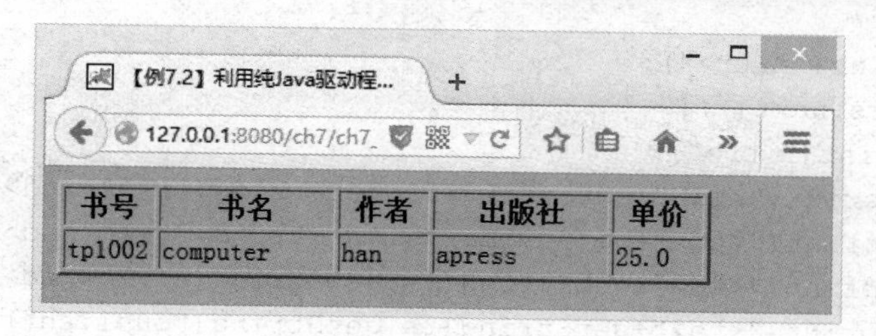

书号	书名	作者	出版社	单价
tp1002	computer	han	apress	25.0

图 7.16 ch7_2.jsp 页面效果

ch7_2.jsp 文件内容如下：

```
<%@ page contentType="text/html;charset=GB2312" %>
<%@ page import="java.sql.*" %>
<html>
<head>
<title>【例 7.2】利用纯 Java 驱动程序连接 mysql 数据库</title>
</head>
<body bgcolor=cyan>
<%!
  public String getString(String s){
    String str=s;
    try{byte bb[]=str.getBytes("ISO-8859-1");
       str=new String(bb);
    }
    catch(Exception e){}
    return str;
  }
%>
<% Connection conn;
   Statement stmt=null;
   ResultSet rs;
   try{ Class.forName("com.mysql.jdbc.Driver");
   }
   catch(ClassNotFoundException e){
      out.print(e);
   }
   try{ //和数据库建立连接
      conn=DriverManager.getConnection(
           "jdbc:mysql://localhost:3306/booklib","root","");
      stmt=conn.createStatement(); //创建 Statement 对象，用于执行 SQL 语句
      rs=stmt.executeQuery("select * from book");
      out.print("<table border=2>");
      out.print("<tr>");
        out.print("<th width=50>"+"书号");
        out.print("<th width=100>"+"书名");
        out.print("<th width=50>"+"作者");
        out.print("<th width=100>"+"出版社");
        out.print("<th width=50>"+"单价");
```

```
            out.print("</tr>");
            while(rs.next()){
             out.print("<tr>");
                out.print("<td>"+getString(rs.getString(1))+"</td>");
                out.print("<td>"+getString(rs.getString(2))+"</td>");
                out.print("<td>"+getString(rs.getString(3))+"</td>");
                out.print("<td>"+getString(rs.getString("publish"))+"</td>");
                out.print("<td>"+rs.getFloat("price")+"</td>");
             out.print("</tr>");
            }
            out.print("</table>");
            stmt.close();  //关闭数据库
            conn.close();
        }
        catch(Exception e){
            out.println(e.toString());
        }
    %>
    </body>
    </html>
```

由例 7.1、例 7.2 可以看出：连接不同的数据库，除 Class.forName()中的驱动程序类及 url 的书写方式不同之外，其他所有部分均相同。

7.3.3 配置和连接不同的数据库

1. 连接 Oracle 数据库

Oracle 数据库是目前常用的网络数据库，具有较强的功能、安全性和不依赖平台的特点，经常应用在较大规模的 Web 应用系统中。如 7.3.1 节和 7.3.2 节所述，可以使用 JDBC-ODBC 桥接器或加载纯 Java 驱动程序与 Oracle 数据库建立连接。下面介绍加载纯 Java 驱动程序的连接方法。

安装完 Oracle 后，jdbc 的驱动可以在 Oracle 11g 的安装目录中找到(本教材目录为：F:\app\admin\product\11.2.0\dbhome_1\jdbc\lib)，找到文件 ojdbc6.jar。将 ojdbc6.jar 复制到 Tomcat 7.0 安装目录的 lib 子目录中(C:\Program Files\Apache Software Foundation\Tomcat 7.0\lib)

将 7.3.1 节中的 2 编写相应的程序步骤中的前两步改为如下两步。

(1) 加载驱动程序。

```
try{
   Class.forName("oracle.jdbc.driver.OracleDriver");
}
catch(Exception e){
   out.print(e.toString());
}
```

(2) 建立连接。

```
try {Connection conn=
   DriverManager.getConnection("jdbc:oracle:thin:@host：端口号：数据库名",
    "用户名","密码");
    …
   conn.close();
  }
catch(Exception e){
   out.print(e.toString());
}
```

2. 连接 SQL Server 数据库

SQL Server 数据库是一个网络数据库，功能强大简单易学，目前在中型的 Web 应用中使用较多。目前 SQL Server 比较流行的有 2008 版本，本教材以 2008 版为例，介绍纯 Java 驱动程序与 SQL Server 连接的方法。

微软提供了 SQL Server 2008 的驱动程序，可以下载 sqljdbc_3.0.1301.101_chs.tar.gz 驱动压缩包。解压后可得到 sqljdbc4.jar 文件，sqljdbc4.jar 类库要求使用 6.0 或更高版本的 Java 运行时环境 (JRE)。在 JRE 1.4 或 5.0 上使用 sqljdbc4.jar 会引发异常。将 sqljdbc4.jar 复制到 Tomcat 7.0 安装目录的 lib 子目录中(C:\Program Files\Apache Software Foundation\Tomcat 7.0\lib)，这样 Tomcat 服务器就可以使用该驱动程序了。同 Oracle 一样，加载驱动和建立连接的方法如下。

(1) 加载驱动程序。

```
try{
   Class.forName("com.microsoft.sqlserver.jdbc.SQLServerDriver");
}
catch(Exception e){
   out.print(e.toString());
}
```

(2) 建立连接。

```
try {Connection conn=
    DriverManager.getConnection(
      "jdbc:sqlserver://127.0.0.1:1443;DatabaseName=数据库名",
     "用户名","密码");
     …
    conn.close();
  }
catch(Exception e){
  out.print(e.toString());
}
```

提示： 如果应用程序和要连接的 SQL Server 2008 服务器在不同的计算机上，可以使用 SQL Server 2008 服务器的 ip 地址，例如 192.168.1.1。

7.4　数据库操作案例

在介绍了与数据库的连接之后，就可以使用 JDBC 应用程序接口和数据库交互信息，例如查询数据、添加数据、修改数据和删除数据等。JDBC 和数据库交互主要是使用 SQL 语句实现，JDBC 将标准的 SQL 语句发送给数据库，数据库将结果集或操作结果返回给应用程序。本节主要使用纯 Java 驱动程序连接 MySQL 数据库，通过几个典型的数据库操作例子，介绍 JSP 中通过 JavaBean 来操作数据库的编程技术。使用前其他驱动程序或连接其他数据库，只需对程序中的连接部分稍做修改即可。

7.4.1　查询数据

1. 顺序查询数据

如 7.3.1 节所述的编程步骤，在得到了数据库连接之后，就可以通过 Statement 对象发送标准的 SQL 语句，查询数据库中表的数据，并将查询的结构放到一个 ResultSet 接口声明的对象中。ResultSet 对象是以统一的行列形式组织数据的。

例如：

ResultSet rs=stmt.executeQuery("select bid,name,author,publish,price from book");

对于结果集 rs 的列数为 5 列，第一列对用 bid，第二列对应 name，第三列对应 author，第四列对应 publish，第五列对应 price；而每一次 rs 只能看到一行，要在看到下一行，必须使用 next()方法移动当前行。ResultSet 对象使用 getXXX()方法获得当前行字段的值，有关方法请参照 7.2.4 节。

在下面的例子中，有一个负责查询数据库的 Bean，该 Bean 加载纯 Java 数据库驱动程序连接数据库，查询表，并负责生成查询结果集显示的 HTML 代码。用户通过 JSP 页面输入用户名和密码，这里的用户名是数据库管理员的名称，然后 JSP 页面调用 Bean 查询数据库，并显示查询结果。

【例 7.3】顺序查询 booklib 库中的 book 表中的数据并显示，相关代码参见 QueryBean.java 和 ch7_3.jsp，页面效果如图 7.17 所示。

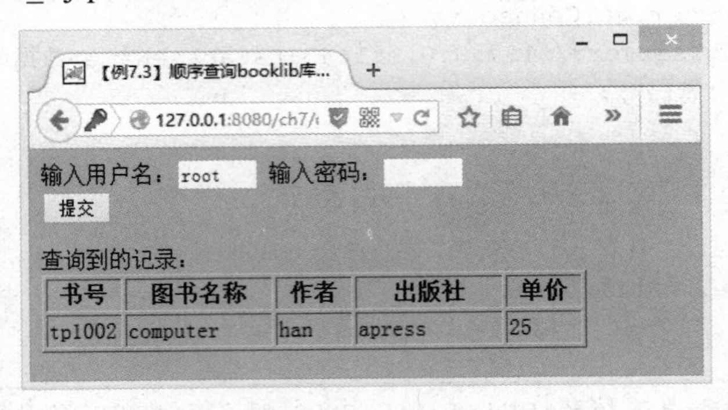

图 7.17　ch7_3.jsp 页面效果

QueryBean.java 文件代码如下：

```
package mybean.database;
import java.sql.*;
public class QueryBean {
    String user="";
    String password="";
    StringBuffer queryResult;
    public QueryBean(){
        queryResult=new StringBuffer();
        try{  Class.forName("com.mysql.jdbc.Driver");
        }
        catch(Exception e){}
    }
    public String  getString(String str){
        String s=str.trim();
        try{byte bb[]=s.getBytes("ISO-8859-1");
            s=new String(bb);
        }
        catch(Exception e){}
        return s;
    }
    public String getUser() {
        return user;
    }
    public void setUser(String user) {
        this.user =getString( user);
        queryResult=new StringBuffer();
    }
    public String getPassword() {
        return password;
    }
    public void setPassword(String password) {
        this.password = getString(password);
        queryResult=new StringBuffer();
    }
    public StringBuffer getQueryResult() {
        Connection conn;
        Statement stmt;
        ResultSet rs;
        try{
            queryResult.append("<table border=1>");
            String uri="jdbc:mysql://localhost:3306/booklib";
            String userid=user;
            String userpwd=password;
            conn=DriverManager.getConnection(uri,userid,userpwd);
            stmt=conn.createStatement();
            rs=stmt.executeQuery("select * from book");
```

```
            queryResult.append("<tr>");
              queryResult.append("<th width=50>书号</th>");
              queryResult.append("<th width=100>图书名称</th>");
              queryResult.append("<th width=50>作者</th>");
              queryResult.append("<th width=100>出版社</th>");
              queryResult.append("<th width=50>单价</th>");
            queryResult.append("</tr>");
            while (rs.next()){
                queryResult.append("<tr>");
                for(int k=1;k<=5;k++){
                    queryResult.append("<td>"+
                            getString(rs.getString(k))+"</td>");
                }
                queryResult.append("</tr>");
            }
            queryResult.append("</table>");
            conn.close();
            }
        catch(Exception e)
        {  queryResult.append("请输入正确的用户名和密码！"+"
注意：<br>用户名为 root，密码为空。");
        }
        return queryResult;
    }
}
```

ch7_3.jsp 文件内容如下：

```
<%@ page contentType="text/html;charset=GB2312" %>
<%@ page import="mybean.database.QueryBean" %>
<jsp:useBean id="query1" class="mybean.database.QueryBean"
  scope="session" ></jsp:useBean>
<html>
<head>
<title>【例 7.3】顺序查询 booklib 库中的 book 表中的数据并显示</title>
</head>
<body bgcolor=cyan>
<form action="" method="post" >
   输入用户名：<input type="text" name="user" size=6 value="root">
   输入密码：<input type="password" name="password" size=6 value="" >
   <br>
  <input type="submit" name="sub" value="提交">
</form>
<jsp:setProperty name="query1" property="*" />
查询到的记录：<br>
<jsp:getProperty name="query1" property="queryResult"/>
</body>
</html>
```

2. 随机查询

前面介绍的顺序查询数据库技术，使用了 Connection 的 createStatement()方法创建了一个 Statement 对象，然后执行 Statement 对象的 executeQuery 方法得到了一个 ResultSet 结果集对象，使用该对象的 next()方法实现了顺序查询数据。如果需要在结果集中前后移动或随机显示某一条记录，这时就必须得到一个可以滚动的结果集。得到滚动结果集的方法如下：

```
Statement stmt=conn.createStatement(int type,int concurrency);
ResultSet rs=stmt.executeQuery(sqlstr);
```

其中 type 的取值决定结果集的滚动方式，可以取下列值。

- ResultSet.TYPE_FORWORD_ONLY：结果集的游标只能向下移动。
- ResultSet.TYPE_SCROLL_INSENSITIVE：结果集的游标可以上下移动，当数据库的数据变化时，当前结果集不变。
- ResultSet.TYPE_SCROLL_SENSITIVE：结果集的游标可以上下移动，当数据库的数据变化时，当前结果集同步变化。

其中 concurrency 的取值决定是否可以使用结果集更新数据库，可以取下列值。

- ResultSet.CONCUR_READ_ONLY：不能用结果集中的数据更新数据库中的表。
- ResultSet.CONCUR_UPDATABLE：能用结果集中的数据更新数据库中的表。

滚动结果集 ResultSet 对象可以使用的方法如下。

- public boolean previous()：将游标向上移动，当移动到结果集第一行之前返回 false。
- public void beforeFirst()：将游标移到结果集第一行之前。
- public void afterLast()：将游标移到结果集最后一行之后。
- public void first()：将游标移到结果集的第一行。
- public void last()：将游标移到结果集的最后一行。
- public boolean isAfterLast()：判断游标是否在结果集的最后一行之后。
- public boolean isBeforeFirst()：判断游标是否在结果集第一行之前。
- public boolean isFirst()：判断游标是否在结果集第一行。
- public boolean isLast()：判断游标是否在结果集的最后一行。
- public int getRow()：得到当前游标所在的行，行号从 1 开始，如果结果集没有行，则返回 0。
- public boolean absolute(int row)：将游标移动到 row 所指定的行，row 取负值则为倒数，如-1，则移动到最后一行。

在下面的例子中得到一个可滚动的结果集，首先将游标移动到最后一行，获取行号得到记录个数，然后逆序显示所有记录，最后显示第 2 条记录。

【例 7.4】随机查询示例，相关文件参见 RandomQuerybean.java 和 ch7_4.jsp，页面效果如图 7.18 所示。

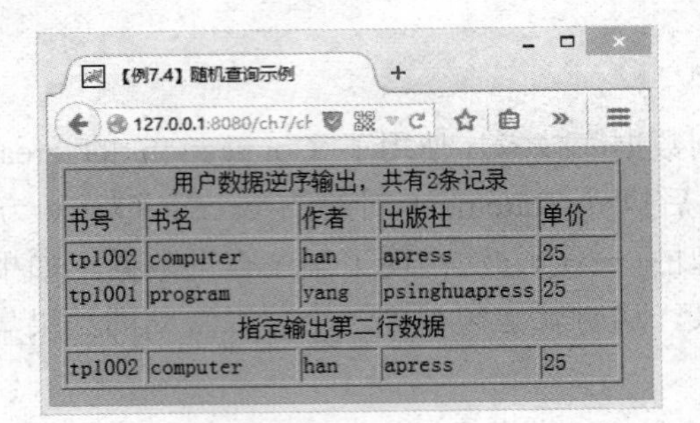

图 7.18　ch7_4.jsp 页面效果

RandomQuerybean.java 文件代码如下：

```
package mybean.database;
import java.sql.*;
public class RandomQuerybean {
    int recordCount;
    StringBuffer queryResult;
    Connection conn;
    Statement stmt;
    ResultSet rs;
    public String handleString(String s){
        String str=s;
        try{
            byte bb[]=str.getBytes("ISO-8859-1");
            str=new String(bb);
            return str;
        }
        catch(Exception e){
            return str;
        }
    }
    public RandomQuerybean(){
        queryResult=new StringBuffer();
        try{
            Class.forName("com.mysql.jdbc.Driver");
        }
        catch(Exception e){}
    }
    public int getRecordCount() {
        try{
            String uri="jdbc:mysql://localhost:3306/booklib";
            String user="root";
            String password="";
            conn=DriverManager.getConnection(uri,user,password);
            stmt=conn.createStatement(ResultSet.TYPE_SCROLL_SENSITIVE,
                    ResultSet.CONCUR_READ_ONLY);
            rs=stmt.executeQuery("select * from book");
```

```
            rs.last();
            recordCount=rs.getRow();
            conn.close();
        }
        catch(Exception e){
            recordCount=-1;
        }
        return recordCount;
    }
    public StringBuffer getQueryResult() {
        try{
            String uri="jdbc:mysql://localhost:3306/booklib";
            String user="root";
            String password="";
            conn=DriverManager.getConnection(uri,user,password);
            stmt=conn.createStatement(ResultSet.TYPE_SCROLL_SENSITIVE,
                    ResultSet.CONCUR_READ_ONLY);
            rs=stmt.executeQuery("select * from book");
            rs.last();
            int rowNumber=rs.getRow();
            queryResult.append("<table border>");
            queryResult.append("<tr><td colspan=5 align=center>"+
"用户数据逆序输出，共有"+rowNumber+"条记录</td></tr>");
            queryResult.append("<tr>");
             queryResult.append("<td width=50>书号</td>");
             queryResult.append("<td width=100>书名</td>");
             queryResult.append("<td width=50>作者</td>");
             queryResult.append("<td width=100>出版社</td>");
             queryResult.append("<td width=50>单价</td>");
            queryResult.append("</tr>");
            rs.afterLast();
            while(rs.previous()){
                queryResult.append("<tr>");
                  queryResult.append("<td>"+rs.getString(1)+"</td>");
                  queryResult.append("<td>"+rs.getString(2)+"</td>");
                  queryResult.append("<td>"+rs.getString(3)+"</td>");
                  queryResult.append("<td>"+rs.getString(4)+"</td>");
                  queryResult.append("<td>"+rs.getString(5)+"</td>");
                queryResult.append("</tr>");
            }
            rs.absolute(2);
            queryResult.append("<tr><td colspan=5 align=center>"+
"指定输出第二行数据</td></tr>");
            queryResult.append("<tr>");
              queryResult.append("<td>"+rs.getString(1)+"</td>");
              queryResult.append("<td>"+rs.getString(2)+"</td>");
              queryResult.append("<td>"+rs.getString(3)+"</td>");
```

```
                queryResult.append("<td>"+rs.getString(4)+"</td>");
                queryResult.append("<td>"+rs.getString(5)+"</td>");
              queryResult.append("</tr>");
              rs.close();
              stmt.close();
              conn.close();
          }
          catch(SQLException e){
              queryResult.append("Sql 异常");
          }
          return queryResult;
      }
}
```

ch7_4.jsp 文件内容如下：

```
<%@ page contentType="text/html;charset=GB2312" %>
<%@ page import="mybean.database.RandomQuerybean" %>
<jsp:useBean id="query1" class="mybean.database.RandomQuerybean"
  scope="request"  ></jsp:useBean>
<html>
<head>
<title>【例 7.4】随机查询示例</title>
</head>
<body bgcolor=cyan>
<jsp:getProperty name="query1" property="queryResult"/>
</body>
</html>
```

3. 条件查询

在顺序查询和随机查询中得到的是表中的全部记录。如果要得到表中满足条件的记录，就要采用条件查询。条件查询是由客户端提供查询条件，即查询的参数，再由这些参数构造 SQL 语句，执行该语句并得到筛选结果的查询。

下面的例子根据用户选择的字段和输入的关键字查询图书信息，如果没有输入关键字则显示所有图书的信息。

【例 7.5】条件查询图书信息，相关文件参见 keyQueryBean.java 和 ch7_5.jsp，页面显示效果如图 7.19 所示。

图 7.19　ch7_5.jsp 页面效果

keyQueryBean.java 文件代码如下：

```
package mybean.database;
import java.sql.*;
public class keyQueryBean {
    String feild="",
        keyWord="";
    StringBuffer queryResult;
    public keyQueryBean(){
        queryResult=new StringBuffer();
        try{ Class.forName("com.mysql.jdbc.Driver");}
        catch(Exception e){}
    }
    public String getString(String str){
        String s=str;
        try{byte bb[]=s.getBytes("ISO-8859-1");
          s=new String(bb);
        }
        catch(Exception e){}
        return  s;
    }
    public void setKeyWord(String s){
        keyWord=getString(s.trim());
    }
    public String getKeyWord(){
        return keyWord;
    }
    public String getFeild()  {
        return feild;
    }
    public void setFeild(String feild)  {
        this.feild = getString(feild.trim());
    }
    public StringBuffer getQueryResult()  {
        String condition="";
        if(feild.equals("") || keyWord.equals("")){//判断是否有值输入
            condition="select * from book ";
        }
        else{
          condition="select * from book where "+
            feild+" like '%"+keyWord+"%'";
        }
        Connection conn;
        Statement stmt;
        ResultSet rs;
        try{
            String uri="jdbc:mysql://localhost:3306/booklib";
```

```
            String user="root";
            String password="";
            conn=DriverManager.getConnection(uri,user,password);
            stmt=conn.createStatement();
            rs=stmt.executeQuery(condition);
            queryResult.append("<table border>");
            queryResult.append("<tr><td colspan=5 align=center>"+
"book 表中数据</td></tr>");
           queryResult.append("<tr>");
             queryResult.append("<td width=50>书号</td>");
             queryResult.append("<td width=100>书名</td>");
             queryResult.append("<td width=50>作者</td>");
             queryResult.append("<td width=100>出版社</td>");
             queryResult.append("<td width=50>单价</td>");
           queryResult.append("</tr>");
           while(rs.next()){
               queryResult.append("<tr>");
                 queryResult.append("<td>"+
                     getString(rs.getString(1))+"</td>");
                 queryResult.append("<td>"+
                     getString(rs.getString(2))+"</td>");
                 queryResult.append("<td>"+
                     getString(rs.getString(3))+"</td>");
                 queryResult.append("<td>"+
                     getString(rs.getString(4))+"</td>");
                 queryResult.append("<td>"+rs.getString(5)+"</td>");
               queryResult.append("</tr>");
            }
            queryResult.append("</table>");
            rs.close();
            stmt.close();
            conn.close();
        }
        catch(Exception e){
            System.out.println(e);
            queryResult.append(e.toString());
        }
        return queryResult;
    }
}
```

ch7_5.jsp 文件内容如下：

```
<%@ page contentType="text/html;charset=GB2312" %>
<%@ page import="mybean.database.keyQueryBean" %>
<jsp:useBean id="query1" class="mybean.database.keyQueryBean"
  scope="request"  ></jsp:useBean>
<jsp:setProperty name="query1" property="feild" param="feild"/>
```

```
<jsp:setProperty name="query1" property="keyWord" param="keyWord"/>
<html>
<head>
<title>【例7.5】条件查询示例</title>
</head>
<body bgcolor=cyan>
<form action="" method="post" name="form1">
  选择<select name="feild">
     <option value="bid" >书号
     <option value="author">作者
     <option value="name">书名
     <option value="publish">出版社
  </select>
  含有<input type="text" name="keyWord">
     <input type="submit" name="g" value="提交">
</form>
<jsp:getProperty name="query1" property="feild"/>含有关键字
<jsp:getProperty name="query1" property="keyWord"/>的记录:
<jsp:getProperty name="query1" property="queryResult"/>
</body>
</html>
```

7.4.2 更新查询

Statement 对象提供了 int executeUpdate(String sqlStatement)方法，用于实现对数据库中数据的添加、删除和更新操作。sqlStatement 参数是由 insert、delete 和 update 等关键字构成的 SQL 语句。函数返回值为查询所影响的行数，失败返回 0。

例如添加记录的 SQL 语句：

```
String sqlStatement="insert book values('tp1004','高数','杨',
         '清华出版社',35)";
    stmt.executeUpdate(sqlStatement);
```

例如更新记录的 SQL 语句：

```
    String sqlStatement="update book set name='新值' where bid='tp1001'";
    stmt.executeUpdate(sqlStatement);
```

例如删除记录的 SQL 语句：

```
    String sqlStatement="delete from book where bid='tp1001' ";
    stmt.executeUpdate(sqlStatement);
```

1. 添加记录

添加记录就是在数据库中表的最后一行追加一条记录。下面的案例使用 addBean 向数据库的表中增加一条记录，使用 7.4.1 节中的 keyQueryBean 来查看追加成功后的表记录。

【例 7.6】向 MySQL 数据库中的表追加记录，相关文件参见 addBean.java 和 ch7_6.jsp，页面效果如图 7.20 所示。

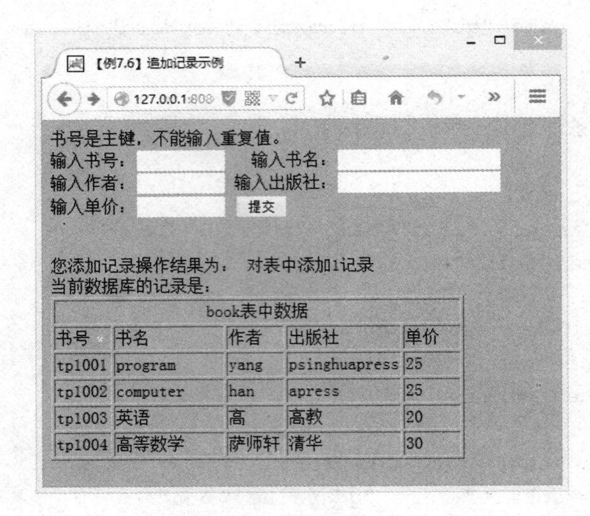

图 7.20　ch7_6.jsp 页面效果

addBean.java 文件代码如下：

```
package mybean.database;
import java.sql.*;
public class addBean {
    String strbid="",
    strname="",strauthor="",strpublish="";
    float fprice=0.0f;
    String addMessage="";
    public addBean(){
        try{Class.forName("com.mysql.jdbc.Driver");
        }
        catch(Exception e){}
    }
    public String getStrbid() {
        return strbid;
    }
    public void setStrbid(String strbid) {
        this.strbid = strbid;
    }
    public String getStrname() {
        return strname;
    }
    public void setStrname(String strname) {
        this.strname = strname.trim();
    }
    public String getStrauthor() {
        return strauthor;
    }
    public void setStrauthor(String strauthor) {
        this.strauthor = strauthor.trim();
    }
```

```
    public String getStrpublish() {
        return strpublish;
    }
    public void setStrpublish(String strpublish) {
        this.strpublish = strpublish.trim();
    }
    public float getFprice() {
        return fprice;
    }
    public void setFprice(float fprice) {
        this.fprice = fprice;
    }
    public String getAddMessage(){
        if(strbid.equals("")){
            addMessage="没有插入记录";
        }
        else{
            String condition="insert book values(?,?,?,?,?)";
            String uri="jdbc:mysql://localhost:3306/booklib";
            String user="root";
            String password="";
            Connection conn;
            PreparedStatement stmt;
            try{conn=DriverManager.getConnection(uri,user,password);
                stmt=conn.prepareStatement(condition);
                stmt.setString(1, strbid);
                stmt.setString(2, strname);
                stmt.setString(3, strauthor);
                stmt.setString(4, strpublish);
                stmt.setFloat(5, fprice);
                int m=stmt.executeUpdate();
                if(m!=0){
                 addMessage="对表中添加"+m+"记录";
                }
                else{
                 addMessage="添加记录失败！";
                }
                stmt.close();
                conn.close();
            }
            catch(Exception e){
                addMessage="输入的书号不能为重复！";
            }
        }
        return addMessage;
    }
}
```

ch7_6.jsp 文件内容如下：

```
<%@ page contentType="text/html;charset=GB2312" %>
<%@ page import="mybean.database.keyQueryBean" %>
<%@ page import="mybean.database.addBean" %>
<jsp:useBean id="query1" class="mybean.database.keyQueryBean"
  scope="request"  ></jsp:useBean>
<jsp:useBean id="update1" class="mybean.database.addBean"
  scope="request" ></jsp:useBean>
<html>
<head>
<title>【例 7.6】追加记录示例</title>
</head>
<body bgcolor=cyan>
<form action="" method="post" name="form1">
   书号是主键，不能输入重复值。
   <br>
   输入书号：<input type="text" name="strbid" size=10>
   输入书名：<input type="text" name="strname" size=20>
  <br>输入作者：<input type="text" name="strauthor" size=10>
  输入出版社：<input type="text" name="strpublish" size=20>
  <br>输入单价：<input type="text" name="fprice" size=10>
  <input type="submit" name="G" value="提交">
</form>
<jsp:setProperty name="update1" property="*"/>
<br>您添加记录操作结果为：
<jsp:getProperty name="update1" property="addMessage" />
<br>当前数据库的记录是：
<jsp:getProperty name="query1" property="queryResult"/>
</body>
</html>
```

2. 删除记录

下面的程序以例 7.3 为基础，在显示 book 表中数据的页面 ch7_7.jsp 中加入了一个删除列，显示一个“删除”链接，通过“删除”链接传递书号(bid)给一个删除页面 ch7_7_delete.jsp。书号为图书的主键，能够唯一标识一条记录。

【例 7.7】删除 book 表中的记录，相关代码参见 listBean.java、deleteBean.java、ch7_7.jsp 和 ch7_7_delete.jsp，页面效果如图 7.21、图 7.22 所示。

listBean.java 文件代码如下：

```
package mybean.database;
import java.sql.Connection;
import java.sql.DriverManager;
import java.sql.ResultSet;
import java.sql.Statement;
public class listBean {
```

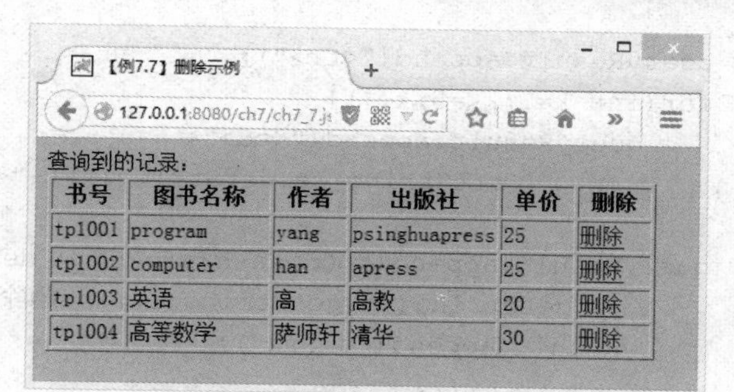

书号	图书名称	作者	出版社	单价	删除
tp1001	program	yang	psinghuapress	25	删除
tp1002	computer	han	apress	25	删除
tp1003	英语	高	高教	20	删除
tp1004	高等数学	萨师轩	清华	30	删除

图 7.21　ch7_7.jsp 页面效果

```
StringBuffer queryResult;
public listBean(){
    queryResult=new StringBuffer();
    try{  Class.forName("com.mysql.jdbc.Driver");
    }
    catch(Exception e){}
}
public String  getString(String str){
    String s=str.trim();
    try{byte bb[]=s.getBytes("ISO-8859-1");
        s=new String(bb);
    }
    catch(Exception e){}
    return s;
}
public StringBuffer getQueryResult() {
    Connection conn;
    Statement stmt;
    ResultSet rs;
    try{
        queryResult.append("<table border=1>");
        String uri="jdbc:mysql://localhost:3306/booklib";
        String userid="root";
        String userpwd="";
        conn=DriverManager.getConnection(uri,userid,userpwd);
        stmt=conn.createStatement();
        rs=stmt.executeQuery("select * from book");
        queryResult.append("<tr>");
          queryResult.append("<th width=50>书号</th>");
          queryResult.append("<th width=100>图书名称</th>");
          queryResult.append("<th width=50>作者</th>");
          queryResult.append("<th width=100>出版社</th>");
          queryResult.append("<th width=50>单价</th>");
          queryResult.append("<th width=50>删除</th>");
      queryResult.append("</tr>");
      while (rs.next()){
```

```
                queryResult.append("<tr>");
                for(int k=1;k<=5;k++){
                    queryResult.append("<td>"+
                          getString(rs.getString(k))+"</td>");
                }
                queryResult.append("<td><a href=ch7_7_delete.jsp?bid="+
                      getString(rs.getString(1))+">删除</a></td>");
                queryResult.append("</tr>");
            }
            queryResult.append("</table>");
            conn.close();
            }
        catch(Exception e)
        {  queryResult.append("请输入正确的用户名和密码！"+
                "注意：<br>用户名为 root，密码为空。");
        }
        return queryResult;
    }
}
```

图 7.22　ch7_7_delete.jsp 页面效果

deleteBean.java 文件代码如下：

```
package mybean.database;
import java.sql.*;
public class deleteBean {
    String strbid="";
    String deleteMessage="";
    public deleteBean(){
        try{Class.forName("com.mysql.jdbc.Driver");
        }
        catch(Exception e){}
    }
    public String getStrbid() {
        return strbid;
    }
    public void setStrbid(String strbid) {
        this.strbid = strbid;
```

```
    }
    public String getDeleteMessage() {
        String condition="";
        if (!"".equals(strbid)){
            condition="delete from book where bid='"+strbid+"'";
            System.out.print(condition);
            Connection conn;
            Statement stmt;
            try{ String uri="jdbc:mysql://localhost:3306/booklib";
            String user="root";
            String password="";
            conn=DriverManager.getConnection(uri,user,password);
            stmt=conn.createStatement();
            int n=stmt.executeUpdate(condition);
            if(n!=0)
            {deleteMessage="删除成功！";}
            else{
                deleteMessage="删除失败!";
            }
            stmt.close();
            conn.close();
            }
            catch(Exception e){
                deleteMessage="删除异常！";
            }
        }
        return deleteMessage;
    }
}
```

ch7_7.jsp 文件内容如下：

```
<%@ page contentType="text/html;charset=GB2312" %>
<%@ page import="mybean.database.listBean" %>
<jsp:useBean id="query1" class="mybean.database.listBean"
  scope="request"  ></jsp:useBean>
<html>
<head>
<title>【例 7.7】删除示例</title>
</head>
<body bgcolor=cyan>
查询到的记录：<br>
<jsp:getProperty name="query1" property="queryResult"/>
</body>
</html>
```

ch7_7_delete.jsp 文件内容如下：

```
<%@ page contentType="text/html;charset=GB2312" %>
<%@ page import="mybean.database.deleteBean" %>
<jsp:useBean id="query1" class="mybean.database.deleteBean"
 scope="request"  ></jsp:useBean>
<% String strbid=null;
  try{
      strbid=request.getParameter("bid");
  }
  catch(Exception e){
      strbid="";
  }
  query1.setStrbid(strbid);
%>
<html>
<head>
<title>【例 7.7】删除示例</title>
</head>
<body bgcolor=cyan>
删除结果：<br>
<jsp:getProperty name="query1" property="deleteMessage"/>
<a href="ch7_7.jsp">返回</a>
</body>
</html>
```

3. 修改记录

修改记录是数据库编程经常使用的技术，下面的例子演示了如何修改 book 表中的数据。第一个页面 ch7_8.jsp 是在顺序查询的基础上增加了一个“修改”列，在“修改”列制作了一个以 bid 为传递参数的链接。单击“修改”链接，将会传递 bid 参数，转到图书数据编辑页面 ch7_8_show.jsp。编辑完图书数据后，单击“提交”提交数据，单击“返回”链接返回到 ch7_8.jsp 页面。在提交数据页面 ch7_8update.jsp 所使用的 Bean 中，调用了 booklib 数据库中的存储过程 update_book_p。

在 MySQL 数据库中创建 update_book_p 存储过程如图 7.23 所示，创建存储过程的 SQL 代码如下：

```
use booklib
delimiter $$
create procedure update_book_p(in p_bid char(20),
    in p_name char(20),
    in p_author char(10),
    in p_publish varchar(50),
    in p_price float)
reads sql data
begin
```

```
    update book set name=p_name,
        author=p_author,
        publish=p_publish,
        price=p_price
    where bid=p_bid;
end $$
delimiter;
```

```
mysql> use booklib
Database changed
mysql> delimiter $$
mysql> create procedure update_book_p(in p_bid char(20),
    ->      in p_name char(20),
    ->      in p_author char(10),
    ->      in p_publish varchar(50),
    ->      in p_price float)
    -> reads sql data
    -> begin
    ->     update book set name=p_name,
    ->         author=p_author,
    ->         publish=p_publish,
    ->         price=p_price
    ->     where bid=p_bid;
    -> end $$
Query OK, 0 rows affected (0.00 sec)

mysql> delimiter ;
```

图 7.23　创建存储过程的 MS-DOS 窗口

调用存储过程的 SQL 语句如下：

```
call update_book_p('tp1007','电子商务','杨梁','清华大学出版社',30);
```

删除存储过程的语句如下：

```
drop procedure update_book_p;
```

查看存储过程的定义的语句如下：

```
show create procedure update_book_p \G;
```

【例 7.8】 修改记录，相关代码参见 update_book_p、showBean.java、ch7_8.jsp、ch7_8_show.jsp、updateBean.java 和 ch7_8_update.jsp，页面效果如图 7.24、图 7.25 所示。

【例7.8】修改示例

127.0.0.1:8080/ch7/ch7_8

查询到的记录：

书号	图书名称	作者	出版社	单价	更新操作
tp1001	program	yang	psinghuapress	25	修改
tp1002	computer	han	apress	25	修改
tp1003	英语	高	高教	20	修改

图 7.24　ch7_8.jsp 页面效果

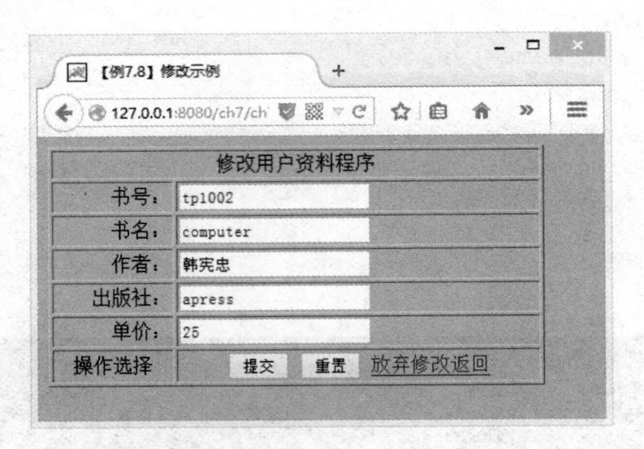

图 7.25　ch7_8_show.jsp 页面效果

第一个页面使用 JavaBean，showBean.java 文件代码如下：

```
package mybean.database;
import java.sql.Connection;
import java.sql.DriverManager;
import java.sql.ResultSet;
import java.sql.Statement;
public class showBean {
    StringBuffer queryResult;
    public showBean(){
        queryResult=new StringBuffer();
        try{  Class.forName("com.mysql.jdbc.Driver");
        }
        catch(Exception e){}
    }
    public String  getString(String str){
        String s=str.trim();
        try{byte bb[]=s.getBytes("ISO-8859-1");
            s=new String(bb);
        }
        catch(Exception e){}
        return s;
    }
    public StringBuffer getQueryResult() {
        Connection conn;
        Statement stmt;
        ResultSet rs;
        try{
            queryResult.append("<table border=1>");
            String uri="jdbc:mysql://localhost:3306/booklib";
            String userid="root";
            String userpwd="";
            conn=DriverManager.getConnection(uri,userid,userpwd);
            stmt=conn.createStatement();
```

```
            rs=stmt.executeQuery("select * from book");
            queryResult.append("<tr>");
              queryResult.append("<th width=50>书号</th>");
              queryResult.append("<th width=100>图书名称</th>");
              queryResult.append("<th width=50>作者</th>");
              queryResult.append("<th width=100>出版社</th>");
              queryResult.append("<th width=50>单价</th>");
              queryResult.append("<th width=100>更新操作</th>");
           queryResult.append("</tr>");
            while (rs.next()){
                 queryResult.append("<tr>");
                 for(int k=1;k<=5;k++){
                      queryResult.append("<td>"+
                                getString(rs.getString(k))+"</td>");
                 }
                 queryResult.append("<td><a href=ch7_8_show.jsp?bid="+
                           getString(rs.getString(1))+">修改</a></td>");
                 queryResult.append("</tr>");
            }
            queryResult.append("</table>");
            conn.close();
            }
        catch(Exception e)
        {  queryResult.append("请输入正确的用户名和密码！"+
                "注意：<br>用户名为 root，密码为空。");
        }
        return queryResult;
    }
}
```

第一个页面 ch7_8.jsp 文件内容如下：

```
<%@ page contentType="text/html;charset=GB2312" %>
<%@ page import="mybean.database.showBean" %>
<jsp:useBean id="query1" class="mybean.database.showBean"
    scope="request"  ></jsp:useBean>
<html>
<head>
<title>【例 7.8】修改示例</title>
</head>
<body bgcolor=cyan>
查询到的记录：<br>
<jsp:getProperty name="query1" property="queryResult"/>
</body>
</html>
```

第二个页面 ch7_8_show.jsp 文件内容：

```
<%@ page contentType="text/html;charset=GB2312" %>
<%@ page import="java.sql.*" %>
<%!
   public String getString(String s){
     String str=s.trim();
     try{byte bb[]=str.getBytes("ISO-8859-1");
        str=new String(bb);
     }
     catch(Exception e){}
     return str;
 }
%>
<% String strbid;
   try{  strbid=request.getParameter("bid");
   }
   catch(Exception e){
       strbid="";
   }
%>
<html>
<head>
<title>【例 7.8】修改示例</title>
</head>
<body bgcolor=cyan>
<%
     if(!"".equals(strbid))
     {
         String condition="select * from book where bid='"+strbid+"'";
         try{
             Connection conn;
             Statement stmt;
             ResultSet rs;
             String uri="jdbc:mysql://localhost:3306/booklib";
             Class.forName("com.mysql.jdbc.Driver");
             conn=DriverManager.getConnection(uri,"root","");
             stmt=conn.createStatement();
             rs=stmt.executeQuery(condition);
             rs.next();
 %>
 <form action="ch7_8_update.jsp" method="post" name="form1">
 <table border="1" width="400">
    <tr>
      <td width="100%" colspan="2" align="center" >修改用户资料程序</td>
    </tr>
    <tr><td width="25%" align="right">书号: </td>
    <td width="75%">
    <input type="text" name="bid"
```

```
        value="<%=getString(rs.getString("bid"))%>">
    </td></tr>
    <tr><td width="25%" align="right">书名：</td>
    <td width="75%">
    <input type="text" name="name"
        value="<%=getString(rs.getString("name"))%>">
    </td></tr>
    <tr><td width="25%" align="right">作者：</td>
    <td width="75%">
    <input type="text" name="author"
        value="<%=getString(rs.getString("author"))%>">
    </td></tr>
    <tr><td width="25%" align="right">出版社：</td>
    <td width="75%">
    <input type="text" name="publish"
        value="<%=getString(rs.getString("publish"))%>">
    </td></tr>
    <tr><td width="25%" align="right">单价：</td>
    <td width="75%">
    <input type="text" name="price"
        value="<%=rs.getString("price")%>">
    </td></tr>
    <tr><td width="25%" align="center">操作选择
      </td>
    <td width="75%" align="center">
      <input type="submit" name="g" value="提交">
      <input type="reset" name="r" value="重置">
      <a href="ch7_8.jsp">放弃修改返回</a>
    </td></tr>
</table>
</form>
<%      rs.close();
        stmt.close();
        conn.close();
         }
         catch(SQLException e){
             out.print("Sql 异常");
         }
     }
%>
</body>
</html>
```

第三个页面的 Javabean，updateBean.java 文件代码如下：

```
package mybean.database;
import java.sql.*;
public class updateBean {
```

```
String bid="",
    name="",
    author="",
    publish="",
   updateMessage="";
float price=0.0f;
public updateBean(){
    updateMessage="没有修改";
    try{
        Class.forName("com.mysql.jdbc.Driver");
    }
    catch(Exception e){}
}
public String getBid() {
    return bid;
}
public void setBid(String bid) {
    this.bid = bid;
}
public String getName() {
    return name;
}
public void setName(String name) {
    this.name = name;
}
public String getAuthor() {
    return author;
}
public void setAuthor(String author) {
    this.author = author;
}
public String getPublish() {
    return publish;
}
public void setPublish(String publish) {
    this.publish = publish;
}
public float getPrice() {
    return price;
}
public void setPrice(float price) {
    this.price = price;
}
public String getUpdateMessage(){
    if(!"".equals(bid))
    {
        try{
```

```
                Connection conn;
                String call_procedure_string="{call update_book_p"+
                  "(?,?,?,?,?)}";
                CallableStatement pstmt;
                String uri="jdbc:mysql://localhost:3306/booklib";
                conn=DriverManager.getConnection(uri,"root","");
                pstmt=conn.prepareCall(call_procedure_string);
                pstmt.setString(1, bid);
                pstmt.setString(2, name);
                pstmt.setString(3, author);
                pstmt.setString(4, publish);
                pstmt.setFloat(5, price);
                pstmt.executeUpdate();
                pstmt.close();
                conn.close();
                updateMessage="成功修改记录";
            }
            catch(SQLException e1){
                updateMessage=e1.toString();
            }
        }
        return updateMessage;
    }
}
```

第三个页面 ch7_8_update.jsp 文件内容如下：

```
<%@ page contentType="text/html;charset=GB2312" %>
<%@ page import="mybean.database.updateBean" %>
<jsp:useBean id="query1" class="mybean.database.updateBean"
 scope="request"  ></jsp:useBean>
<%
   String bid=request.getParameter("bid");
   String name=request.getParameter("name");
   String author=request.getParameter("author");
   String publish=request.getParameter("publish");
   String strprice=request.getParameter("price");
   float price;
   try{price=Float.parseFloat(strprice);
   }
   catch(Exception e){
       price=0.0f;
   }
   query1.setAuthor(author);
   query1.setBid(bid);
   query1.setName(name);
   query1.setPublish(publish);
   query1.setPrice(price);
```

```
%>
<html>
<head>
<title>【例 7.8】修改示例</title>
</head>
<body bgcolor=cyan>
修改记录的结果：<br>
<jsp:getProperty name="query1" property="updateMessage"/>
<br><a href="ch7_8.jsp">返回</a>
</body>
</html>
```

7.4.3 分页查询

用户在客户端查询数据时，如果查询结果较多，不能够在一页将信息全部显示完，这时就需要数据分页显示。ResultSet 对象与 Connection 对象是紧密相连的，如果 Connection 对象关闭，ResultSet 对象中的数据也会随之消失，因此使用 ResultSet 对象实现分页显示就必须保持与数据库的连接或者多次访问数据库。一般情况下，数据库服务器的连接资源是有限的，因此当多个用户分页显示数据库数据时应避免长时间占用数据库的连接资源。为了解决这个问题，在分页显示程序中可以使用 CachedRowSetImpl 类，CachedRowSetImpl 对象可以保存 ResultSet 对象中的数据，它不依赖与 Connection 对象，并且继承了 ResultSet 的所有方法。CachedRowSetImpl 类在 com.sun.rowset 包中。

使用 CachedRowSetImpl 对象的代码如下：

```
import com.sun.rowse.*;
…
ResultSet rs=stmt.executeQuery("select * from book");
CachedRowSetImpl rowSet=new CachedRowSetImpl();
rowSet.populate(rs);
while(rowSet.next())
{
   …
   rowSet.getString(1);
}
```

下面的例子介绍了数据分页显示。例子使用了一个有效范围为 session 的 Bean，该 Bean 负责连接数据库并得到 CachedRowSetImpl 对象 rowSet，得到 rowSet 对象后就关闭与数据库的连接，通过 rowSet 对象的 absolute 方法定位要显示的记录，实现记录分页显示。

【例 7.9】数据库记录的分页显示，相关文件参见 showbypageBean.java 和 ch7_9.jsp，页面效果如图 7.26 所示。

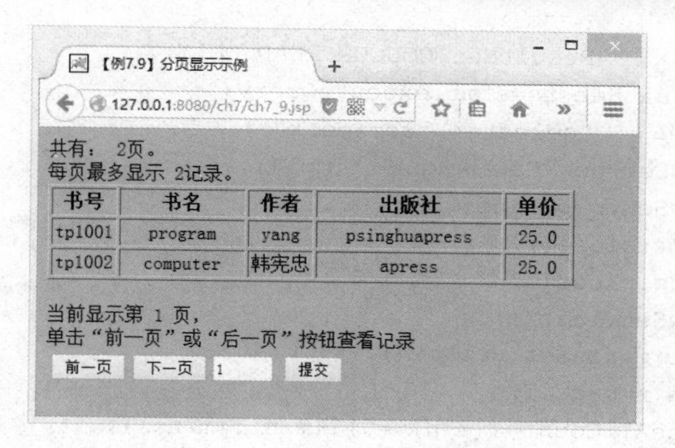

图 7.26　ch7_9.jsp 页面效果

showbypageBean.java 文件代码如下：

```
package mybean.database;
import java.sql.*;
import com.sun.rowset.*;
public class showbypageBean {
    int pageSize=0;
    int pageAllCount=0;
    int showPage=1;
    StringBuffer presentPageResult;
    CachedRowSetImpl rowSet;
    public showbypageBean(){
        presentPageResult=new StringBuffer();
        try{Class.forName("com.mysql.jdbc.Driver").newInstance();
        }
        catch(Exception e){}
    }
    public String getString(String s){
        String str=s;
        try{
            byte bb[]=str.getBytes("ISO-8859-1");
            str=new String(bb);
        }
        catch(Exception e){}
        return str;
    }
    public void setPageSize(int size){
        pageSize=size;
        String uri="jdbc:mysql://localhost:3306/booklib" ;
        try{
            System.out.println(uri);
            Connection conn=DriverManager.getConnection(uri,"root","");
            Statement stmt=conn.createStatement(
                    ResultSet.TYPE_SCROLL_SENSITIVE,
```

```
                ResultSet.CONCUR_READ_ONLY);
            ResultSet rs=stmt.executeQuery(
                "select * from book");
            rowSet=new CachedRowSetImpl();
            rowSet.populate(rs);
            stmt.close();
            conn.close();
            rowSet.last();
            int m=rowSet.getRow();
            int n=pageSize;
            pageAllCount=((m%n)==0)?(m/n):(m/n+1);
        }
        catch(Exception exp){}
    }
    public int getPageSize(){
        return pageSize;
    }
    public int getPageAllCount(){
        return pageAllCount;
    }
    public void setShowPage(int n){
        showPage=n;
    }
    public int  getShowPage(){
        return showPage;
    }
    public StringBuffer getPresentPageResult(){
        if(showPage>pageAllCount)
            showPage=1;
        if(showPage<=0)
            showPage=pageAllCount;
        presentPageResult=showRecord(showPage);
        return presentPageResult;
    }
    public StringBuffer showRecord(int page){
        StringBuffer showMessage=new StringBuffer();
        showMessage.append("<table border=1>");
        showMessage.append("<tr>");
          showMessage.append("<th width=50 align=center>书号</th>");
          showMessage.append("<th width=100 align=center>书名</th>");
          showMessage.append("<th width=50 align=center>作者</th>");
          showMessage.append("<th width=150 align=center>出版社</th>");
          showMessage.append("<th width=50 align=center>单价</th>");
        showMessage.append("</tr>");
        try{
            rowSet.absolute((page-1)*pageSize+1);
            for(int i=1;i<=pageSize;i++)
```

```
            {
                showMessage.append("<tr>");
                  showMessage.append("<td width=50 align=center>"+
                          getString(rowSet.getString(1))+"</td>");
                  showMessage.append("<td width=100 align=center>"+
                          getString(rowSet.getString(2))+"</td>");
                  showMessage.append("<td width=50 align=center>"+
                          getString(rowSet.getString(3))+"</td>");
                  showMessage.append("<td width=150 align=center>"+
                          getString(rowSet.getString(4))+"</td>");
                  showMessage.append("<td width=50 align=center>"+
                          rowSet.getString(5)+"</td>");
                showMessage.append("</tr>");
                rowSet.next();
            }
        }
        catch(SQLException exp){}
        showMessage.append("</table>");
        return showMessage;
    }
}
```

ch7_9.jsp 文件内容如下：

```
<%@ page contentType="text/html;charset=GB2312" %>
<%@ page import="mybean.database.showbypageBean" %>
<jsp:useBean id="query1" class="mybean.database.showbypageBean"
  scope="session"  ></jsp:useBean>
<html>
<head>
<title>【例 7.9】分页显示示例</title>
</head>
<body bgcolor=cyan>
<jsp:setProperty name="query1" property="pageSize" value="2" />
共有:
   <jsp:getProperty name="query1" property="pageAllCount"/>页。<br>
每页最多显示
   <jsp:getProperty name="query1" property="pageSize"/>记录。<br>
<jsp:setProperty name="query1" property="showPage"/>
<!-- 显示表中记录 -->
  <jsp:getProperty name="query1" property="presentPageResult"/>
  <br>
当前显示第
   <jsp:getProperty name="query1" property="showPage" /> 页,
   <br>
单击"前一页"或"后一页"按钮查看记录
<table>
  <tr><td>
```

```
  <form action="">
  <input type="hidden" name="showPage"
    value="<%= query1.getShowPage()-1 %>">
  <input type="submit" name="g" value="前一页">
  </form>
</td>
<td>
  <form action="">
    <input type="hidden" name="showPage"
     value="<%=query1.getShowPage()+1 %>">
    <input type="submit" name="g" value="下一页">
  </form>
</td>
<td>
   <form action="">
     <input type="text" name="showPage"
        value="1" size=5>
     <input type="submit" name="g" value="提交">
   </form>
</td></tr>
</table>
</body>
</html>
```

7.4.4 使用连接池

所谓数据库连接池就是为数据库连接建立一个“存储池”。Tomcat 服务器预先在“连接池”中放入一定量的连接，当程序需要建立连接时，只需从“连接池”中申请一个连接，使用完毕之后再将该连接放入“连接池”中，以便供其他程序申请使用。这种预先建好连接，然后申请分配的连接池机制，提高了数据库处理连接的速度；同时，还可以通过连接池管理和监视数据库的连接数量，为系统开发、测试及性能调整提供依据。不同的数据库的连接池配置方法会有所不同，本书以 MySQL 为例介绍数据库连接池的配置和使用。

1. 在 Tomcat 7.0 上配置连接池

用任意一种文本编辑器打开 Tomcat 7.0 安装目录下 conf 子目录中的 server.xml 文件，在文档内容中的</host>之前加入如下配置：

```
<Context path="/ch7" docBase="E:\programJsp\ch7" debug="0"
   reloadable="true">
   <Resource name="jdbc/dataBook" auth="Container"
        type="javax.sql.DataSource"
        driverClassName="org.gjt.mm.mysql.Driver"
        url="jdbc:mysql://localhost:3306/booklib"
        username="root"  password="" maxActive="5000"
```

```
        maxIdle="10" maxWait="-1" />
</Context>
```

上面代码中，<Resource/>项表示数据库的连接资源，它有几个属性 name、auth、type、driverClassName、url、username、password 等属性。其中：

- name 项是 JNDI 的名称定义，程序通过 JNDI 才能找到此对象，这里取名为“jdbc/dataBook”。
- auth 项即连接池管理权限，这里取值为 Container，声明为容器管理，此连接池为容器管理池，对应 ch7 服务目录。
- type 项即对象类型，这里取值为 javax.sql.DataSource，声明为数据库连接池。
- driverClassName 项为数据库驱动程序。url 项为数据库连接地址。username 为连接用户名。password 为连接用户密码。maxActive 为最大连接数，maxIdle 为最小连接数。

2. 使用连接池示例

配置好 Tomcat 的数据库连接池后，下面的例子在例 7.2 的基础上完成。程序首先通过 JNDI 找到 jdbc/dataBook 对象，使用 Object obj=(Object) ctx.lookup("jdbc/dataBook")；然后将得到的 obj 对象转换成 DataSource 类型的 ds 对象，使用 ds 得到数据库连接 conn 对象。得到数据库连接对象之后，就可以用前面所讲的方法操作数据库，这里不再赘述。

【例 7.10】使用连接池操作数据库(其文件参见 ch7_10.jsp)，页面效果如图 7.27 所示。

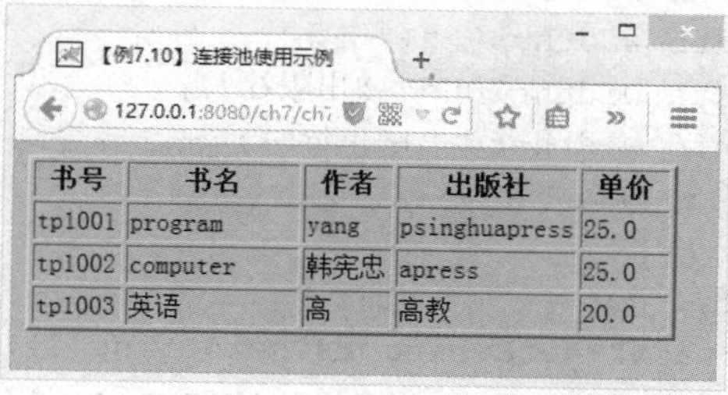

【例7.10】连接池使用示例

127.0.0.1:8080/ch7/ch7

书号	书名	作者	出版社	单价
tp1001	program	yang	psinghuapress	25.0
tp1002	computer	韩宪忠	apress	25.0
tp1003	英语	高	高教	20.0

图 7.27　ch7_10.jsp 页面效果

ch7_10.jsp 文件内容如下：

```
<%@ page contentType="text/html;charset=GB2312" %>
<%@ page import="java.sql.*" %>
<%@ page import="javax.naming.*" %>
<html>
<head>
<title>【例 7.10】连接池使用示例</title>
</head>
<body bgcolor=cyan>
<%!
  public String getString(String s){
    String str=s;
    try{byte bb[]=str.getBytes("ISO-8859-1");
       str=new String(bb);
```

```
    }
    catch(Exception e){}
    return str;
  }
%>
<% Connection conn;
   Statement stmt=null;
   ResultSet rs;
   try{
       Context initCtx=new InitialContext();
       Context ctx=(Context)initCtx.lookup("java:comp/env");
       //获取连接池对象
       Object obj=(Object)ctx.lookup("jdbc/dataBook");
       //类型转换
       javax.sql.DataSource ds=(javax.sql.DataSource)obj;
       //得到连接
       conn=ds.getConnection();
       stmt=conn.createStatement(); //创建 Statement 对象，用于执行 SQL 语句
       rs=stmt.executeQuery("select * from book");
       out.print("<table border=2>");
       out.print("<tr>");
         out.print("<th width=50>"+"书号");
         out.print("<th width=100>"+"书名");
         out.print("<th width=50>"+"作者");
         out.print("<th width=100>"+"出版社");
         out.print("<th width=50>"+"单价");
       out.print("</tr>");
       while(rs.next()){
        out.print("<tr>");
          out.print("<td>"+getString(rs.getString(1))+"</td>");
          out.print("<td>"+getString(rs.getString(2))+"</td>");
          out.print("<td>"+getString(rs.getString(3))+"</td>");
          out.print("<td>"+getString(rs.getString("publish"))+"</td>");
          out.print("<td>"+rs.getFloat("price")+"</td>");
        out.print("</tr>");
       }
       out.print("</table>");
       stmt.close();  //关闭数据库
       conn.close();
   }
   catch(Exception e)
   {
       out.println(e.toString());
   }
%>
</body>
</html>
```

7.5　上机实训

实训目的

- 理解JDBC编程接口。
- 掌握利用JDBC访问数据库的方法。
- 掌握对表中数据的增、删、改、查编程技术。
- 掌握连接池的使用技术。
- 掌握查询结果的分页显示技术。

实训内容

实训　编写一个简单的图书管理系统。

要求：

(1) 建立一个图书管理数据库booklib，库中有一个book表。

(2) 完成一个管理系统主界面，主界面功能包括添加图书、修改图书、删除图书和查询图书等功能的链接。

(3) 参照本章示例，完成图书管理系统各链接的功能。

实训总结

通过本章的上机实训，学员应该能够理解JDBC接口；掌握JDBC编程的基本技术，掌握数据的增加、删除、修改与查找的编程方法；掌握查询结果的分页显示技术。掌握使用数据库连接池的编程技术。

7.6　本章习题

思考题

(1) 什么是JDBC编程接口？

(2) 什么是数据库连接池？如何使用？

(3) 使用纯Java驱动操作MySQL数据库的步骤？

(4) 加载SQL Server 2000纯Java驱动程序的代码是什么？

(5) 使用预处理语句和存储过程有什么好处？

(6) 使用CachedRowSetImpl类有什么好处？

(7) 如何使用滚动的结果集？

拓展实践题

(1) 如何使用Connection的事务机制？

(2) 如何使用元数据来提高程序的适应性？

第 8 章　Servlet 技术

学习目的与要求：

Servlet 拥有面向对象语言 Java 的所有优势，是开发 Web 程序的主要技术。本章主要学习 Servlet 的基本概念；Servlet 技术原理，包括 Servlet 的生命周期、结构等；开发 Servlet 的常用类；编写、配置和调用 Servlet 的方法；Servlet 的典型应用。通过本章的学习，学员要理解 Servlet 的概念和特点，掌握编写、编译、调试、配置和调用 Servlet 的方法；掌握 HttpServlet API 的常用接口和类；掌握 Servlet 的典型应用：读入表单数据，读取 cookie，读取 session 和读取请求头信息等应用。

8.1　Servlet 介绍

8.1.1　什么是 Servlet

Servlet 是使用 Java Servlet API 所定义的相关类和方法的 Java 程序，它运行在启用 Java 的 Web 服务器或应用服务器端，用于扩展该服务器的能力。

Servlet 与 Applet 相对应，Applet 是运行在客户端浏览器上的程序，而 Servlet 是运行在 Web 服务器端程序。它们都是字节码对象，可以动态地从网络加载，Applet 称为客户端小程序，所以人们又将 Servlet 称为“服务器端小程序”。

Servlet 功能强大，体系结构先进，但它在表示层的实现上存在一些缺陷，它输出 HTML 语句还是使用传统 CGI 的方法，即使用 print 语句一句一句的输出，编写复杂的表示层显得很烦琐，这使得它的应用受到了很大的限制。JSP 技术，采用 Java 语言和 HTML 语言镶嵌的形式，表示层的 HTML、脚本语言和 Java 语言可以混合地编写在一起，这极大地方便了网页的设计和修改，但也带来了另外一个问题。由于一个 HTML、Java 和脚本语言混合在一起的程序可读性较差，维护起来较困难。

JSP 技术是在 Servlet 之后产生的，它以 Servlet 为核心技术，是 Servlet 技术的一个成功应用，当 JSP 页面被请求时，JSP 页面会被 JSP 引擎翻译成 Servlet 字节码执行。基于以上的 Servlet 和 JSP 各自的优点，一般用 JSP 来实现页面，用 Servlet 来处理业务逻辑，数据的持久层则采用 JavaBean 或者专门的持久层框架来实现。

8.1.2　Servlet 的功能

Servlet 可以在服务器端完成对数据库的访问、调用 JavaBean、响应浏览器的各种请求、向客户端发送页面等，总结起来，Servlet 具有如下功能。

- Servlet 可以同其他资源交互，例如文件、数据库、Applet、Java 应用程序等资源，并能控制外部用户的访问数量及访问性质。

- 创建并返回一个包含基于客户端请求性质的动态的完整 HTML 页面，也可以创建嵌入现有 HTML 页面中的 HTML 片段。
- 与多个客户机处理连接，同时处理多个浏览器的请求，并在各浏览器间通信，例如 Servlet 可以是多个用户参与的游戏服务器。
- 与 Applet 通信。Servlet 可以建立服务器与 Applet 的新连接，并将该连接保持在打开状态。
- 对客户端提交的特殊类型数据进行过滤，例如 Servlet 处理文件上传、图像转换等。
- Servlet 可被连接。Servlet 可以调用另一个或一系列 Servlet，即成为它的客户端。

8.1.3 Servlet 技术的特点

Servlet 程序在服务器端运行，动态地生成 Web 页面，与传统的 CGI 程序相比有许多优势，主要包括以下特点。

- 高效。在传统的 CGI 中，每个请求都要启动一个新的进程，而 Servlet 中，每个请求由一个线程来处理，因此效率较高。
- 使用方便。Servlet 提供的大量的实用工具程序，自动解析和解码 HTML 表单数据、读取和设置 HTTP 头、处理 Cookie、跟踪会话等，编程很方便。
- 功能强大。Servlet 能够直接与服务器交互，能够在各个应用程序之间共享数据。
- 可移植性好。Servlet 用 Java 语言编写，具有 Java 语言的优点，可以“一次编写，到处运行”。Servlet API 具有完善的标准，所有 Servlet 可以运行的 Tomcat、Apache 等 Web 服务器上。

8.2 Servlet 技术原理

对于有一定 Java 基础的用户来讲，编写一个 Servlet 并不难，因为编写 Servlet 就是编写一个特殊的 Java 类，这个特殊的 Java 类与其他的 Java 类编写相似，只是它必须直接或间接实现 Servlet 接口，该接口定义了 Servlet 的生命周期方法。

8.2.1 Servlet 的生命周期

用户所编写的 Servlet 是如何执行的？它的生存期有多长？当服务器收到对某一个 Servlet 请求时，它会检查该 Servlet 类的实例是否存在，如果不存在就会创建这个 Servlet 的实例，这个过程称为载入 Servlet，如果存在就会直接调用该 Servlet 的实例。Servlet 对象创建之后，服务器就可以调用该实例响应客户的请求了，当多个客户请求一个 Servlet 时，服务器为每一个客户启动一个线程而不是启动一个进程，这个线程调用内存中的 Servlet 实例的 Service 方法响应客户的请求。当服务器关闭或者卸载应用程序时，关闭该 Servlet 实例，释放 Servlet 所占的资源，这就是一个 Servlet 的生命周期。

javax.servlet.Servlet 接口定义了三个用于 Servlet 生命周期的方法，任何一个 Servlet 都会直接或间接地实现这三个方法。这三个用于 Servlet 生命周期的方法如下。

(1) public void init(ServletConfig config) throws ServletException 方法。该方法在 Servlet 载入时执行，且只执行一次，对 Servlet 进行初始化，例如读入配置信息等。ServletConfig 对象保存这服务器的一些设置信息。如果初始化失败就会发生 ServletException。

(2) public void service(ServletRequest req, ServletResponse res)方法。该方法用来为请求服务，在 Servlet 生命周期中，Servlet 每被请求一次它就会被调用一次。服务器将两个参数传递给该方法，即 ServletRequest 类型和 ServletResponse 类型的对象。

(3) public void destroy()方法。当服务器关闭时，调用 destroy 方法释放 Servlet 所占的资源。一般用户编写程序不会调用该方法。

对于读者来讲，一般编程情况下最关心的是 Service 方法。它通过 ServletRequest 对象可以使 Servlet 得到关于请求的所有信息，通过 ServletResponse 对象可以产生响应用户的消息，例如设定首部、状态码等。

8.2.2 Servlet 的结构

学习 Servlet 编程的关键是学习 Servlet 的接口，Servlet 接口定义了基本的方法来管理服务器小程序以及该程序与客户端的通信，当用户开发一个 Servlet 时，必须直接或间接实现 Servlet 接口所定义的方法，例如通过继承 HttpServlet 类就可以间接地实现 Servlet 的生命周期方法。在编写 Servlet 时，通常会提供部分或者全部的方法。

Servelt 所适用的网络协议可以是多种多样的，比如 HTTP，FTP，SMTP 等，但目前来讲只有 HTTP 已经形成了标准的 Java 组件，即 javax.servlet.http 和 javax.servlet.jsp 软件包，它们分别对应的通常所说的 Servlet 和 JSP 编程。本书介绍的 Servlet 编程就是指的针对 HTTP 的 Servlet 编程，用到的主要是 HttpServlet 类。下面通过一个简单 Servlet 程序来介绍 Servlet 的结构。

exampleServlet.java 文件中的代码如下：

```
package myservlet.example;
import java.io.*;
import javax.servlet.*;
import javax.servlet.http.*;
public class FirstServlet extends HttpServlet {
    public void init(ServletConfig config)throws ServletException{
        super.init(config);
    }
    protected void service(HttpServletRequest request,
            HttpServletResponse response)throws
            ServletException,IOException {
        response.setContentType("text/html;charset=GB2312");
        PrintWriter out=response.getWriter();
        out.println("<html><body><h3>Welcome to the " +
                "First Servlet!</h3></body></html>");
        out.flush();
    }
}
```

由此例可以看出，一般编写一个 Servlet 类首先要引入 java.io 包、javax.servlet 包和 javax.servlet.http 包，然后就是编写一个 HttpServlet 类的子类，这个子类重载了 HttpServlet 父类 GenericServlet 的 init 方法和 HttpServlet 类的 service 方法。init 方法用来初始化 Servlet 实例，将服务器的配置信息传给 Servlet 对象。service 方法用来响应用户的请求，当一个 Servlet 接收来自客户端的请求时，它同时接收两个对象，一个是 HttpServletRequest 对象，另一个是 HttpServletResponse 对象。HttpServletRequest 对象中保存着客户端传递的信息，如通过客户端传递的参数名称，客户端使用的协议，还有远程提出的请求，远程主机名称以及远程数据输入输出流，通过输入流 Servlet 可以得到客户端的数据。HttpServletResponse 对象则提供了返回给客户端的信息，例如使用 HttpServletResponse 对象的方法可以设置响应的长度和种类，还可以得到输出流，包括小程序的输出流和一个 Writer 对象，通过 Writer 对象发送给远程客户端所请求的数据。

8.2.3 Servlet 常用类与接口的层次关系

Java API 提供了编写 Servlet 的接口和类，这些接口和类存放在了 javax.servlet 和 javax.servlet.http 包中。javax.servlet 包中有下列主要接口和类。

- javax.servlet.Servlet 接口；
- javax.servlet.ServletConfig 接口；
- javax.servlet.GenericServlet 类；
- javax.servlet.ServletRequest 接口；
- javax.servlet.ServletResponse 接口；
- javax.servlet.RequestDispatcher 接口。

javax.servlet.http 包中有下列主要接口和类。

- javax.servlet.http.HttpServlet 类；
- javax.servlet.http.HttpServletRequest 接口；
- javax.servlet.http.HttpServletResponse 接口；
- javax.servlet.http.HttpSession 接口。

编写 Servlet 用到的主要接口和类的层次结构如图 8.1 所示。

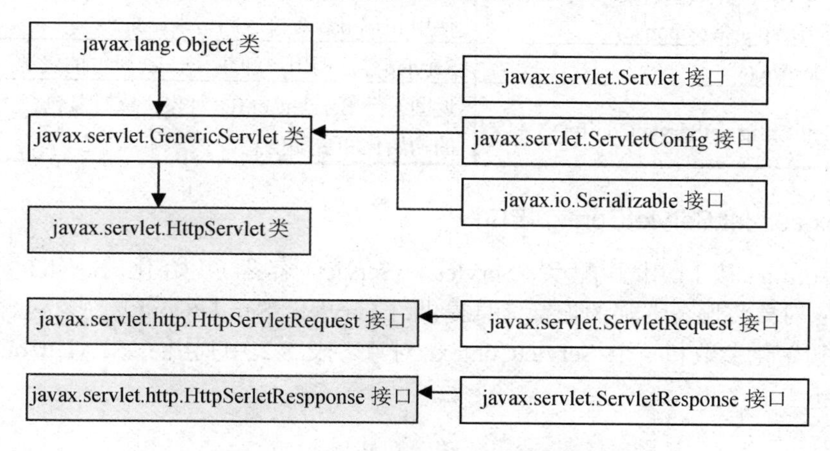

图 8.1　Servlet 常用类与接口层次图

javax.servlet.GenericServlet 类是 javax.lang.Object 类的子类，它实现了 javax.servlet.Servlet 接口、javax.servlet.ServletConifg 和 javax.io.Serializable 接口。HttpServlet 类是 javax.servlet.GenericServlet 类的子类。在多数应用中，用户编写的程序多数继承 javax.servlet.httpServlet 类来创建自己的 Servlet。

8.3 Servlet 的常用类、接口及其方法

使用 Java API 进行 Servlet 开发只需要程序员做很少量的工作，一般都是重写 HttpServlet 类的 doPost 和 doGet 方法，其他工作由 Java API 来完成。因此，熟悉用于开发 Servlet 的 API 就显得非常重要，这里介绍的是 Java API 常用接口、类及其方法，详细的内容请参照 Java API 文档。

8.3.1 javax.servlet 包

javax.servlet 包中的类和接口说明如下。

1. javax.servlet.Servlet 接口

javax.servlet.Servlet 接口用于开发 Servlet，所有的 Servlet 对要直接或间接地实现这个接口，这个接口定义了 Servlet 生命周期的方法。一般不用直接实现该接口，可以扩展 javax.servlet.GenericServlet 来实现一般协议的 Servlet，也可扩展 javax.servlet.http.HttpServlet 来实现 HTTP 的 Servlet。表 8.1 中列出了 Servlet 接口的常用方法。

表 8.1　Servlet 接口的常用方法

序号	方法名称	功　能
1	void init(ServletConfig config)	Servlet 容器调用，用于初始化 Servlet，用 ServletConfig 对象参数来启动配置，只执行一次
2	void service(ServletRequest request, ServletResponse response)	Servlet 容器调用，处理 request 对象中描述的请求，使用 response 对象返回请求结果
3	void destroy()	Servlet 容器调用，卸载 Servlet 所占的资源
4	ServletConfig getServletConifg()	返回一个 ServletConfig 对象，该对象包含当前 Servlet 的初始化和启动信息

2. javax.servlet.ServletConfig 接口

ServletConfig 接口用于配置 Servlet。Servlet 容器初始化 Servlet 时，使用 ServletConfig 对象将服务器端的配置信息提供给 Servlet，ServletConifg 对象包含一组由名/值对形式的初始化参数和一个 ServletContext 对象。表 8.2 中列出了 ServletConfig 的主要方法。

表 8.2 ServletConfig 接口的常用方法

序号	方法名称	功 能
1	String getInitParameter(String)	返回一个字符串，包含参数指定的初始化参数值
2	Enumaration getInitParameterNames()	以 String 对象的枚举形式，返回当前 Servlet 所有初始化参数名字，如果没有初始化参数，返回一个空枚举
3	ServletContext getServletContext()	返回当前 Servlet 正在执行的上下文 ServletContext 对象
4	String getServletName()	返回当前 Servlet 的实例名字

3. javax.servlet.ServletContext 接口

javax.servlet.ServletContext 是 Servlet 的环境上下文接口，每个 Servlet 在 Java 虚拟机内都有一个环境上下文实例。当 Servlet 初始化时，Web 容器将初始化一个 ServletContext 环境上下文对象，并将其包含在当前的 ServletConfig 对象中，通过 ServletConfig 对象给 Servlet。表 8.3 中列出了 ServletContext 接口的主要方法。

表 8.3 ServletContext 接口的常用方法

序号	方法名称	功 能
1	Java.lang.Object getAttribute(String)	返回一个由参数给定名字的属性对象
2	Enumaration getAttributeNames()	以 String 对象的枚举形式，返回当前 ServletContext 对象所有有效属性名字，如果没有有效属性，返回一个空枚举
3	ServletContext getContext(String uriPath)	返回服务器端的一个与 uriPath 相符的 ServletContext 对象
4	Enumaration getInitParameterNames()	以 String 对象的枚举形式，返回当前上下文中所有初始化参数名字，如果没有初始化参数，返回一个空枚举
5	Java.lang.String getInitParameter(String name)	返回一个以 name 为名字的初始化参数的值，这个值的类型为 String 类型
6	void setAttribute(String ,Object)	绑定一个对象到当前的 ServletContext 上下文中

4. javax.servlet.GenericServlet 类

javax.servlet.GenericServlet 类实现了 Servlet、SerlvetConfig 和 javax.io.Serializable 接口，定义了一个与协议无关的 Servlet。GenericServlet 提供了简单版本的生命周期方法和 destroy 以及 ServletConfig 接口中的方法，它可以直接扩展开发 Servlet，用户只需覆盖抽象方法 service()。当开发用于 Web 的 HTTP 的 Servlet 时，一般都是用 GenericServlet 类的子类 HttpServlet，而不是直接使用它，这样可以提高开发效率。HttpServlet 类继承了 GenericServlet 的方法，所以通过 GenericServlet 类的常用方法用户可以了解 HttpServlet 类，表 8.4 中列出了 GenericServlet 类的常用方法。

表 8.4　GenericServlet 类的常用方法

序号	方法名称	功　能
1	ServletConfig getServletConfig()	返回当前 Servlet 的 ServletConfig 对象
2	ServletContext getContext()	返回当前 Servlet 正在执行的上下文 ServletContext 上下文的引用
3	Enumaration getInitParameterNames()	以 String 对象的枚举形式，返回当前 Servlet 所有初始化参数名字，如果没有初始化参数，返回一个空枚举
4	String getServletInfo()	返回当前 Servlet 的有关信息，如作者版本等信息
5	String getServletName()	返回当前 Servlet 实例的名字
6	void init()	初始化 Servlet，是一个生命周期方法，可以被覆盖
7	void init(ServletConfig config)	Servlet 容器调用，指示当前 Servlet 放入服务器栈并使用 ServletConfig 配置
8	abstract void service(ServletRequest request, ServletResponse response)	Servlet 容器调用该方法，用来响应客户请求，开发 Servlet 必须直接或间接实现该方法

5. javax.servlet.ServletRequest 接口

javax.servlet.ServletRequest 接口用于定义封装客户端请求信息的“请求对象”，它与协议无关，并有一个指定 HTTP 的子接口。每次 Servlet 被请求，Servlet 容器创建一个 ServletRequest 对象并把它作为 Service 方法的一个参数。ServletRequest 对象封装了客户端的请求，例如参数名/值和输入流。表 8.5 中列出了 javax.servlet.ServletRequest 接口的常用方法。

表 8.5　ServletRequest 接口的方法

序号	方法名称	功　能
1	Enumaration getAttributeNames()	返回当前所有属性的名字的枚举，如果没有属性，返回一个空枚举
2	javax.lang.object getAttribute()	返回一个给出名字的属性值的对象
3	String getCharacterEncoding()	返回当前请求体中字符编码方式的名字
4	int getContentLength()	返回当前请求体的长度
5	String getContentType()	返回当前请求的 MIME 类型
6	ServletInputStream getInputStream()	获得请求体的输入流
7	String[] getParameterValues()	返回所有参数值的一个 String 数组
8	RequestDispatcher getRequestDispatcher()	返回一个 RequestDispatcher 对象
9	void setAttributer(String ,Object)	设置一个属性
10	Object getAttribute(String)	从当前请求中得到一个给定名字的属性值
11	String getProtocol()	返回当前请求使用的协议名字和版本
12	BufferedReader getReader()	获得请求体字符数据流
13	String getRemoteAddr()	获得发送请求的客户端 IP 地址
14	String getRemoteHost()	获得发送请求的客户端全名或者 IP 地址

6. javax.servlet.ServletResponse 接口

javax.servlet.ServletResponse 接口用于定义发送给客户端信息的“响应对象”，它与协议无关，并有一个指定的 HTTP 子接口。每次 Servlet 被请求，Servlet 容器创建一个 ServletResponse 对象并把它作为 Service 方法的一个参数。ServletResponse 对象可以调用本身的方法发送给客户端信息。表 8.6 中列出了 javax.servlet.ServletResponse 的接口常用方法。

表 8.6 ServletResponse 的常用方法

序号	方法名称	功 能
1	void flushBuffer()	将缓冲区的内容输出到客户端
2	int getBufferSize()	返回当前缓冲区的大小
3	String getCharacterEncoding()	返回当前 MIME 中字符编码的名字
4	ServletOutputStream getOutputStream()	返回一个输出流对象
5	PrintStream getWriter()	返回一个 PrinterWriter 对象，用于发送字符文本到客户端
6	boolean isCommited()	当前响应是否已经提交
7	void reset()	清除缓冲区的数据
8	void setBufferSize(int size)	设置响应体缓冲区的大小
9	void setContentLength(int len)	设置响应体的内容长度
10	void setContentType(String type)	设置发送到客户端的内容类型

7. javax.servlet.RequestDispatcher 接口

javax.servlet.RequestDispatcher 接口用于定义“请求转发”对象，该对象(RequestDispatcher)可以把用户对当前页面或 Servlet 的请求和响应转发给另一个 JSP 页面或 Servlet，所转发到的目的页面或 Servlet 可以使用 ServletRequest 对象获取用户提交的数据。Servlet 容器创建 RequestDispatcher 对象，该对象封装了转发的目的路径。表 8.7 中列出了 RequestDispatcher 接口的常用方法。

表 8.7 RequestDispatcher 接口的常用方法

序号	方法名称	功 能
1	void forward(ServletRequest req, ServletResponse res)	转发一个来自 Servlet 的请求到另外一个资源
2	void include(HttpServletRequest reg, HttpServletResponse res)	在响应中包含另外一个服务器资源

8.3.2 javax.servlet.http 包

javax.servlet.http 包中的类和接口说明如下。

1. javax.servlet.http.HttpServlet 类

javax.servlet.http.HttpServlet 是一个抽象类，它继承自 javax.servlet.GenericServlet 类，提供了一个处理 HTTP 的框架，用来处理客户端的 HTTP 请求。这个类中的 service 方法支持 GET、POST、PUT、DELETE 这些标准的 HTTP 请求类型，service 方法为每个 HTTP 请求类型调用相应的 doXXX()方法来处理，如 POST 请求调用 doPost()方法来处理。开发 HTTP 的 Servlet，只需实现 doXXX()方法即可，一般不用重写 service 方法。表 8.8 中列出了 HttpServlet 类的常用方法。

表 8.8　HttpServlet 类常用方法

序号	方法名称	功　能
1	void doGet(HttpServletRequest req, HttpServletResponse res)	由 service 方法调用，用于处理 GET 请求
2	void doPost(HttpServletRequest req, HttpServletResponse res)	由 service 方法调用，用于处理 POST 请求
3	void doPut(HttpServletRequest req, HttpServletResponse res)	由 service 方法调用，用于处理 PUT 请求
4	void doDelete(HttpServletRequest req, HttpServletResponse res)	由 service 方法调用，用户处理 DELETE 请求
5	void service(HttpServletRequest req, HttpServletResponse res)	接收 HTTP 的标准请求，并将它分配给响应的 doXXX()方法，一般不重载此方法
6	void init(ServletConfig config)	初始化 HttpServlet

2. javax.servlet.http.httpServletRequest 接口

javax.servlet.http.httpServletRequest 接口继承了 ServletRequest 接口，它用于定义封装客户端 HTTP 请求的“请求对象”。Servlet 容器创建一个 HttpServletRequest 对象，并将它作为一个参数传递给 service、doPost、doGet 等方法。表 8.9 中列出了 HttpServletRequest 接口特有的常用方法，继承自 ServletRequest 接口的方法在这里没有列出。

表 8.9　HttpServletRequest 接口的常用方法

序号	方法名称	功　能
1	String getContextPath()	返回指定 Servlet 上下文的 URL 的前缀
2	Cookie[] getCookies()	返回与请求相关的 Cookie 的一个数组
3	String getHeader(String name)	返回指定的 HTTP 头
4	String getMethod()	返回 HTTP 请求方法(如 GET、POST)
5	String getQueryString()	返回查询字符串，即 URL 中的“?”后面的部分
6	String getRequestedSessionId()	返回客户端的会话 ID
7	String getRequestURI()	返回 URL 中的一部分，从“/”开始，包括上下文，但不包括任意查询字符串

续表

序号	方法名称	功 能
8	String getServletPath()	返回URL上下文后的子串
9	HttpSession getSession(boolean create)	返回当前HTTP会话，如果不存在，则创建一个新的会话，create参数为true
10	boolean isRequestedSessionIdValid()	如果客户端返回的会话ID仍然有效，则返回true

3. javax.servlet.http.HttpServletResponse接口

javax.servlet.http.HttpServletResponse接口继承了ServletResponse接口，它用于定义使用HTTP响应客户端的“响应对象”。Servlet容器创建HttpServletResponse对象，该对象允许Servlet的service、doPost、doGet等方法使用HTTP头部信息域，并把数据发送给客户端。表8.10中列出了HttpServletResponse接口特有的常用方法，继承自ServletResponse接口的方法在此表中没有列出。

表8.10　HttpServletResponse接口的常用方法

序号	方法名称	功 能
1	void addCookie(Cookie cookie)	将一个Set-Cookie头标加入响应中
2	void addDateHeader(String name,long date)	使用指定日期值加入带有指定名字的响应头部
3	void setHeader(String name,String value)	设置具有指定名字和取值的一个响应头部
4	boolean containsHeader(String name)	判断响应是否包含指定名字的头部
5	void setStatus(int status)	设置响应使用字符编码的名称

4. javax.servlet.http.HttpSession接口

javax.servlet.http.HttpSession接口用于定义一个会话对象，该会话对象能够从页面的多个请求中甚至整个Web站点的范围内识别一个用户，并且能够存储该用户的有关信息。Servlet容器使用该接口在HTTP客户端和HTTP服务器端之间建立一个session(会话)。一个session可以有多种方法维持，如cookie或url重写等。表8.11中列出了javax.servlet.http.HttpSession接口的常用方法。

表8.11　javax.servlet.http.Httpsession接口的常用方法

序号	方法名称	功 能
1	Object getAttribute(String name)	返回当前session中指定名字的对象
2	Enumeration getAttributeNames()	返回一个所有属性名字的String对象的枚举变量
3	void invalidate()	是当前session失效，并将绑定的对象解除
4	boolean isNews()	客户端不知道当前session或不加入当前session则返回true
5	void removeAttribute(String name)	从当前session中删除指定名字的绑定对象
6	void setAttribute(String name,Object value)	使用指定的名字绑定一个对象到当前的session
7	void setStatus(int sc,String msg)	给当前的响应设定状态码和信息

8.4 编写、配置和调用 Servlet

在实际的 Web 编程中，一般编写一个 Servlet 就是编写一个 HttpServlet 的子类，该类实现响应用户的 POST、GET、PUT 等请求的方法，这些方法是 doPost、doGet 和 doPut 等 doXXX 方法。本节以例 8.1 为主线，介绍基于 HTTP 协议的 Servlet 编写、配置和调用的完整过程和方法。

8.4.1 编写第一个 Servlet

可以使用任意的文本编辑器编写 Servlet 类，编写方法类似于 JavaBean。Tomcat 7.0 要求 Servlet 必须有包名。下面是第一个 Servlet 示例。

【例 8.1】编写第一个 Servlet(相关文件参见 FirstServlet.java)，当客户端请求它时向用户问好。

FirstServlet.java 文件中的代码如下：

```
package myservlet.example;                                //01 行
import java.io.*;
import javax.servlet.*;
import javax.servlet.http.*;
public class FirstServlet extends HttpServlet {
    public void init(ServletConfig config)throws ServletException{//06 行
        super.init(config);
    }
    protected void doGet(HttpServletRequest request,//09 行
            HttpServletResponse response)throws
            ServletException,IOException {
        response.setContentType("text/html;charset=GB2312");
        PrintWriter out=response.getWriter();
        out.println("<html><body><h3>欢迎你使用第一 " +
                "Servlet!</h3></body></html>");
        out.flush();                                    //16 行
    }
    protected void doPost(HttpServletRequest request,//18 行
            HttpServletResponse response) throws
            ServletException,IOException{
        doGet(request,response);                        //21 行
    }
}
```

第 1 行将 Servlet 类放在了 myservlet.example 包中。第 2～4 行引入了编写 Servlet 所需的包。第 6～8 行重载了 init 方法。第 9~16 行重载了 doGet 方法，实现了 Servlet 响应用户请求的功能。第 18～21 行重写了 doPost 方法，调用了 doGet 方法，很方便地实现了响应 POST 请求。

8.4.2 配置 Servlet

1. 保存 Servlet 字节码文件

首先将 Servlet 类编译生成字节码文件(以 class 为扩展名的文件)，例如 FirstSerlvet.java 编译成 FirstServlet.class 文件。如果不使用 MyEclipse 等开发工具，为了编译 Servlet 源文件，需要将 HttpServlet、HttpServletRequest 等类的类文件包 servlet-api.jar 所在目录 C:\Program Files\Apache Software Foundation\Tomcat 7.0\lib\servlet-api.jar 添加到 CLASSPATH 环境变量中。为了让 Tomcat 服务器能够使用 FirstServlet.class 字节码文件创建一个 Servlet 对象，需要将 Servlet 字节码文件复制到 Tomcat 服务器的 Web 服务目录中的特定子目录下。本章使用的 Web 服务目录是 ch8，ch8 是用户创建的 Web 虚拟服务目录，它的物理路径是 E:\programJsp\ch8，创建虚拟目录的方法请参见本书第 1 章 1.3 节。使用 Servlet 需要在 ch8 的实际目录中创建以下特定的目录结构：

E:\programJsp\ch8\WEB-INF\classes

然后根据 Servlet 的包名在 classes 目录下创建相应的子目录，这里与 JavaBean 相似。对于 FirstServlet 创建的目录为：

E:\programJsp\ch8\WEB-INF\classes\myservlet\example

用户要注意，这里的目录结构区分大小写。目录创建好后将 Servlet 的字节码文件(如 FirstServlet.class)，复制到 classes\myservlet\example 目录中，重新启动 Tomcat 服务器，Tomcat 服务器就会启用上述目录。

2. 编写 Servlet 部署文件

为了能让 Tomcat 服务器使用 Servlet 字节码文件创建一个 Servlet 对象，必须为 Tomcat 编写一个部署文件。这个部署文件是一个 XML 文件，文件名为 web.xml，它保存在 Web 服务目录的 WEB-INF 子目录中，例如本章的 web.xml 文件保存在 E:\programJsp\ch8\WEB-INF 目录中，使用纯文本编辑器编写 web.xml 文件，部署 FirstServlet 的文件内容如下：

```
<?xml version="1.0" encoding="UTF-8"?>
<web-app>
  <welcome-file-list>
    <welcome-file>index.jsp</welcome-file>
  </welcome-file-list>
  <servlet>
      <servlet-name>Hello</servlet-name>
      <servlet-class>myservlet.example.FirstServlet</servlet-class>
  </servlet>
  <servlet-mapping>
      <servlet-name>Hello</servlet-name>
      <url-pattern>/helpHello</url-pattern>
  </servlet-mapping>
</web-app>
```

一个 XML 文件应当以 xml 声明作为文件的第一行，在其前面不能有空白。xml 声明以“<?xml”开始，以“?>”结束。例如<?xml version="1.0" encoding="UTF-8" ?>，XML 文件保存的编码必须与 encoding 指定的编码相同。有关 XML 的更多内容请参见第 10 章 10.1 节。Servlet 部署文件 web.xml 中使用的标记含义如表 8.12 所示。

表 8.12 Servlet 部署文件中的标记

序号	标记名称	功能描述
1	<web-app>	web.xml 的根标记
2	<servlet>	该标记标识 Servlet，由服务器处理；一个部署文件中可以有若干个该标记
3	<servlet-mapping>	与<servlet>标记对应出现，用来将 Servlet 映射到一个 URL
4	<servlet-name>	<servlet>标记和<servlet-mapping>的子标记，标识 Servlet 的正式名字，名字必须唯一
5	<servlet-class>	<servlet>标记的子标记，标识 Servlet 的带包类名
6	<url-pattern>	<servlet-mapping>标记的子标记，用来将 Servlet 映射到一个 url
7	init-param	Servlet 可用的初始化参数，其后跟着参数的名字
8	param-name	参数名
9	param-value	参数值

一个 Web 服务目录下的 web.xml 部署文件负责管理该目录下的所有 Servlet，当需要更多的 Servlet 时，只需在 web.xml 文件中增加<servlet>和<servlet-mapping>子标记即可。

8.4.3 调用 Servlet

当用户请求一个 Servlet 时，服务器会检查这个 Servlet 的实例是否存在，如果不存在，服务器就会根据 web.xml 部署文件和 Servlet 字节码文件创建一个 servlet 对象，如果存在就会直接调用这个 Servlet 的实例。Serlvet 类可以使用 getServletNmae()方法返回部署文件中的 Servlet 名字，例如 FirstServlet 返回的名字为 Hello。

用户可以有多种方式请求 Servlet，如浏览器直接调用，页面 form 中提交调用、超级链接调用、Servlet 调用等。

1. 在浏览器地址栏中直接调用 Servlet

在浏览器地址栏中直接输入 URL 地址：http://127.0.0.1:8080/ch8/helpHello，客户端访问 Servlet 得到的页面如图 8.2 所示。

图 8.2 FirstServlet 的运行效果

2. 在页面的 form 中调用 Servlet

Web 服务目录的 JSP 页面可以通过表单请求该 Web 服务目录下的某个 Servlet。在表单中的 action 属性中指定被调用 Servlet 的 URL，在表单的 method 属性中指定调用方式为 post 或 get 等，提交表单后，Web 服务器就会根据提交的方式(如 post 或 get)来调用 Servlet 响应的方法来响应该请求。下面的代码演示了表单调用 Servlet。

ch8_1.jsp 文件中的代码如下：

```
<%@ page contentType="text/html;charset=GB2312" %>
<%@ page import="java.sql.*" %>
<html>
<body bgcolor=cyan>
<form action="helpHello" method="get">
    <input type="submit" name="g" value="提交">
</form>
</body>
</html>
```

ch8_1.jsp 的页面效果如图 8.3 所示。

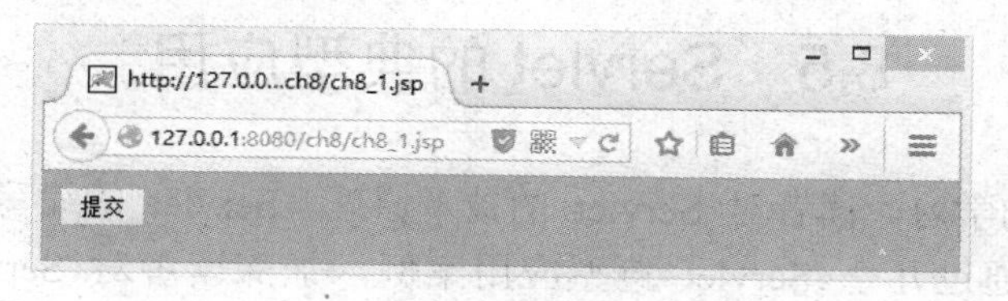

图 8.3　ch8_1.jsp 的页面效果

3. 页面超级链接调用 Servlet

Web 服务目录的 JSP 页面可以通过超级链接请求该 Web 服务目录下的某个 Servlet。在页面中使用<a href="helpHello"></a>标记来调用 FirstServlet。例如创建一个 ch8_1_link.jsp 页面效果如图 8.4 所示，文件内容如下：

```
<%@ page contentType="text/html;charset=GB2312" %>
<%@ page import="java.sql.*" %>
<html><body bgcolor=cyan>
<a href="helpHello">访问 FirstServlet</a>
</body>
</html>
```

图 8.4　ch8_1_link.jsp 的页面效果

4. 使用<jsp:forward>标签请求转发到 Servlet

在 JSP 页面中可以使用<jsp:forward>标记将请求转发到 Servlet，使用格式为：

```
<jsp:forward page="Servlet 映射名" />
```

或者：

```
<jsp:forward page="Servlet 映射名">
    <jsp:param name="nameid" value=""/>
    <jsp:param name="nameid" value=""/>
</jsp:forward>
```

例如创建一个 ch8_1_forward.jsp 页面，文件内容如下：

```
<%@ page contentType="text/html;charset=GB2312" %>
<%@ page import="java.sql.*" %>
<html>
<body bgcolor=cyan>
<jsp:forward page="helpHello"/>
</body>
</html>
```

8.5 Servlet 的典型应用

通过前面几节的学习，读者对 Servlet 的技术原理、Servlet 的编写、配置和调用有了一定的了解，本节通过几个 Servlet 典型应用案例，加深读者对 Servlet 的理解，提高 Servlet 的应用能力。

8.5.1 读取表单数据

第 4 章 4.3 节介绍了在 JSP 页面中使用 request 对象获取 HTML 表单数据的方法，在 Servlet 中同样也可以通过 HttpServletRequest 对象来读取表单数据。HttpServletRequest 对象提供了 getParameter、getParameterNames 和 getParameterValues 方法。getParameter (String name)方法的返回值是一个字符串，如果 name 变量不存在则返回 null。getParameterValues(String name)的返回值是一个字符串数组。getParameterNames 返回的是一个 Enumeration。下面是一个访问指定表单参数的例子。

【例 8.2】读取表单指定参数值(相关代码参见 LoginServlet.java、ch8_2.jsp 以及 web.xml)，页面效果如图 8.5 和图 8.6 所示。

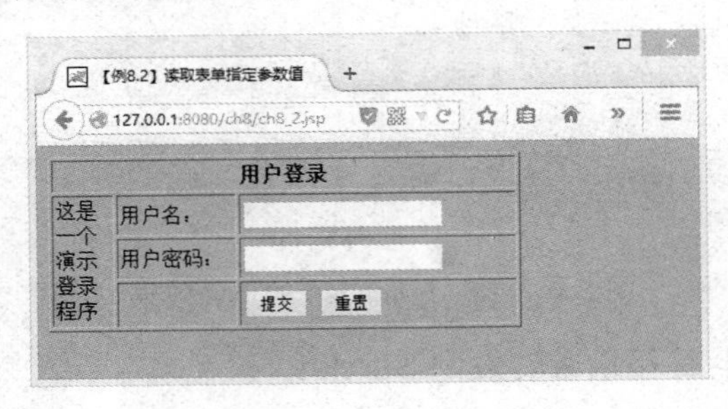

图 8.5　ch8_2.jsp 页面效果

图 8.6　LoginServlet 运行效果

LoginServlet.java 文件中的代码如下：

```
package myservlet.example;
import java.io.*;
import javax.servlet.*;
import javax.servlet.http.*;
public class LoginServlet extends HttpServlet {
    String name;
    String password;
    public void init(ServletConfig config)throws ServletException {
        super.init(config);
    }
    public void doPost(HttpServletRequest request,
            HttpServletResponse response)throws ServletException,
            IOException{
        name=request.getParameter("name");
        password=request.getParameter("password");
        response.setContentType("text/html;charset=GB2312");
        PrintWriter out=response.getWriter();
        out.println("<html>");
        out.println("<head><title>【例 8.2】获得表单参数</title></head>");
        out.println("<body>");
        out.println("<h3>您输入的参数是：</h3>");
        out.println("<li>用户名(name)： "+name);
        out.println("<li>用户密码(password)： "+password);
        out.println("</body></html>");
    }
}
```

ch8_2.jsp 文件中的代码如下：

```
<%@ page contentType="text/html;charset=GB2312" %>
<html>
<head>
<title>【例 8.2】读取表单指定参数值</title>
</head>
<body bgcolor=cyan>
<form action="helpLogin" method="post">
```

```
  <table width="366" border="1" cellspacing="2" cellpadding="2">
    <tr>
      <th colspan="3" scope="col">用户登录</th>
    </tr>
    <tr>
      <td width="48" rowspan="3">这是一个演示登录程序</td>
      <td width="102">用户名：</td>
      <td width="222"><label>
        <input type="text" name="name" id="name">
      </label></td>
    </tr>
    <tr>
      <td>用户密码：</td>
      <td><label>
        <input type="text" name="password" id="password">
      </label></td>
    </tr>
    <tr>
      <td> </td>
      <td><input type="submit" name="g" value="提交">
      <label>
      <input type="reset" name="reset" id="reset" value="重置">
      </label></td>
    </tr>
  </table>
</form>
</body>
</html>
```

在 web.xml 文件的适当位置，填写如下内容：

```
<servlet>
     <servlet-name>Login</servlet-name>
     <servlet-class>myservlet.example.LoginServlet</servlet-class>
</servlet>
<servlet-mapping>
     <servlet-name>Login</servlet-name>
     <url-pattern>/helpLogin</url-pattern>
</servlet-mapping>
```

下面是一个访问表单所有参数的例子。

【例 8.3】 访问表单所有参数的例子(相关代码参见 RegisterServlet.java、ch8_3.jsp 以及 web.xml)，如图 8.7 和图 8.8 所示。

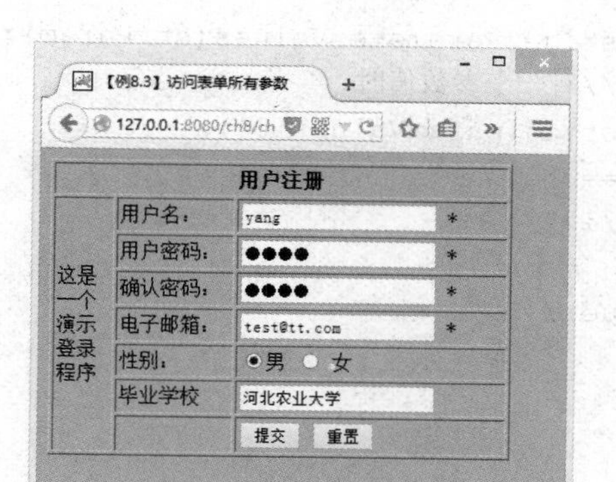

图 8.7　ch8_3.jsp 的页面效果

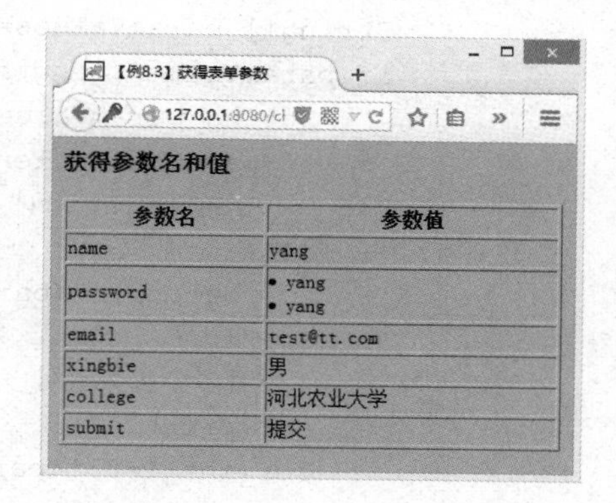

图 8.8　helpRegister 的页面效果

RegisterServlet.java 文件中的代码如下：

```
package myservlet.example;
import java.io.*;
import java.util.Enumeration;
import javax.servlet.*;
import javax.servlet.http.*;
public class RegisterServlet extends HttpServlet{
    public void init(ServletConfig config)throws ServletException{
        super.init(config);
    }
    public String getString(String s){//处理汉字
        String str=s.trim();
        try{
            byte bb[]=str.getBytes("iso-8859-1");
            str=new String(bb);
        }
        catch(Exception e){}
        return str;
    }
    public void doPost(HttpServletRequest request,
            HttpServletResponse response)throws ServletException,
            IOException{
        Enumeration enuNames=request.getParameterNames();
        response.setContentType("text/html;charset=GB2312");
        PrintWriter out=response.getWriter();
        out.println("<html>");
        out.println("<head><title>【例 8.3】获得表单参数</title></head>");
        out.println("<body bgcolor=cyan>");
        out.println("<h3>获得参数名和值</h3>");
        out.println("<table width=400 border=1>");
        out.println("<tr><th>参数名</th><th>参数值</th></tr>");
        while(enuNames.hasMoreElements()){
            String strParam=(String)enuNames.nextElement();
            out.println("<tr><td>"+strParam+"</td>\n<td>");
```

```
            String[] paramValues=request.getParameterValues(strParam);
            if(paramValues.length==1){//有一个参数值时
                String paramValue=getString(paramValues[0]);
                if(paramValue.length()==0)
                    out.println("<i>空</i>");
                else
                    out.print(paramValue);
            }
            else{//有多个参数值时
                out.println();
                for(int i=0;i<paramValues.length;i++){
                    out.println("<li>"+getString(paramValues[i]));
                }
                out.println("</ul>");
            }
            out.println("</td>");
        }
        out.println("</table>\n");
        out.println("</body>\n</html>");
    }
}
```

ch8_3.jsp 文件中的代码如下：

```
<%@ page contentType="text/html;charset=GB2312" %>
<html>
<head>
<title>【例 8.3】访问表单所有参数</title>
</head>
<body bgcolor=cyan>
<form action="helpRegister" method="post">
  <table width="366" border="1" cellspacing="2" cellpadding="2">
    <tr>
      <th colspan="3" scope="col">用户注册</th>
    </tr>
    <tr>
      <td width="48" rowspan="7">这是一个演示登录程序</td>
      <td width="102">用户名: </td>
      <td width="222"><label>
        <input type="text" name="name">
      *</label></td>
    </tr>
    <tr>
      <td>用户密码: </td>
      <td><label>
        <input type="password" name="password" >
      *</label></td>
    </tr>
```

```
    <tr>
      <td>确认密码：</td>
      <td><label>
        <input type="password" name="password" >
      *</label></td>
    </tr>
    <tr>
      <td>电子邮箱：</td>
      <td><input type="text" name="email" >
        *</td>
    </tr>
    <tr>
      <td>性别：</td>
      <td>
        <input name="xingbie" type="radio"  value="男" checked
        >男
        <input name="xingbie" type="radio"  value="女"> 女
        </td>
    </tr>
    <tr>
      <td>毕业学校</td>
      <td><input type="text" name="college"></td>
    </tr>
    <tr>
      <td> </td>
      <td><input type="submit" name="submit" value="提交">
        <label>
        <input type="reset" name="reset"  value="重置">
        </label></td>
    </tr>
  </table>
</form>
</body>
</html>
```

在web.xml文件的适当位置，填写如下内容：

```
<servlet>
   <servlet-name>Register</servlet-name>
   <servlet-class>myservlet.example.RegisterServlet</servlet-class>
</servlet>
<servlet-mapping>
   <servlet-name>Register</servlet-name>
   <url-pattern>/helpRegister</url-pattern>
</servlet-mapping>
```

8.5.2 读取 cookie 数据

cookie 是 Web 服务器保存用户硬盘上的一段文本，这段文本是以“关键词/值对”的格式保存的。cookie 允许一个 Web 站点在用户的计算机上保存信息并且随后再取回它。使用 cookie 的基本步骤如下。

(1) 创建 cookie 对象：

```
Cookie c=new Cookie("name","value");
```

(2) 传送 cookie 对象：

```
response.addCookie( c);
```

(3) 读取 cookie 对象：

```
Cookie[] cs=request.getCookies();
cs[i].getName();
cs[i].getValue();
```

(4) 设置 cookie 对象的有效时间：

```
c.setMaxage(int age);
```

age 为 cookie 保存的最大时间，为 0 时删除 cookie，为负值时关闭浏览器后删除 cookie。

下面介绍一个访问计数 cookie 的 Servlet。

【例 8.4】读取 cookie 数据的示例(其代码参见 CookieServlet.java)，Servlet 的运行效果如图 8.9 所示。

图 8.9 访问 CookieServlet 的页面效果

CookieServlet.java 文件中的代码如下：

```
package myservlet.example;
import java.io.*;
import javax.servlet.*;
import javax.servlet.http.*;
public class CookieServlet extends HttpServlet{
    public void init(ServletConfig config)throws
          ServletException{
        super.init(config);
    }
    public void service(HttpServletRequest request,
```

```
            HttpServletResponse response) throws IOException{
        boolean flag=false;//是否存在计数 cookie
        Cookie cookieLogin=null;
        Cookie[] cookies=request.getCookies();
        response.setContentType("text/html;charset=GB2312");
        PrintWriter out=response.getWriter();
out.println("<html><head><title>【例 8.4】Cookie 示例</title></head>");
        out.println("<body bgcolor=cyan>");
        out.println("<h3>您的登录信息如下：</h3>");
        if(cookies!=null){
            for(int i=0;i<cookies.length;i++){
                //判断是 cookie 中是否有 countNum 项
                if(cookies[i].getName().equals("countNum")){
                    flag=true;
                    cookieLogin=cookies[i];
                }
            }
        }
        if(flag){
        int tempCount=Integer.parseInt(cookieLogin.getValue());
        tempCount++;
        out.println("这是您第"+String.valueOf(tempCount)+"次访问该网页！");
        cookieLogin.setValue(String.valueOf(tempCount));
        cookieLogin.setMaxAge(108000*24);//设置 cookie 生存时间
        response.addCookie(cookieLogin);//将 cookie 加入的响应中
        }
        else{
            int tempCount=1;
            out.println("这是您第一次访问该页！");
           cookieLogin=new  Cookie("countNum",String.valueOf(tempCount));
            cookieLogin.setMaxAge(108000*24);
            response.addCookie(cookieLogin);
        }
        out.println("</body>");
        out.println("</html>");
    }
}
```

在 web.xml 文件的适当位置，增加如下内容：

```
<servlet>
    <servlet-name>Count</servlet-name>
    <servlet-class>myservlet.example.CookieServlet</servlet-class>
</servlet>
<servlet-mapping>
    <servlet-name>Count</servlet-name>
    <url-pattern>/helpCount</url-pattern>
</servlet-mapping>
```

8.5.3 读取 session 数据

HTTP 是一种无状态协议。JSP 内置了 Session 对象用来管理用户与服务器的会话，第 4 章 4.5 节对其做了介绍。Servlet 提供了 HttpSession API。HttpSession API 可以用来管理 Session，如 8.3 节所述。HttpSession API 是一个基于 Cookie 或者 URL 重写机制的高级会话管理接口。如果浏览器支持 Cookie 则使用 Cookie，如果不支持 Cookie 则自动采用 URL 重写，程序员不必关心细节，API 自动为 Servlet 开发者提供一个可以方便存储会话信息的地方。

在 Servlet 中管理 Session，通常采用如下步骤。

(1) 调用 HttpServletRequest 的 getSession 方法得到一个会话对象(Session)。

```
HttpSession session=request.getSession(true);
```

(2) 查看和会话有关的信息。

```
Integer oldAccessCount=
    (Integer)session.getAttribute("accessCount");
if(oldAccessCount!=null){
    accessCount=
        new Integer(oldAccessCount.intValue()+1);
}
```

(3) 在会话中保存数据。

```
session.setAttribute("accessCount", accessCount);
```

下面介绍一个读取 Session 数据的 Servlet。

【例 8.5】读取 Session 中数据的示例(其代码参见 SessionServlet.java)，Servlet 运行效果如图 8.10 所示。

图 8.10 helpSessionServlet 的页面效果

SessionServlet.java 文件中的代码如下：

```
package myservlet.example;
import java.io.*;
import javax.servlet.*;
import javax.servlet.http.*;
```

```
import java.util.*;
public class SessionServlet extends HttpServlet{
    public void init(ServletConfig config)throws ServletException{
        super.init(config);
    }
    public void doGet(HttpServletRequest request,
            HttpServletResponse response) throws
            ServletException,IOException{
        //得到一个Session对象
        HttpSession session=request.getSession(true);
        response.setContentType("text/html;charset=gb2312");
        PrintWriter out=response.getWriter();
        String heading;
        String title="Servlet中的会话管理";
        Integer accessCount=new Integer(0);
        if(session.isNew()){
            heading="欢迎，新客户！";
        }
        else{
            heading="欢迎您回来！";
        }
        //获得session中的数据
        Integer oldAccessCount=
            (Integer)session.getAttribute("accessCount");
        if(oldAccessCount!=null){
            accessCount=
                new Integer(oldAccessCount.intValue()+1);
        }
        //保存数据到session中
        session.setAttribute("accessCount", accessCount);
        out.println("<html>");
        out.println("<head><title>"+title+"</title></head>");
        out.println("<body bgcolor=#FDF5E6>");
        out.println("<h2>"+heading+"</h2>");
        out.println("<h2>您的会话(session)中的信息</h2>");
        out.println("<table border=1 align=center");
        out.println("<th>类型</th><th>值</th>");
        out.println("<tr><td>Session ID</td><td>");
        out.println(session.getId()+"</td></tr>");
        out.println("<tr><td>您访问该网站的次数</td><td>");
        out.println(accessCount+"</td></tr>");
        out.println("<tr><td>会话创建时间</td><td>");
        out.println(new Date(session.getCreationTime())+"</td></tr>");
        out.println("<tr><td>上次访问时间</td><td>");
        out.println(new
        Date(session.getLastAccessedTime())+"</td></tr>");
        out.println("</table>");
```

```
        out.println("</body></html>");
    }
    public void doPost(HttpServletRequest request,
            HttpServletResponse response)throws
                ServletException,IOException{
        doGet(request,response);
    }
}
```

在 web.xml 文件的适当位置，增加如下内容：

```
<servlet>
   <servlet-name>mySession</servlet-name>
   <servlet-class>myservlet.example.SessionServlet</servlet-class>
</servlet>
<servlet-mapping>
   <servlet-name>mySession</servlet-name>
   <url-pattern>/helpSessionSerlvet</url-pattern>
</servlet-mapping>
```

8.5.4 读取 HTTP 请求头数据

第 4 章 4.1 节介绍了 HTTP 的请求消息格式。浏览器向服务器发送请求的时候必须指明请求类型(一般是 post 或 get)。请求消息中包含许多的请求头，表 8.13 所示为常见的 HTTP 请求头。

表 8.13 常见 HTTP 头信息

序号	名 称	说 明
1	Accept	浏览器可以接受的 MIME 类型
2	Accept-Charset	浏览器可以接受的字符集
3	Accept-Encoding	浏览器能够进行解码的数据编码方式
4	Accept-Language	浏览器所希望的语言种类
5	Authorization	授权信息
6	Connection	表示是否需要持久连接
7	Content-Length	表示请求消息的长度
8	Cookie	保存客户信息
9	From	请求发送者的 E-Mail 地址
10	Host	初始 URL 中的主机地址和端口
11	Pragma	指定 no-cache 值表示服务器必须返回一个刷新后的文档
12	User-Agent	浏览器类型

在 Servlet 中读取 HTTP 头信息非常容易，只需调用 HttpServletRequest 的 getHeader 等方法即可。如果要读取指定的头，如 cookie 可以使用 getCookies 方法。除了读取指定头之外，程序员还可以使用 getHeaderNames 方法得到请求头中所有头名字的一个

Enumeration 对象，通过该对象的 hasMoreElements 方法遍历所有的头。

下面是在 Servlet 中读取 HTTP 请求头信息的例子。

【例 8.6】读取 HTTP 请求头信息的示例(其代码参见 HeaderServlet.java)，Servlet 执行效果如图 8.11 所示。

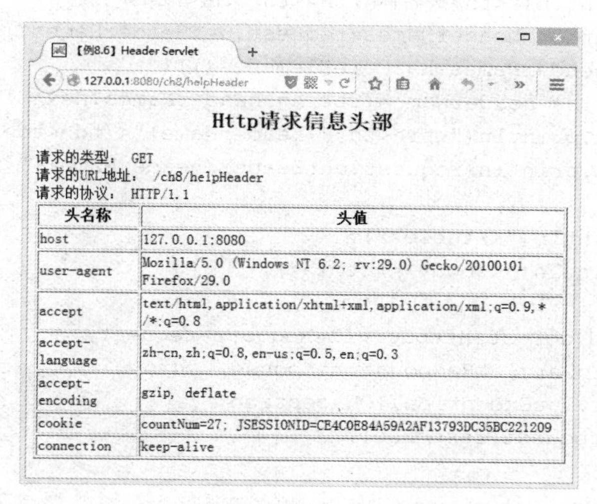

头名称	头值
host	127.0.0.1:8080
user-agent	Mozilla/5.0 (Windows NT 6.2; rv:29.0) Gecko/20100101 Firefox/29.0
accept	text/html,application/xhtml+xml,application/xml;q=0.9,*/*;q=0.8
accept-language	zh-cn,zh;q=0.8,en-us;q=0.5,en;q=0.3
accept-encoding	gzip, deflate
cookie	countNum=27; JSESSIONID=CE4C0E84A59A2AF13793DC35BC221209
connection	keep-alive

图 8.11　HeaderServlet 的执行效果

HeaderServlet.java 文件中的代码如下：

```
package myservlet.example;
import java.io.*;
import javax.servlet.*;
import javax.servlet.http.*;
import java.util.*;

public class HeaderServlet extends HttpServlet{
    public void init(ServletConfig config)throws ServletException
    {
        super.init(config);
    }
    public void doGet(HttpServletRequest request,
            HttpServletResponse response) throws
            ServletException,IOException{
        response.setContentType("text/html;charset=gb2312");
        PrintWriter out=response.getWriter();
        out.println("<html>");
        out.println("<head><title>【例 8.6】Header
Servlet</title></head>");
        out.println("<body bgcolor=#FDF5E6>");
        out.println("<h2 align=center >Http 请求信息头部</h2>");
        out.println("<p>请求的类型：");
        out.println(request.getMethod());
        out.println("<br>请求的 URL 地址：");
```

```
        out.println(request.getRequestURI());
        out.println("<br>请求的协议：");
        out.println(request.getProtocol());
        out.println("<table border=1 align=center");
        out.println("<th>头名称</th><th>头值</th>");
        Enumeration headerNames=request.getHeaderNames();
        while(headerNames.hasMoreElements()){
            String headerName=(String)headerNames.nextElement();
            out.println("<tr><td>"+headerName+"</td><td> ");
            out.println(request.getHeader(headerName)+"</td></tr>");
        }
        out.println("</table>");
        out.println("</body></html>");
    }
    public void doPost(HttpServletRequest request,
            HttpServletResponse response) throws
            ServletException,IOException{
        doGet(request,response);
    }
}
```

在 web.xml 文件的适当位置，增加如下内容：

```
<servlet>
    <servlet-name>myHeader</servlet-name>
    <servlet-class>myservlet.example.HeaderServlet</servlet-class>
</servlet>
<servlet-mapping>
    <servlet-name>myHeader</servlet-name>
    <url-pattern>/helpHeader</url-pattern>
</servlet-mapping>
```

8.6 上 机 实 训

实训目的

- 理解 Servlet 的概念和特性。
- 掌握编写、配置和使用 Servlet 的方法。
- 掌握编写 Servlet 的常用类和接口。
- 掌握 Servlet 获取表单数据的技术。
- 掌握 Servlet 读取 Session 数据的技术。
- 掌握 Servlet 读取 Cookie 的数据。

实训内容

实训 1 编写一个读取 Session 的 Servlet 并将其配置好之后执行该 Servlet。

要求：

获得 Session 并保存用户数据。

实训 2 编写一个读取 cookie 的 Servlet 并将其配置好之后执行该 Servlet。

要求：

(1) 创建一个 accessCount 对象并保存在 cookie 中。

(2) accessCount 对象记录用户重复访问次数。

实训 3 将一个使用 JavaBean 访问数据库的例子改为使用 Servlet 来实现。

要求：

(1) 不改变页面的风格。

(2) 配置并执行 Servlet。

实训总结

通过本章的上机实训，学员应该能够理解 Servlet 的概念和特点；掌握 Servlet 的编写、编译、布置和使用方法；掌握 Servlet 获得表单数据，读取 cookie 和 session 的方法。

8.7 本章习题

思考题

(1) 试述 Servlet 的生命周期。

(2) 如何使 Servlet 既能处理 GET 请求，又能处理 POST 请求？

(3) 获取表单数据的基本方法有哪些？

(4) HttpServletResponse 接口有哪些用处？

(5) Servlet 处理表单提交比起 JSP 页面处理表单提交有哪些优点？

(6) 是否一定要重写 Service 方法？重写了 Servlet 的 doPost 和 doGet 方法如何被调用？

(7) Servlet 对象如何获取用户的会话对象？

(8) Servlet 如何与 Servlet 或者 JSP 进行通信？

(9) 如何编写、编译、调试和配置 Servlet？

拓展实践题

重写通过 JavaBean 访问数据库的例子，使用 Servlet 技术实现对数据库的操作。

第9章　基于Servlet的MVC模式

学习目的与要求：

MVC模式的核心思想是有效地组合“视图”、“模型”和“控制器”，实现良好地面向对象Web程序设计，以便更好地复用组件，提高程序的可维护性和扩展性。本章主要学习MVC设计模式的思想以及JSP中的两种规范模式，重点介绍基于Servlet的MVC模式，也就是JSP模式2，阐述MVC中模型、视图和控制器各部分的功能，介绍模型的生命周期、视图的更新和控制器的重定向与转发技术。最后给出几个基于MVC模式的典型Web应用。通过本章的学习，读者要理解MVC模式概念和特点，掌握编写、编译、调试、基于Servlet的MVC模式Web应用程序方法和技术。

9.1　MVC模式介绍

面向对象技术的出现和应用极大地提高了软件复用和质量。采用什么样的交互软件设计模式，才能使编程既有对具体问题的针对性，又有对将来问题和需求的通用性？另外用户需要保持交互操作界面的稳定性，也需要根据需求的变化调整显示内容和形式，如何使数据的显示和处理之间松耦合，便于系统的维护和扩展？要解决这些问题，关键是采用显示部分和计算模型部分相互独立的软件组织结构。

9.1.1　MVC设计模式

MVC(Model-View-Controller)是20世纪80年代为编程语言Smalltalk-80发明的一种软件设计模式。MVC模式的结构如图9.1所示。MVC模式将交互式应用分成模型(Model)、视图(View)和控制器(Controller)三部分。模型是指从现实世界中抽象出来的对象模型，是应用逻辑的反映。模型封装了数据和对数据的操作，是实际进行数据处理计算的地方。视图是应用和用户之间的接口，它负责将应用显现给用户和显示模型的状态。控制器负责视图和模型之间的交互，控制对用户输入的响应、响应方式和流程，它主要负责两方面的动作：把用户的请求分发到相应的模型；将模型的改变及时反映到视图上。

MVC是一种理想的设计模式，它将这些对象、显示和控制分离以提高软件的灵活性和复用性。从面向对象的角度分析，MVC结构可以使程序具有对象化特性，也更容易维护。在设计程序时一般将某个对象看成“模型”，然后为“模型”提供合适的显示组件(可视对象)，即“视图”。

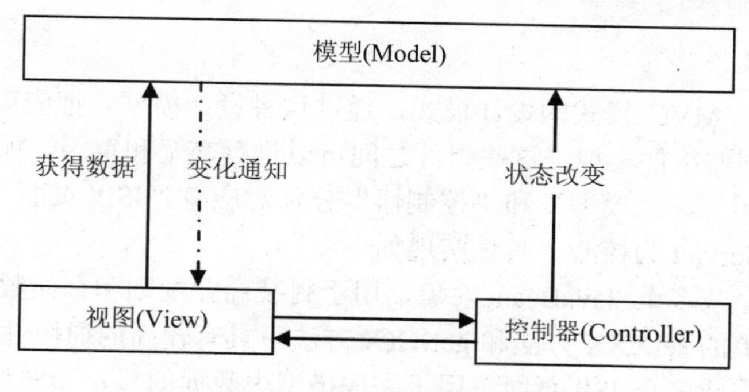

图 9.1 MVC 设计模式的结构

9.1.2 JSP 中的 MVC 模式

经典 MVC 模式在桌面程序中发挥了很重要的作用，随着软件规模的不断扩大，MVC 设计模式正在应用到各种应用程序的设计中。如何将 MVC 模式映射到 Web 应用中，MVC 各个部分对应着 Web 应用的哪个部分？对 MVC 模式又做了哪些改进？

Sun 公司的 JSP 规范提出了两种用 JSP 技术建立应用程序的方式，分别称为 JSP Model 1(模式 1)和 JSP Model 2(模式 2)。

1. JSP Model 1

JSP Model 1 是以 JSP 为中心的开发模型，如图 9.2 所示。JSP 页面独自响应请求并将处理结果返回客户，数据读取都是由 JavaBean 完成的。这种方式中 JSP 页面中同时实现业务逻辑、数据显示和流程控制，在开发小规模的 Web 应用程序时，有非常大的优势。但从工程化的角度考虑，这种模式也有一些不足。

- 应用是基于过程的，一组 JSP 页面实现一个业务流程，如果用户需求变化，软件维护和扩展困难。
- 业务逻辑和表示逻辑混合在 JSP 页面中没有实现抽象和分离，不利于应用系统业务的重用和改动。

因此，它适合简单应用的需要，不能满足复杂的大型应用程序的实现。

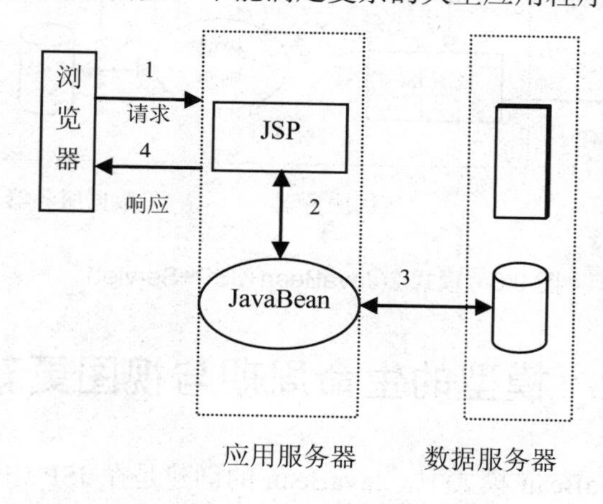

图 9.2 模式 1(JSP+JavaBean)

2. JSP Model 2

Model 2 是基于 MVC 模式的设计模式，通过这种设计模型，把应用逻辑、处理过程和显示逻辑分成不同的组件实现，这些组件之间可以进行交互和重用，如图 9.3 所示。在 Model 2 中，“视图”、“模型”和“控制器”分别对应着“JSP 页面”、“JavaBean”和“Serlvet”，以 Servlet 为核心，具体实现如下。

- 模型：一个或多个 JavaBean 对象，用于封装商业规则和存储数据，JavaBean 主要提供简单的 setXXX 方法和 getXXX 方法，具有较高的通用性。
- 视图：一个或多个 JSP 页面，用于完成模型中数据的显示和向控制器提交必要的数据和请求。主要使用 HTML 标记和 JavaBean 标记，在静态模板中显示存储在 JavaBean 中的数据。
- 控制器：一个或多个 Servlet 对象，负责创建 JavaBean 对象，根据视图提交的请求进行数据处理操作，完成处理逻辑，并将结果存储到 JavaBean 中，然后 Servlet 使用重定向方式请求视图中的某个 JSP 页面更新显示，即让 JSP 页面通过 JavaBean 标记显示控制器存储在 JavaBean 中的数据。

模式 2 发挥了 JSP 和 Servlet 的技术特长，JSP 页面擅长数据的显示，适合做视图，应尽量避免在 JSP 中大量使用 Java 代码来处理数据；Servlet 擅长数据处理，应避免在其中有大量的 HTML 标记输出。模式 2 清晰地分离了视图层和业务层，使视图、模式和控制之间具有较低的耦合性，有利于软件的扩展和维护。从软件工程化的角度来讲，有利于开发团队各司其职，便于软件开发的工程化管理。Struts、WebWork 和 Struts2 是基于 MVC 模式的开源框架，它们能够使混乱的东西结构化，是 MVC 发展的产物，学习本章能为今后使用这些框架打下坚实的设计模式基础。

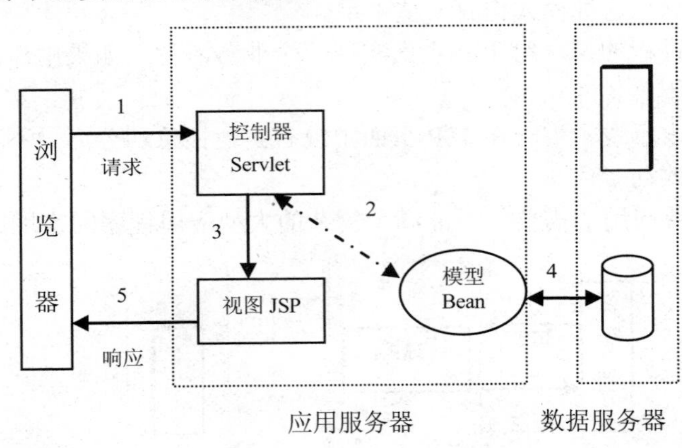

图 9.3　模式 2(JavaBean+JSP+Servlet)

9.2　模型的生命周期与视图更新

前面学习的 JSP+JavaBean 模式中，JavaBean 的创建是在 JSP 中使用 JavaBean 标记完成的，具体方法参见第 5 章。

在MVC模式中，JavaBean的创建是由Servlet负责的。Servlet创建Bean并将有关的数据存储在所创建的Bean中，然后Servlet再请求某个JSP页面，JSP页面通过<jsp:getProperty>标记显示Bean中的内容。在MVC模式中，因为Servlet负责创建JavaBean，所以JavaBean的构造函数可以带有参数，除了保留get和set规则外，还可以有其他功能的函数。Servlet创建JavaBean也涉及生命周期，它可以创建request、session和application周期的JavaBean。下面讨论这三种周期的JavaBean创建和对应视图更新方法。为了讲述方便，本章假设Servlet要创建Bean的类为JavaBeanClass，该类的包名为mybean.mvc。

9.2.1　requst周期的JavaBean与视图更新

1. 创建JavaBean

Servlet创建和使用request生命周期的JavaBean的步骤分两步，第一步使用JavaBeanClass类的构造方法创建Bean对象，例如：

```
JavaBeanClass bean=new JavaBeanClass(parameter);
```

第二步将创建的bean对象存放到HttpServletRequest的对象(request)中，并指定该bean对象的id，该步骤决定了bean的生命周期为request。例如：

```
request.setAttribute("keyword",bean);
```

2. 视图更新

当Servlet向某个页面发出请求时，Servlet创建的request周期的bean对象只对所请求的JSP页面有效。Servlet所请求的JSP页面使用<jsp:useBean>标记获得Servlet所创建的bean的引用(JSP页面不负责创建bean)，使用<jsp:getProperty>标记显示bean中的数据，该页面结束后，bean释放所占内存空间结束生命周期。下面是JSP页面中获得bean和显示bean中数据的方法：

```
<jsp:useBean
id="keyword" type="mybean.mvc.JavaBeanClass" scope="request"/>
<jsp:getProperty name="keyword" property="bean的变量名"/>
```

9.2.2　session周期的JavaBean与视图更新

1. 创建JavaBean

Servlet创建和使用session生命周期的JavaBean的步骤分两步，第一步使用JavaBeanClass类的构造方法创建bean对象，例如：

```
JavaBeanClass bean=new JavaBeanClass(parameter);
```

第二步将创建JavaBean对象存放在HttpServletSession的对象session中，并指定该bean对象的id，该步骤决定了JavaBean的生命周期为session。例如：

```
HttpSession session=request.getSession(true);
session.setAttribute("keyword",bean);
```

2. 视图更新

只要用户会话没有消失，Servlet 创建的 session 周期的 bean 对象就会一直存在，它对 Web 服务目录中的各个 JSP 页面都有效。Servlet 所请求的 JSP 页面都可以使用<jsp:useBean>标记获得 Servlet 所创建的 bean 的引用，使用<jsp:getProperty>标记显示 bean 中的数据，只有该会话结束后，bean 释放所占内存空间结束生命周期。

下面是 JSP 页面中获得 bean 和显示 bean 中数据的方法：

```
<jsp:useBean
id="keyword" type="mybean.mvc.JavaBeanClass" scope="session"/>
<jsp:getProperty name="keyword" property="bean 的变量名"/>
```

需要注意：不同用户的 session 周期的 bean 是互不相同的，即占有不同的内存空间。

9.2.3 application 周期

1. 创建 JavaBean

Servlet 创建和使用 application 生命周期的 JavaBean 的步骤分两步，第一步使用 JavaBeanClass 类的构造方法创建 bean 对象，例如：

```
JavaBeanClass bean=new JavaBeanClass(parameter);
```

第二步将创建的 JavaBean 对象存放在服务器创建的 ServletContext 对象中，并指定该 bean 的关键字。ServletContext 对象可以使用 getServletContext()方法得到其引用。该步骤决定了创建的 bean 的生命周期为 application。例如：

```
getServletContext().setAttribute("keyword",bean);或者
application.setAttribute("keyword",bean);
```

2. 视图更新

只要该 Web 应用程序不结束，Servlet 创建的 application 周期的 bean 对象就会一直存在，它对 Web 服务目录中的各个 JSP 页面都有效。Servlet 所请求的 JSP 页面都可以使用<jsp:useBean>标记获得 Servlet 所创建的 bean 的引用，使用<jsp:getProperty>标记显示 bean 中的数据，只有该 Web 应用程序关闭，bean 才释放所占内存空间结束生命周期。下面是 JSP 页面中获得 bean 和显示 bean 中数据的方法：

```
<jsp:useBean
id="keyword" type="mybean.mvc.JavaBeanClass" scope="application"/>
<jsp:getProperty name="keyword" property="bean 的变量名"/>
```

需要注意：所有用户的 application 周期的 bean 是相同的，即占有相同的内存空间。

9.3　控制器的重定向与转发

接受用户的请求，并根据数据处理的结果决定显示的视图是模式 2 中控制器完成的主要功能之一。

Servlet 如何实现页面流程控制？它主要是使用重定向与转发技术实现页面流程控制的。重定向功能是将用户从当前页面或 Servlet 定向到另一个 JSP 页面或 Servlet；转发的功能是将用户对当前 JSP 页面或 Servlet 的请求转发给另一个 JSP 页面或 Servlet。在 Servlet 中，主要使用 HttpServletResponse 类的重定向方法 sendRedirect 方法实现重定向，以及使用 RequestDispatcher 类的转发方法 forward 方法实现转发功能。下面讨论重定向与转发。

9.3.1　重定向

HttpServletResponse 类中提供了一个 sendRedirect(java.lang.String.location)方法，该方法是一个重定向方法。当用户请求一个 Servlet 时，该 Servlet 处理完成数据后，可以将用户重新定向到一个 JSP 页面或另一个 Servlet。重定向仅仅是将用户定向到其他的 JSP 页面或 Servlet，而不能将 HttpServletRequest 对象转发给所指向的资源。换一句话来讲，就是重定向的目标页面或 Servlet 不能使用 request 对象获取用户提交的数据。重定向到目的页面，在用户浏览器的地址栏中会见到重定向目标页面的 URL 地址。

9.3.2　转发

RequestDispatcher 类提供了一个 forward(ServletRequest request, ServletResponse response)方法，该方法可以把用户对当前 JSP 页面或 Servlet 的请求转发给另一个 JSP 页面或 Servlet，并且能够将用户对当前 JSP 页面或 Servlet 的请求对象(HttpServletRequest 对象)和响应对象(HttpServletResponse 对象)传递给目标页面或 Servlet。换一句话来说，就是目标 JSP 页面或 Servlet 可以使用 request 对象获得用户的请求数据。

实现转发需要两个步骤，首先在 Servlet 中要得到 RequestDispatcher 对象，然后在调用该对象的 forward 方法实现转发。

1. 得到 RequestDispatcher 对象

调用 HttpServletRequest 对象 request 的 getRequestDispatcher(java.lang.String path)方法可以得到一个 RequestDispathcer 对象，例如：

```
RequestDispatcher dispatcher=request.getRequestDispatcher("a.jsp");
```

其中："a.jsp"是要转发的 JSP 页面或者 Servlet 的地址。request 是 HttpServletRequest 类的对象。

2. 转发到目的页面

调用第一步中得到的 RequestDispatcher 对象 dispatcher 的 forward 方法，将用户对当

前 JSP 页面或 Servlet 的请求转发给 RequestDispatcher 对象所指定的 JSP 页面或 Servlet。例如：

```
dispatcher.forward(request,response);
```

其中：request 对象和 response 对象分别为当前页面或 Servlet 的 HttpServletRequest 对象和 HttpServletResponse 对象；dispatcher 对象为第一步得到的 RequestDispatcher 对象。该语句的含义是将用户对当前页面或 Servlet 的请求转发给“a.jsp”页面。转发与重定向是不同的，用户不能在浏览器的地址栏中见到转发的页面或 Servlet 地址，只能见到目的页面的运行效果，浏览器地址栏中显示的是原来的 URL 地址。

9.4 MVC 模式的分析

前面几节介绍了 MVC 模式的一些基本知识，以下通过几个简单的例子演示 MVC 三个部分的设计与实现。本章 JavaBean 的包名为 mybean.mvc，Servlet 类的包名为 myservlet.mvc。需要按照有关 Java 类的编译方法对 Servlet 类和 JavaBean 类进行编译。本章约定将编写的 JavaBean.java 类文件存放在 E:\user\myservlet\mvc\mybean\mvc 目录中，将 Servlet.java 文件存放在 E:\user\myservlet\mvc 目录中，使用 JavaJDK 的 javac 命令编译类文件。为了正确的运行程序，将通过编译的 Servlet 字节码复制到 ch9\WEB-INF\classes\myservlet\mvc 目录中，JavaBean 字节码复制到 ch9\WEB-INF\classes\mybean\mvc 目录中。

9.4.1 用户登录

设计一个简单的用户登录 Web 应用，该例中有一个 Servlet、一个 JavaBean 和三个页面。Login.jsp 页面负责收集登录用户名和密码，并将请求交给 loginServlet，Login.jsp 页面效果如图 9.4 所示；名字为 loginBean 的 JavaBean 负责保存用户登录信息；loginServlet 接受用户的请求并处理登录操作，根据用户输入的用户名和密码判断是否为合法用户，登录成功则转发到 LoginSuccess.jsp 页面，如图 9.5 所示，登录失败则重定向到 LoginFalse.jsp 页面，如图 9.6 所示。Servlet 创建一个 session 周期的 Bean，根据登录是否成功设置 Bean 的 flag 属性，其他页面根据 session 中 Bean 的 flag 值判断是否为合法用户。

本章使用的 Web 服务目录是 ch9，ch9 是用户创建的 Web 虚拟服务目录，它的物理路径是 E:\programJsp\ch9，创建虚拟目录的方法请参见有关章节。另外，需要在 ch9 中创建如下的目录结构：

```
ch9\WEB-INF\classes
```

然后根据 Servlet 和 JavaBean 的包名，在 classes 下再建立相应的子目录。本例 Servlet 类的包名为 myservlet.mvc，JavaBean 类的包名为 mybean.mvc，则在 classes 目录下创建：

```
ch9\WEB-INF\classes\myservlet\mvc
ch9\WEB-INF\classes\mybean\mvc
```

为了让 Tomcat 服务器启用上述目录，需要重新启动服务器。有关 JavaBean 和 Servlet 的编写和配置详细内容，请参考有第 5 章和第 8 章相关内容。

为了让 Servlet 运行，需要编写 web.xml 配置文件，并保存到 Web 服务目录的 WEB-INF 子目录中，即 ch9\WEB-INF 中，web.xml 文件的内容如下：

```
<?xml version="1.0" encoding="UTF-8"?>
<web-app>
  <welcome-file-list>
    <welcome-file>index.jsp</welcome-file>
  </welcome-file-list>
  <servlet>
      <servlet-name>LoginServlet</servlet-name>
      <servlet-class>myservlet.mvc.loginServlet</servlet-class>
  </servlet>
  <servlet-mapping>
     <servlet-name>LoginServlet</servlet-name>
     <url-pattern>/helpLoginServlet</url-pattern>
  </servlet-mapping>
</web-app>
```

1. 模型

如果使用 MyEclipse 等开发工具，工具会自动编译 Servlet 和 Bean，如果用户使用 Java JDK 手动编译 Servlet 类和 JavaBean，本章约定 Servlet 放到 E:\user\myservlet\mvc 目录；将 JavaBean 放到 E:\user\myservlet\mvc\mybean\mvc 目录。编译 Servelt 和 JavaBean 的方法如下：

```
E:\user\myservlet\mvc  javac myservlet.java
```

JavaBean 和 Servlet 编译后，用户需要将其字节码复制到虚拟目录 ch9 的对应物理目录中，本章 Servlet 的目录为：

```
E:\programJsp\ch9\WEB-INF\classes\myservelt\mvc
```

JavaBean 的目录为：

```
E:\programJsp\ch9\WEB-INF\classes\mybean\mvc
```

模型 loginBean 中设置了用户名、密码和登录是否成功标志等，并不涉及登录的细节，这样可以提高模型的通用性，便于组件复用。

loginBean.java 文件代码如下：

```
package mybean.mvc;
import java.io.*;
import java.util.*;

public class loginBean {
    String name=null;
    String password=null;
```

```
    boolean flag=false;
    public String getName() {
        return name;
    }
    public void setName(String name) {
        this.name = name;
    }
    public String getPassword() {
        return password;
    }
    public void setPassword(String password) {
        this.password = password;
    }
    public boolean isFlag() {
        return flag;
    }
    public void setFlag(boolean flag) {
        this.flag = flag;
    }
}
```

2. 视图

接收用户名和密码的界面 Login.jsp 页面效果如图 9.4 所示。登录成功后的页面 LoginSuccess.jsp 页面效果如图 9.5 所示。登录失败后转向的页面如图 9.6 所示。

Login.jsp 文件代码如下：

```
<%@ page contentType="text/html;charset=GB2312" %>
<html>
<body bgcolor="#33CCFF">
<form action="helpLoginServlet" method="post" name="loginform">
 <font size="2">
  输入登录名字：<input type="text" name="loginname" >
  <br>输入登录密码：<input type="password" name="password">
  <br><input type="submit" value="提交">
  </font>
</form>
</body>
</html>
```

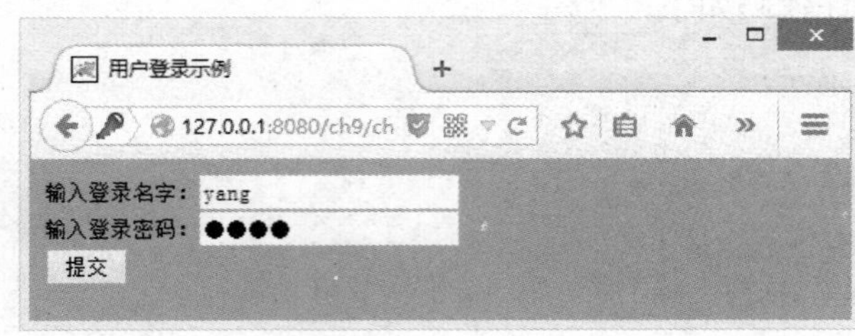

图 9.4　Login.jsp 页面效果

LoginSuccess.jsp 文件代码如下：

```
<%@ page contentType="text/html;charset=GB2312" %>
<!-- 获得 bean 的引用 -->
<jsp:useBean id="loginbean" type="mybean.mvc.loginBean"
 scope="session"/>
<html>
<head><title>这是登录成功页面</title>
</head>
<body bgcolor="#33CCFF">
 <font size="2">
  您登录的用户名是：
  <!-- 得到 bean 里的数据 -->
  <jsp:getProperty name="loginbean" property="name"/>
  <br>您输入的密码是：
  <jsp:getProperty name="loginbean" property="password"/>
  <br>您登录状态是：
  <jsp:getProperty name="loginbean" property="flag"/>
  <br><a href="Login.jsp" target="_self">返回登录界面：</a>
  </font>
</body>
</html>
```

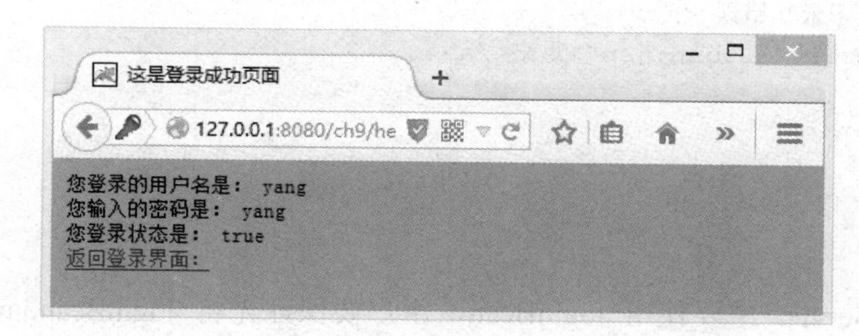

图 9.5　LoginSuccess.jsp 页面效果

LoginFalse.jsp 文件代码如下：

```
<%@ page contentType="text/html;charset=GB2312" %>
<!-- 获得 bean 的引用 -->
<jsp:useBean id="loginbean" type="mybean.mvc.loginBean"
 scope="session"/>
<html>
<head><title>这是登录失败页面</title>
</head>
<body bgcolor="#33CCFF">
 <font size="2">
  您登录的用户名是：
  <!--得到 bean 中的数据-->
  <jsp:getProperty name="loginbean" property="name"/>
  <br>您输入的密码是：
```

```
  <jsp:getProperty name="loginbean" property="password"/>
  <br>您登录状态是:
  <jsp:getProperty name="loginbean" property="flag"/>
  <br><a href="Login.jsp" target="_self">返回登录界面: </a>
  </font>
</body>
</html>
```

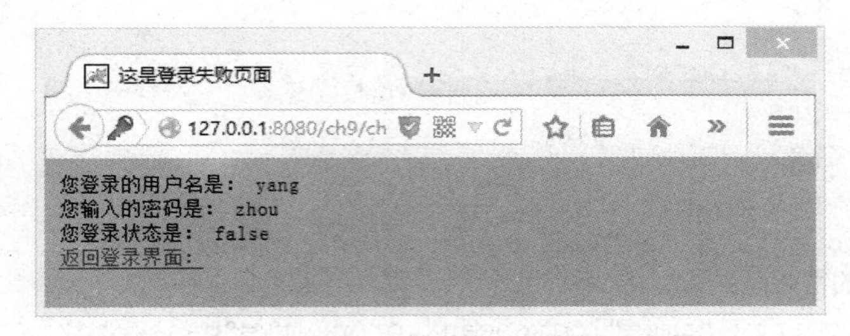

图 9.6　LoginFalse.jsp 页面效果

error1.html 文件中的代码如下：

```
<html>
<body>
<h1>
对不起，发生了未知错误 <br>
请试着 <a href="/Login.jsp">重来</a>.
</h1>
</body>
</html>
```

3. 控制器

由于 loginServlet 中要使用 loginBean，用户要保证先将 loginBean.java 文件保存到 E:\user\myservlet\mvc\mybean\mvc 目录中，然后在使用 javac 命令编译 loginServlet.java 文件，将得到的 loginServlet.class 和 loginBean.class 文件复制到 ch9\WEB-INF\classess 对应的目录中。

loginServlet.java 文件中的代码如下：

```
package myservlet.mvc;
import java.io.*;
import javax.servlet.*;
import javax.servlet.http.*;
import mybean.mvc.loginBean;
public class loginServlet extends HttpServlet{
    public void init(ServletConfig conf) throws ServletException {
        super.init(conf);
    }
    public void doPost(HttpServletRequest req,
            HttpServletResponse res) throws
```

```
ServletException,IOException{
        HttpSession session = req.getSession(false);
        if (session == null) {//如果会话为空，重定向到错误页面
            res.sendRedirect("http://localhost:8080/ch9/error.html");
        }
        loginBean lbean=null;
        lbean=getLBean(req);//得到登录 bean
        session.setAttribute("loginbean", lbean);//将 bean 保存到 session 中
        if(lbean.isFlag()){//登录成功转发页面到 LoginSuccess.jsp 页面
            //保存 bean 到会话中
        session.setAttribute("loginbean", lbean);
        String url="/LoginSuccess.jsp";
        ServletContext sc = getServletContext();
        RequestDispatcher rd = sc.getRequestDispatcher(url);
        rd.forward(req, res);
        }
        else{//登录失败，重定向到 LoginFailse.jsp 页面
            //保存 bean 到会话中
            session.setAttribute("loginbean", lbean);
            String url="LoginFalse.jsp";
          res.sendRedirect(url);
        }
    }
    private loginBean getLBean(HttpServletRequest req) {
        //预设登录用户名和账户密码
        String loginNames[]={"yang","zhou","zhang","wang"};
        String passWords[]={"yang","zhou","zhang","wang"};
        //创建 bean
        loginBean lbean=new loginBean();
        String name=null;
        String password=null;
        name=(String)req.getParameter("loginname");
        if(name!=null){
            name=name.trim();
            }
        password=(String)req.getParameter("password");
        if(password!=null){
            password=password.trim();
        }
        for(int i=0;i<4;i++){
            if(loginNames[i].equals(name)&&
                    passWords[i].equals(password))
            {
                lbean.setFlag(true);
                break;
            }
        }
```

```
        lbean.setName(name);
        lbean.setPassword(password);
        return lbean;
    }
}
```

提示：① 在实际的登录系统中，登录成功则调转到系统的主面，并提供与该用户权限相符的菜单，登录错误则提示登录错误，并跳转到登录窗口。

② 本例演示了模式 2 中视图、模型和控制器的功能、分工与协作，以及控制器根据数据处理逻辑实现重定向和转发技术。

9.4.2 留言板

设计一个使用 application 对象的留言板，该留言板提供一个界面，如图 9.7 所示。页面可以输入用户的名字、留言的标题和留言的内容。提交后在该页面的下部显示所有用户的留言。留言的显示顺序按逆序显示(后发布的内容先显示)，最新留言放在顶部，便于用户查看。

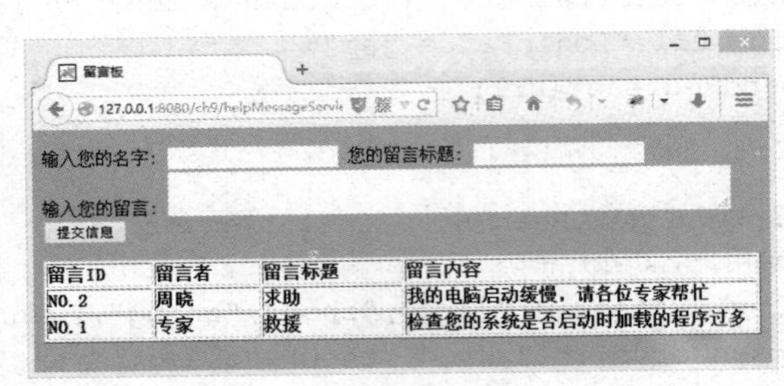

图 9.7 留言板页面效果

该 Web 应用提供一个 submit.jsp 页面和一个 messagePane.jsp 文件，负责留言信息的录入和显示；提供一个名字为 messageServlet 的 Servlet 负责接受和处理用户的留言，并通知 submit.jsp 页面更新显示；名字为 Message 的 Bean 负责保存用户的留言信息。

为了使用 Servlet，需要编写 web.xml 文件，并保存到 Web 服务目录的 WEB-INF 子目录中，即 ch9/WEB-INF 目录中。Web.xml 文件增加的内容为：

```
<servlet>
<servlet-name>MessageServlet</servlet-name>
    <servlet-class>myservlet.mvc.messageServlet</servlet-class>
</servlet>
<servlet-mapping>
    <servlet-name>MessageServlet</servlet-name>
    <url-pattern>/helpMessageServlet</url-pattern>
</servlet-mapping>
```

1. 模型(JavaBean)

由于 messageServlet 类中要使用 Message 类，为了能顺利地编译 messageServlet 类和

Message 类，需要 messageServlet.java 和 Message.java 按照一定的目录结构存放。按照本章的约定，Servlet 类的包名为 myservlet.mvc，JavaBean 类的包名为 mybean.mvc，用户需要将 Message.java 存放在 E:\user\myservlet\mvc\mybean.mvc 目录中，将 messageServlet.java 放到 E:\user\myservlet\mvc 目录中。进入 E:\user\myservlet\mvc\mybean\mvc 目录，首先编译 message.java 文件，然后将编译得到的该类字节码 Message.class 文件复制到 E:\programJsp\ch9\WEB-INF\classes\mybean\mvc 目录中。

模型 Message.java 中的 getXXX 和 setXXX 方法不涉及对数据的具体处理细节，只是仅仅给 XXX 变量赋值和得到 XXX 属性的值，这样可以增强模型的通用性。

Message.java 文件中的代码如下：

```
package mybean.mvc;
public class Message {
    String title=null;//留言标题
    String name=null;//留言者
    String mess=null;//留言内容
    String mID;//留言 ID
    public Message(){//构造函数
        title="";
        name="";
        mess="";
        mID="";
    }
    public String getTitle() {
        return title;
    }
    public void setTitle(String title) {
        this.title = title;
    }
    public String getName() {
        return name;
    }
    public void setName(String name) {
        this.name = name;
    }
    public String getMess() {
        return mess;
    }
    public void setMess(String mess) {
        this.mess = mess;
    }
    public String getMID() {
        return mID;
    }
    public void setMID(String mid) {
        mID = mid;
    }
}
```

2. 视图

submit.jsp 页面效果如图 9.7 所示。

submit.jsp 文件中的代码如下：

```
<%@ page contentType="text/html;charset=GB2312" %>
<html>
<head><title>留言板</title>
</head>
<body bgcolor="#33CCFF">
<form action="helpMessageServlet" method="post" name="form">
   <p><font size="3">输入您的名字：
   <input type="text" name="pepoleName">

    您的留言标题：
   <input type="text" name="messTitle">
     <br>
     输入您的留言：
   <textarea name="message" rows="1" cols="59"
     ></textarea>
     <br>
   </font><font size="3">
   <input type="submit" value="提交信息" name="submit">
   </font></p>
</form>
<jsp:include page="messagePane.jsp" flush="true" />
</body>
</html>
```

messagePane.jsp 文件中的代码如下：

```
<%@ page contentType="text/html;charset=GB2312" %>
<%@ page import="java.util.*, mybean.mvc.Message" %>
<%
Vector<Message> messlist = (Vector) application.getAttribute("message");
if (messlist != null && (messlist.size() > 0)) {
%>
    <center>
    <table border="1" cellpadding="0" width="100%" bgcolor="#FFFFFF">
    <tr>
    <td width="15%"><b>留言 ID</b></td>
    <td width="15%"><b>留言者</b></td>
    <td width="20%"><b>留言标题</b></td>
    <td width="50%"><b>留言内容</b></td>
    </tr>
    <%
    for (int index=messlist.size()-1;index>=0;index-- ) {
        Message anMess = (Message) messlist.elementAt(index);
```

```
        %>
        <tr>
        <td><b><%= anMess.getMID() %></b></td>
        <td><b><%= anMess.getName() %></b></td>
        <td><b><%= anMess.getTitle() %></b></td>
        <td><b><%= anMess.getMess() %></b></td>
        </tr>
    <% } %>
    </table>
    <center>
    <% } %>
```

messagePane.jsp 处理基于 application 的 Bean，它指定了 MVC 结构中的模型。观察 messagePane.jsp 开头这一段脚本：

```
<%
Vector<Message> messlist = (Vector) application.getAttribute("message");
if (messlist != null && (messlist.size() > 0)) {
%>
```

这段脚本从 application 中得到了一个 Vector 的实例 messlist。如果 messlist 为空或者还未创建，messagePane.jsp 不会显示任何留言；如果 messlist 不是空的，那么已经保存的留言会一次一个地从 messlist 中被提出，下面的脚本完成留言的提出：

```
<%
for (int index=messlist.size-1; index>=0; index-- ) {
Message anMessage = (Message) messlist.elementAt(index);
%>
```

一旦描述留言的变量 anMessage 已创建，它们就简单地被 JSP 表达式插入到静态 HTML 模板中去。图 9.7 显示了用户已经发布了一些留言，这里体现了模型与视图的关系。

3. 控制器

messageServlet.java 要使用 Message.java 类，本章将 messageServlet.java 放到 E:\user\myservlet\mvc 目录中，Message.java 放到 E:\user\myservlet\mvc\mybean\mvc 目录中。进入 E:\user\myservlet\mvc 目录编译 messageServlet 类，并将得到的字节码文件 messageServlet.class 复制到 E:\programJsp\ch9\WEB-INF\classes\myservlet\mvc 中。

控制器 messageServlet 实现了对留言动作的处理，每次用户在 submit.jsp 中发表一个留言，请求都被发送到这个控制器 messageServlet。它处理相应要添加的留言的请求参数，实例化一个新的 Message 的 bean(anMessage)，这个 anMessage 代表这个发布的留言，接着更新 messlist，然后把包含有这个 anMessage 的 messlist 放进 application。完成这些处理后，控制器转发请求到 submit.jsp 页面，通知 submit.jsp 页面更新显示。详细内容见 messageServlet.java 代码。

messageServlet.java 文件中的代码如下：

```
package myservlet.mvc;
import java.util.*;
```

```
import java.io.*;
import javax.servlet.*;
import javax.servlet.http.*;
import mybean.mvc.Message;
public class messageServlet extends HttpServlet{
    private Integer i=0;
    public void init(ServletConfig conf) throws ServletException {
        super.init(conf);
    }
    public
    synchronized void sendMessage(HttpServletRequest req){
        Vector<Message> messlist=null;
       ServletContext application;
       application=getServletContext();
       if(null!=application.getAttribute("message"))
        messlist=(Vector)application.getAttribute("message");
    Message amess = getMessage(req);
       if (messlist==null)
           messlist = new Vector<Message>(); //first order
        messlist.addElement(amess);
    application.setAttribute("message", messlist);
    }
    public String codeString(String str){
        String s=str;
        try{
            byte b[]=s.getBytes("ISO-8859-1");
            s=new String(b);
        }
        catch(Exception e){}
        return s;
    }
    public void doPost (HttpServletRequest req, HttpServletResponse res)
    throws ServletException, IOException {
       sendMessage(req);
       String url="/submit.jsp";
       ServletContext sc = getServletContext();
       RequestDispatcher rd = sc.getRequestDispatcher(url);
       rd.forward(req, res);
    }
    private Message getMessage(HttpServletRequest req) {
       i++;
       String mID ="NO."+i.toString();
       String name = req.getParameter("pepoleName");
       System.out.print(name);
       String title=req.getParameter("messTitle");
       String message=req.getParameter("message");
       Message Mess = new Message();
```

```
        Mess.setMID(codeString(mID));
        Mess.setName(codeString(name));
        Mess.setTitle(codeString(title));
        Mess.setMess(codeString(message));
        return Mess;
    }
}
```

9.4.3 访问数据库

设计一个访问数据库的 Web 应用，该应用有 showDatabase.jsp 和 showRecord.jsp 两个页面，一个名字为 databaseServlet 的 Servlet，一个名字为 showRecPage.java 的 JavaBean。showDatabase.jsp 页面效果如图 9.8 所示，用于接收用户输入的数据库名和数据库中所包含的表名，并将请求发送给控制器 databaseServlet。showRecord.jsp 页面负责显示查询记录，并提供分页显示，该页面效果如图 9.9 所示。控制器 databaseServlet 根据用户的请求查询数据库，如果数据库和表不存在则重定向到 showDatabase.jsp 页面，如果查询成功则转发到 showRecord.jsp 记录显示页面。模型 showRecPage 负责保存查询记录数据和分页显示数据。

用户需要配置 ch9\WEB-INF 目录中的 web.xml 文件，文件增加的内容如下：

```
<servlet>
<servlet-name>DatabaseServlet</servlet-name>
    <servlet-class>myservlet.mvc.databaseServlet</servlet-class>
</servlet>
<servlet-mapping>
    <servlet-name>DatabaseServlet</servlet-name>
    <url-pattern>/helpDatabaseServlet</url-pattern>
</servlet-mapping>
```

1. 模型

showRecPage 中使用了 CachedRowSetImpl 对象 rowSet 和 feildSet，作为 bean 的属性用于保存数据库查询的结果集，feildSet 保存表的字段名结果集，rowSet 保存查询记录结果集。CachedRowSetImpl 对象可以在连接对象关闭后使用，这样就节省了数据库的连接资源，减少了数据库的访问次数。showRecPage 对象由 Servlet 创建，并保存在 session 对象中。

用户需要将 showRecPage.java 保存在 E:\user\myservlet\mvc\mybean\mvc 目录中，编译后将字节码 showRecPage.class 文件复制到 ch9\WEB-INF\classes 对应目录中。

showRecPage.java 文件中的代码如下：

```
package mybean.mvc;
import com.sun.rowset.*;
public class showRecPage {
    CachedRowSetImpl rowSet=null;//存储表中全部记录的行集对象
    CachedRowSetImpl feildSet=null;//存储字段的行集对象
    int pageSize;//每页显示的记录数
```

```
int pageAllCount;//分页后的总页数
int showPage;//当前显示页
String databaseName=null;//数据库的名字
String tableName=null;//表的名字
//构造函数
public showRecPage() {
    super();
    this.rowSet = null;
    this.feildSet = null;
    this.pageSize = 10;
    this.pageAllCount =0;
    this.showPage =1;
    this.databaseName =null;
    this.tableName = null;
}
public CachedRowSetImpl getRowSet() {
    return rowSet;
}
public void setRowSet(CachedRowSetImpl rowSet) {
    this.rowSet = rowSet;
}
public CachedRowSetImpl getFeildSet() {
    return feildSet;
}
public void setFeildSet(CachedRowSetImpl feildSet) {
    this.feildSet = feildSet;
}
public int getPageSize() {
    return pageSize;
}
public void setPageSize(int pageSize) {
    this.pageSize = pageSize;
}
public int getPageAllCount() {
    return pageAllCount;
}
public void setPageAllCount(int pageAllCount) {
    this.pageAllCount = pageAllCount;
}
public int getShowPage() {
    return showPage;
}
public void setShowPage(int showPage) {
    this.showPage = showPage;
}
public String getDatabaseName() {
    return databaseName;
```

```
    }
    public void setDatabaseName(String databaseName) {
        this.databaseName = databaseName;
    }
    public String getTableName() {
        return tableName;
    }
    public void setTableName(String tableName) {
        this.tableName = tableName;
    }
}
```

2. 视图

showDatabase.jsp 页面效果如图 9.8 所示。文本框 databaseName、tableName 和 pageSize 指定了默认 value。提交后将用户输入的请求发给 databaseServlet 控制器处理。

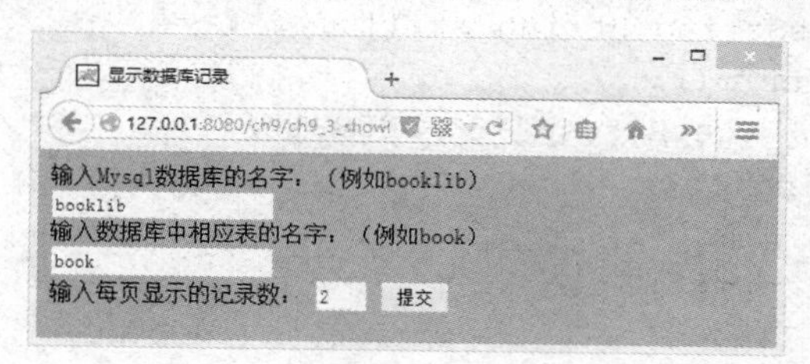

图 9.8　showDatabase.jsp 页面效果

showDatabase.jsp 文件中的代码如下：

```
<%@ page contentType="text/html;charset=GB2312" %>
<html>
<head><title>显示数据库记录</title></head>
<body bgcolor="#33CCFF">
<form action="helpDatabaseServlet" method="post" name="form">
  输入 MySQL 数据库的名字：(例如 booklib)
  <br><input type="text" name="databaseName" value="booklib">
  <br>输入数据库中相应表的名字：(例如 book)
  <br><input type="text" name="tableName" value="book">
  <br>输入每页显示的记录数：
  <input type="text" value="2" name="pageSize" size="3">
  <input type="hidden" value="query" name="action" >
  <input type="submit" value="提交" name="submit">
</form>
</body>
</html>
```

showRecord.jsp 页面效果如图 9.9 所示，该页面实现了分页功能，单击“上一页”或“下一页”，则将当前页码和 action 动作指令提交给 databaseServlet，databaseServlet 根据 action 的值修改 showRecPage 模型中的数据，并通知 showRecord.jsp 页面更新显示。

图 9.9 showRecord.jsp 页面效果

showRecord.jsp 文件中的代码如下:

```
<%@ page contentType="text/html;charset=GB2312" %>
<%@ page import="com.sun.rowset.*" %>
<%@ page import="mybean.mvc.showRecPage" %>
<%!
public String  getString(String str){
    String s=str.trim();
    try{byte bb[]=s.getBytes("ISO-8859-1");
       s=new String(bb);
    }
    catch(Exception e){}
    return s;
}
%>
<html>
<body bgcolor="#33CCFF">
<jsp:useBean id="database" type="mybean.mvc.showRecPage"
    scope="session" />
您查询的数据库:<jsp:getProperty name="database"
    property="databaseName"/>
,查询的表:<jsp:getProperty name="database"
        property="tableName"/>
<br>记录分<jsp:getProperty name="database"
    property="pageAllCount"/>
页,每页最多显示<jsp:getProperty name="database"
        property="pageSize"/>
条记录,目前显示第<jsp:getProperty name="database"
    property="showPage"/>页
<font size="3" >
<%
int feildCount=0;
int k=0,i=0;
CachedRowSetImpl rowSet=null;
CachedRowSetImpl feildSet=null;
feildSet=database.getFeildSet();
if(feildSet!=null&&rowSet!=null);{%>
```

```
<table border=1>
<%
String feilds[]=new String[100];//保存表的字段名
feildSet.beforeFirst();//字段名结果集游标定位
while(feildSet.next()){//统计字段个数
   feildCount++;
   feilds[k]=feildSet.getString(4);//得到字段名
   k++;
}
%>
<tr>
<%
for(k=0;k<feildCount;k++){//输出字段名%>
    <th><%= getString(feilds[k]) %></th>
<%}%>
</tr>
<%
}
rowSet=database.getRowSet();//得到结果集
int pageCount=database.getShowPage();//得到当前页号
int pageSize=database.getPageSize();//得到每页显示记录数
if(rowSet!=null){
    try{rowSet.absolute((pageCount-1)*pageSize+1);//定位记录
        for(k=1;k<=pageSize;k++){//显示一页%>
        <tr>
        <%  for(i=1;i<=feildCount;i++){//显示每一行%>
            <td><%= getString(rowSet.getString(i))%></td>
        <%}%>
    </tr>
        <%  rowSet.next();//移动游标
        }
    }
    catch(Exception e){
    }
%>
</table>
<table>
<tr><td>
<form action="helpDatabaseServlet" method="post" name="form" >
<input type="hidden" value="previousPage" name="action" >
<input type="submit" value="上一页" name="submit">
</form>
</td><td>
<form action="helpDatabaseServlet" method="post" name="form">
<input type="hidden" value="nextPage" name="action">
<input type="submit" value="下一页" name="submit" >
</form>
```

```
    </td></tr>
    </table>
<%
}
%>
<a href="showDatabase.jsp" >返回到查询界面</a>
</font>
</body>
</html>
```

3. 控制器

控制器 databaseServlet 处理来自 showDatabase.jsp 页面和 showRecord.jsp 页面的请求，负责修改模型中的数据并通知视图更新，完成数据库操作。databaseServlet 根据上述两个页面传递的 action 指令值决定执行何种操作，action 为“query”时根据用户提供的数据库和表名查询数据库，并将得到的 ResultSet 对象保存到 showRecPage 对象中；action 为“previousPage”时显示上一页，action 为“nextPage”时显示下一页。databaseServlet 根据 action 指令值来修改模型 showRecPage 中的数据，并重定向到 showDatabase.jsp 或者转发通知 showRecord.jsp 视图更新。下面是 databaseServlet 控制器的代码。

databaseServlet.java 文件中的代码如下：

```
package myservlet.mvc;
import mybean.mvc.showRecPage;
import com.sun.rowset.*;
import java.sql.*;
import java.io.*;
import javax.servlet.*;
import javax.servlet.http.*;
public class databaseServlet extends HttpServlet {
    public void doGet(HttpServletRequest request,
            HttpServletResponse response)
            throws ServletException, IOException {
        doPost(request,response);
    }
    public void doPost(HttpServletRequest request,
            HttpServletResponse response)
            throws ServletException, IOException {
        HttpSession session = request.getSession(false);
        if (session == null) {
            response.sendRedirect("error.html");
        }
        showRecPage databaseBean=null;
        if(null!=session.getAttribute("database"))
        databaseBean=(showRecPage)session.getAttribute("database");
        String action = request.getParameter("action");
    if(action.equals("query")){
```

```
    try{
        CachedRowSetImpl rowSet=null;
        CachedRowSetImpl feildSet=null;
        databaseBean=new showRecPage();
        //得到数据库名和表名
        String databaseName=request.getParameter("databaseName");
        String tableName=request.getParameter("tableName");
        databaseBean.setDatabaseName(databaseName);
        databaseBean.setTableName(tableName);
        //计算每页显示记录数
        String ps=request.getParameter("pageSize");
        if(ps!=null){
            try{
                int mm=Integer.parseInt(ps);
                databaseBean.setPageSize(mm);
            }
            catch(NumberFormatException exp){
                databaseBean.setPageSize(3);
            }
        }
        //得到字段结果集和记录结果集
        rowSet=getRowSet(databaseBean);
        feildSet=getFeildSet(databaseBean);
        //设置 bean 的属性值
        databaseBean.setRowSet(rowSet);
        databaseBean.setFeildSet(feildSet);
        //统计表记录个数
        rowSet.last();
        int m=rowSet.getRow();
        int n=databaseBean.getPageSize();
        //计算总页数
        int pageAllCount=((m%n)==0)?(m/n):(m/n+1);
        databaseBean.setPageAllCount(pageAllCount);
        session.setAttribute("database", databaseBean);
        RequestDispatcher dispatcher=
            request.getRequestDispatcher("showRecord.jsp");
        dispatcher.forward(request, response);
    }
    catch(Exception e){//发生错误后，定向到原页面
        response.sendRedirect("showDatabase.jsp");
    }
  }
  else if(action.equals("previousPage")){
    int showPage=databaseBean.getShowPage();
    //计算显示的页号，小于第一页页号，则从最后一页显示
    if(showPage-1<=0){
        showPage=databaseBean.getPageAllCount();
```

```
        }
        else {
            showPage-=1;
        }
        databaseBean.setShowPage(showPage);
            session.setAttribute("database", databaseBean);
        RequestDispatcher dispatcher=
            request.getRequestDispatcher("showRecord.jsp");
        dispatcher.forward(request, response);
        }
        else if(action.equals("nextPage")){//单击了下一页
        int showPage=databaseBean.getShowPage();
        //计算显示的页号，大于最后一页页号，则从第一页显示
        if(showPage+1>databaseBean.getPageAllCount()){
            showPage=1;
        }
        else {
            showPage+=1;
        }
        databaseBean.setShowPage(showPage);
            session.setAttribute("database", databaseBean);
        RequestDispatcher dispatcher=
            request.getRequestDispatcher("showRecord.jsp");
        dispatcher.forward(request, response);
        }
    }
    public void init(ServletConfig config) throws ServletException {
        super.init(config);
        try{
            Class.forName("com.mysql.jdbc.Driver");
        }
        catch(Exception e){}
    }
    public CachedRowSetImpl getRowSet(showRecPage databaseBean){
        CachedRowSetImpl rowSet=null;//保存 rs 对象中的数据，不依赖 conn
        Connection conn=null;
        Statement stmt=null;
        ResultSet rs=null;//结果集
        String uri="jdbc:mysql://localhost:3306/"+
            databaseBean.getDatabaseName();
        String userid="root";
        String userpwd="";
        try{
            conn=DriverManager.getConnection(uri,userid,userpwd);
            //得到游标可前后移动的，只读的结果集
            stmt=conn.createStatement(ResultSet.TYPE_SCROLL_SENSITIVE,
                    ResultSet.CONCUR_READ_ONLY);
```

```
            rs=stmt.executeQuery("select * from "+
                    databaseBean.getTableName());
            rowSet =new CachedRowSetImpl();
            //将rs中的数据保存到rowSet对象中
            rowSet.populate(rs);
            rs.close();
            stmt.close();
            conn.close();
        }
        catch(Exception e){
            //System.out.println("表查询错误! "+e.toString());
        }
        return rowSet;//返回结果集
    }
    public CachedRowSetImpl getFeildSet(showRecPage databaseBean){
        CachedRowSetImpl feildSet=null;//保存所查询表的表头信息
        Connection conn=null;
        ResultSet rs=null;
        String uri="jdbc:mysql://localhost:3306/"+
           databaseBean.getDatabaseName().trim();
        String uid="root";
        String upw="";
        try{
            conn=DriverManager.getConnection(uri,uid,upw);
            //得到数据库的元数据;
            DatabaseMetaData metadata=conn.getMetaData();
            /**得到某个表的字段信息的结果集(n行,4列),例如:
             * booklib dbo book bid
             * booklib dbo book bookname
             * booklib dbo book author
             * booklib dbo book publish
             * booklib dbo book price
             */
            rs=metadata.getColumns(null, null,
                    databaseBean.getTableName(), null);
            feildSet=new CachedRowSetImpl();
            feildSet.populate(rs);
            rs.close();
            conn.close();
        }
        catch(Exception e){
            //System.out.println("取表头错误! "+e.toString());
        }
        return feildSet;
    }
}
```

注意：该例的 showdatabase.jsp 页面设置了数据库名称和表名的默认值，如果用户输入的数据库名和表名错误，也就是在 MySQL 中没有此数据库或表，数据库访问就会发生异常，程序将重新定向到 showDatabase.jsp 页面。
用户也可以思考一下，程序如何给使用者提供更明确的错误信息。

9.5 上机实训

实训目的

- 理解 MVC 模式的核心思想。
- 理解 MVC 模式中模型、视图与视图的各自功能。
- 掌握各种生命周期 Bean 的创建及其数据修改操作。
- 掌握控制器对数据的处理操作，对各种生命周期 Bean 的保存操作，以及对页面流程控制技术。
- 掌握视图更新操作。
- 掌握数据库访问及记录集的分页显示技术。

实训内容

实训 1 使用基于 Servlet 的 MVC 模式，编写一个计算三角形或矩形面积和周长的 Web 应用。

要求：

(1) 用户通过页面输入三角形或矩形的参数。

(2) 控制器计算三角形和矩形的面积、周长，模型保存计算的周长或面积。

实训 2 参照第 11 章，采用 MVC 模式，完成网上报名系统登录 JavaBean、Servlet 和视图的设计与实现。

要求：

(1) 登录页面名称为 login.jsp，主页面为 index.jsp。

(2) JavaBean 存储用户名和密码以及登录状态。

(3) 控制器根据用户名和密码验证结果调转页面，成功到 index.jsp 页面，失败到 error.jsp 页面。

实训 3 采用 MVC 模式，实现图书购物车。

要求：

(1) 购物车保存在 Session 对象中，结账后会话结束。

(2) 由一个 Servlet 完成商品选购，购物车查看和结账操作。

实训总结

通过本章的上机实训，学员应该能够理解 MVC 模式的设计思想，理解 MVC 模式中视图、模型和控制器的功能；掌握 MVC 模式 Web 应用设计与程序编写的常用技术。

9.6 本章习题

思考题

(1) MVC 模式的核心思想是什么？
(2) 谁来担当 MVC 模式的视图、模型和控制器角色？
(3) 使用 MVC 模式有什么好处？
(4) MVC 模式中的 Bean 由谁来创建？都有哪些生命周期类型？
(5) 控制器如何控制页面的流转。
(6) 怎样编写、编译、调试和布置 MVC 模式中的 Servlet 和 bean？

拓展实践题

将 5.4.2 节购物车实例修改为 MVC 模式。

第 10 章　JSP 中使用 XML

学习目的与要求：

XML 是为互联网而诞生的可扩展置标语言。相对于 HTML 来讲，使用 XML 文件可以按照一定的标准来组织数据，使界面和结构化数据相分离。本章主要学习 XML 文件的结构、声明、元素、标记等基本知识，DOM 解析器的基本概念、基本工作原理，使用 DOM 解析器读取 XML 文档的基本编程技术，SAX 解析器的基本概念、基本工作原理，使用 SAX 解析器解析 XML 文件的基本编程技术等内容。通过本章的学习，学员要理解 XML 的基本概念及简单 XML 文件编写；理解 DOM 解析器和 SAX 解析器的工作原理，掌握其常用接口的编程技术；掌握使用 DOM 和 SAX 读写 XML 文件的基本编程技术。

10.1　XML 简 介

XML(eXtensible Markup Language，可扩展置标语言)是由万维网联盟(World Wide Web Consortium，W3C)定义的一种语言，是表示结构化数据的行业标准。XML 是专门为因特网定制的，它克服了 HTML 一些固有的缺陷，例如结构复杂混乱、内容与形式无法分离、维护困难等，具有自由与开放、超越固有格式、严格的语法校验、便于不同系统之间信息的传输等优点。XML 最重要的功能是用来创建、描述和存储信息，以便各种 Web 应用可以基于 XML 进行更方便的数据交换。本节对 XML 做简单的介绍，有关 XML 详细内容请读者学习 XML 教程。

10.1.1　XML 文件的结构

XML 是在文本处理系统的发展过程中产生的。当用户需要查询文件中的某些内容时，显然希望这种文件具有某种特殊形式的结构，即文件按照一定的标准来组织数据。XML 是一个源描述性语言，可以看作是用来产生标记的工具。因此，XML 并没有预定义一个特定的标记集，而是描述了一个用来定义标记集的方法。当用户用 XML 规范定义一个标记集，并根据这些 XML 的规定填入文本内容，这些标记和纯文本就一起构成了一个 XML 文件。

尽管 XML 允许用户定义自己的标记集，但 XML 文件必须遵守 XML1.0 规范中的语法规则。以下是一个简单的 XML 文件。

【例 10.1】简单 XML 文件示例，其代码参见 resume.xml 文件。

resume.xml 文件中的代码如下：

```
<?xml version="1.0" encoding="UTF-8" ?>
<简历>
```

```
    <学生 ID="001">
      <姓名>张三</姓名>
       <性别>男</性别>
       <年龄>19</年龄>
       <联系电话>0312123456</联系电话>
       <email>zhang@163.com</email>
    </学生>
    <学生 ID="002">
      <姓名>李四</姓名>
       <性别>女</性别>
       <年龄>20</年龄>
       <联系电话>0123567789</联系电话>
       <email>li@163.com</email>
    </学生>
    <学生 ID="003">
      <姓名>王五</姓名>
       <性别>男</性别>
       <年龄>20</年龄>
       <联系电话>03128383737</联系电话>
       <email>wangwu@163.com</email>
    </学生>
</简历>
```

该 XML 文件描述三个学生的简历情况，包括姓名、性别、年龄等各项元素。读者可以看出 XML 文件是采用树形结构来描述数据的。在这种树形结构中，各节点间层次关系非常清晰。下面分析一下这个简单的 XML 文件结构，以便用户更好地理解每行代码的含义。

1. XML 文件声明

resume.xml 文件的第一行：

```
<?xml  version="1.0" encoding="UTF-8" ?>
```

是 XML 声明且必须是 XML 声明，XML 声明指明了 XML 的版本和编码方式等属性。

2. 根标记

resume.xml 文件的<简历>是根标记的开始标记，</简历>是根标记的结束标记。XML 文件有且只有一个根标记，其他标记都必须封装在根标记中，根标记又称为文件标记。

3. 元素

<姓名>张三</姓名>是一个元素，元素是 XML 文档内容的基本单元。一个元素包含一个起始标记、一个结束标记以及标记之间的数据内容。其形式一般为：

```
<元素标记>数据内容</元素标记>
```

4. 树形结构

在根标记<简历></简历>之间有若干<学生></学生>子标记，<学生></学生>子标记之间有<姓名>、<性别>子标记。XML 文件的标记必须形成树形结构，也就是标记之间不允许交叉。

10.1.2 XML 声明

一个规范的 XML 文件第一行是它的声明部分，声明的作用就是告诉 XML 处理程序："下面的这个文件是一个按照 XML 文件的标准对数据进行置标的文件"。下面就是一个简单的 XML 文件的声明部分：

```
<?xml  version="1.0" standalone="yes" encoding="UTF-8" ?>
```

声明部分以"<?xml"开始，以"?>"部分结束，"<?xml"字符之间没有空格。声明部分要求必须指定 version 属性，encoding 和 standalone 属性可选。

1. version 属性

version 为版本属性，指明 XML 文件的版本号，采用"属性="值""对的形式赋值。在一个 XML 文件的声明中必须包含版本属性，并且需要放到属性列表的第一位。目前的版本为"1.0"，所以用户编写的 XML 文件的版本号都是"1.0"。

2. encoding 属性

encoding 为编码属性，该属性规定 XML 文件采用哪种字符集进行编码。如果在声明中没有指定 encoding 属性，默认为 UTF-8 字符集。XML 规范中能够使用中文的字符集有 GB2312(简体中文码)、BIG5(繁体中文码)和 UTF-8。使用 UTF-8、GB2312 和 BIG5 编码，在 XML 文件中标记和标记的内容就可以使用汉字了。

用户在使用记事本等文本编辑工具或者 XML 编辑工具时，要注意声明的字符集编码类型与保存文件的字符编码类型相同，例如：resume.xml 文件的声明为 UTF-8 字符集，则在文件的保存类型必须选择 UTF-8 编码，如图 10.1 所示。

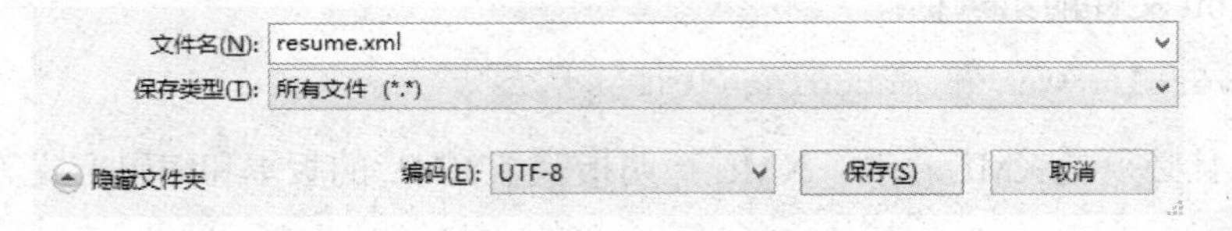

图 10.1　encoding 值与文件保存类型

3. standalone 属性

standalone 属性表明该 XML 文件是否和一个独立的"置标声明"文件配套使用。如果该属性设置为"yes"，则没有配套的 DTD 文件进行置标声明，如果这个属性设置为"no"，则该文件可能有配套的置标声明文件。

10.1.3 XML 元素

元素是 XML 文件内容的基本单元。从语法上讲，一个元素包含一个起始标记、一个结束标记以及标记之间的数据内容。其形式为：

```
<标记>数据内容</标记>
```

<标记>数据内容</标记>的形式在 HTML 中经常见到，其实用户并不陌生。例如在 resume.xml 文件中，<简历>...</简历>以及它们之间包含的所有数据信息是根元素，在一个 XML 文件中有且只能有一个根元素。另外，元素中还可以嵌套别的元素。例如<简历>根元素中还有<学生>元素，<学生>元素中还有<姓名>、<性别>、<年龄>等元素。

XML 元素与 HTML 中的元素一样可以有自己的属性。元素的每一个属性是一个名称=数值对，名称和数值为一个字符串，一元素不能有两个名称相同的属性。例如：

```
<学生 ID="A1001">
...
</学生>
```

<学生>元素有一个属性，属性名为 ID，值为“A1001”，它们以 ID=“A1001”的形式出现。为了描述的方便，一般把“< >”内的文字称为标记，XML 中开始标记和结束标记之间的文字称为“字符数据”。例如：

```
<姓名>张三</姓名>
```

在这个元素中“姓名”称为标记，“张三”称为字符数据。接下来介绍标记的一些规则。

10.1.4　XML 标记

XML 文件是由标记和字符数据组成的文件，置标是 XML 语言的精髓，占有举足轻重的位置。除了注释和 CDATA 部分以外，所有符号“<”和“>”之间的内容都称为标记。其一般形式为：

```
<标记名　属性 1="属性取值 1"　属性 2="属性取值 2"　...　>字符数据</标记名>
```

XML 标记和 HTML 的标记形式上大体相同，但 XML 的标记语法要严格得多。下面是对标记的一些语法规定。

(1) XML 标记的名称可以由字母、数字、下划线、点或连字符组成，但必须是字母或下划线开头。如果 XML 文件的编码是 UTF-8，则汉字、日文片假名以及朝鲜文等其他语言文字都可以作为标记使用。

(2) 标记区分大小写，例如<HELLO>和<hello>是两个不同的标记。标记要有结束标记例如<HELLO>标记的结束标记为</HELLO>。

(3) 标记要求正确的配对，不能像 HTML 中的某些标记，例如
，只有
是不合法的，必须有</br>标记。

(4) 标记要正确嵌套，标记之间不能交叉。例如下面的标记嵌套是正确的：

```
<学生>
        <姓名>张三</姓名>
</学生>
```

下面的标记嵌套是错误的：

```
<学生>
        <姓名>张三</学生>
</姓名>
```

(5) 标记必须具有有效的使用属性。标记中可以包含任意多个属性，在标记中属性是以名称/值对的形式出现的，属性名不能重复，名称和取值之间用“=”分隔，且取值用双引号引起来。例如：

```
<学生 ID="A1001" >
```

1. 空标记

不含有子标记或文本内容的标记称为空标记。因为空标记不含有内容，所以空标记不需要结束标记。空标记的一般形式为：

```
<标记名  属性列表 />
```

或者：

```
<标记名 />
```

例如：

```
<desktop  width="150" heigh="75" />
```

2. 非空标记

相对于空标记来讲，非空标记就是开始标记和结束标记之间含有数据内容的标记。非空标记的一般形式为：

```
<标记名  属性列表>文本内容</标记名>
```

或者：

```
<标记名  属性列表>
    文本内容
</标记名>
```

注意：“<”或者“</”和标记名之间不能有空格。上述的两种非空标记一般形式，所含的文本内容是不同的，第二个标记含有两个换行符。标记之间的内容会忠实地传给处理程序，这一点不同于 HTML。

3. CDATA 节

<、>、&、"、'等符号在 XML 中有特殊用途，如果用户想要在标记内容中使用<、>、&、"、'等符号，可以使用 CDATA 节。在标记 CDATA 下，所有的标记、实体引用都被忽略，<、>、&、"、' 等这些符号被当作普通字符处理。CDATA 的形式如下：

```
<![CDATA[文本内容]]>
```

CDATA 节以“<![CDATA[”开始，以“]]>”结束，在文本内容中不能有“]]>”。

【例 10.2】带有 CDATA 节的 XML 文件示例(其代码参见 noun_xml)。

noun.xml 文件中的代码如下：

```
<?xml version="1.0" encoding="UTF-8" standalone="no" ?>
<专有名词列表>
```

```
    <专有名词>
        <名词>XML</名词>
        <解释>xml是一种可扩展的标记语言</解释>
        <示例>
            <!--一个xml的例子-->
            <![CDATA[
              <联系人>
                  <姓名>张三</姓名>
                  <email>zhang@aaa.com</email>
              </联系人>
            ]]>
        </示例>
    </专有名词>
</专有名词列表>
```

4. XML 注释

在 XML 中，注释的方法与 HTML 完全相同，用“<!--”和“-->”将注释文本括起来。用户需要注意在注释中不能出现“--”，并且注释不能嵌套。

> **提示：** 用户可以使用记事本编辑 XML 文件，也可以使用 XML 文件编辑软件，常见的 XML spy、XML writer 等。在保存 XML 文件时，保存文件的字符编码应与 XML 文件声明中的字符集类型一致。另外，用户还可以使用高版本的浏览器对 XML 文件进行语法检查，如果 XML 文件符合语法规则，就会在浏览器中正确显示。

10.2 DOM 解析器

10.2.1 什么是 DOM 解析器

当一个文本文件一旦经过 XML 结构化后，就称其为 XML 文件。XML 文件的制作、存取、应用都能够用计算机程序进行有效的处理。开发基于 XML 的程序，必须通过能识别 XML 语法的分析器来实现。XML 语法分析器实际上就是一个能对 XML 文档进行语法分析的程序，它读入一个 XML 文件，确认这个文件具有正确的格式，然后将其分解成各种元素，使得程序员能够得到这些元素。用户开发的应用程序正是通过分析器的接口，实现对 XML 文档的识别和访问。如果不同的分析器各自定义不同的接口，就会给 XML 应用程序的开发带来很大的不便。为了使不同的 XML 应用程序可以方便地任意选择、更换合适的分析器，W3C 以及 XMLDEV 邮件列表的成员分别提出了两个标准的应用程序接口：文档对象模型(Document Object Model, DOM)和简单应用程序编程接口(Simple API for XML, SAX)。DOM 和 SAX 接口在 Web 应用开发中的作用如图 10.2 所示。

从图 10.2 中可以看出，应用程序不直接对 XML 文档进行读取操作，而是通过 DOM 或 SAX 接口，对 XML 分析器分析 XML 文档所得结果进行操作，从而间接地实现了对 XML 文档的访问。

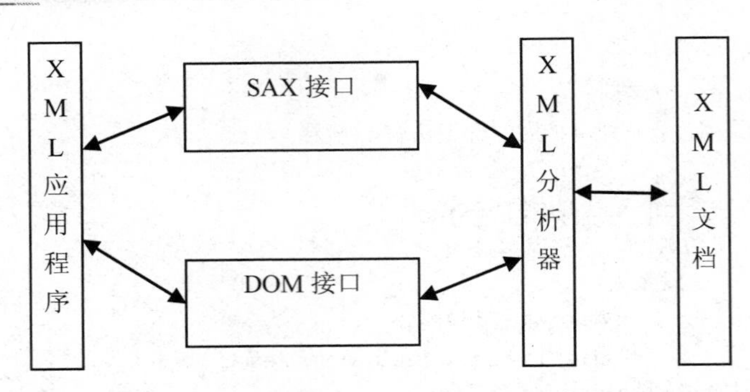

图 10.2　DOM 和 SAX 在应用程序开发中的位置

10.2.2　JAXP 简介

Sun 公司发布的 1.5 以后版本的 SDK 都提供了解析 XML 文件的 JAXP(Java API for XML Processing)库，其中也包含了自己的 DOM 解析器，实现了 W3C 标准化的 DOM 接口。JAXP 可以插入自己的 DOM 解析器，也可以和其他的 DOM 解析器配合。

在 DOM 接口规范中，有四个基本的接口，Document、Node、NodeList 以及 NamedNodeMap。在这四个基本接口中，Document 接口是对文档进行操作的入口，NodeList 接口是一个节点的集合，它包含了某个节点中的所有子节点，NamedNodeMap 接口也是一个节点的集合，通过该接口，可以建立节点名和节点之间的一一映射关系，从而利用节点名可以直接访问特定的节点。如图 10.3 所示 Node 接口是其他大多数接口的父类，像 Documet、Element、Attribute、Text、Comment 等接口都是从 Node 接口继承过来的。本节主要介绍 JAXP 如何使用 DOM 方式解析 XML。

1. 使用 DOM 解析器的步骤

在 JAXP 中，javax.xml.parsers 包中的 DocumentBuilder 类的一个实例就是 DOM 解析器，它由 javax.xml.parsers 包中的 DocumentBuilderFactory 类负责创建，创建的步骤如下。

(1) 调用 DocumentBuilderFactory 类的 newInstance()方法，得到该类的一个对象。

```
DocumentBuilderFactory factory=DocumentBuilderFactory.newInstance();
```

(2) 调用 DocumentBuilderFactory 对象 factory 的 newDocumentBuilder()方法得到一个 DocumentBuilder 对象，该对象称为 DOM 解析器。

```
DocumentBuilder builder=factory.newDocumentBuilder();
```

(3) 调用 DocumentBuilder 对象 builder 的 parse(File f)方法得到一个 Document 的一个实例，这个实例称为 Document 对象。Document 接口的定义在 org.w3c.dom 包中。

```
Document doc=builder.parse(new File("score.xml"));
```

2. Document 对象的结构

Document 对象是 XML 文档的树形结构在内存中的表现，它由实现 Node 接口及其多个子接口的类的对象构成，这些实例又称为 Document 对象中的节点。XML 文件中的标记都和 Document 对象中的节点相对应。XML 文件中的根标记和 Document 对象中的根节点相对应。

Node 接口是整个 DOM 接口中最重要的一个接口，因为 DOM 接口中很大一部分都是从 Node 接口继承而来的。Node 代表了 Document 对象节点树中的一个节点。Node 与子接口之间的关系如图 10.3 所示。

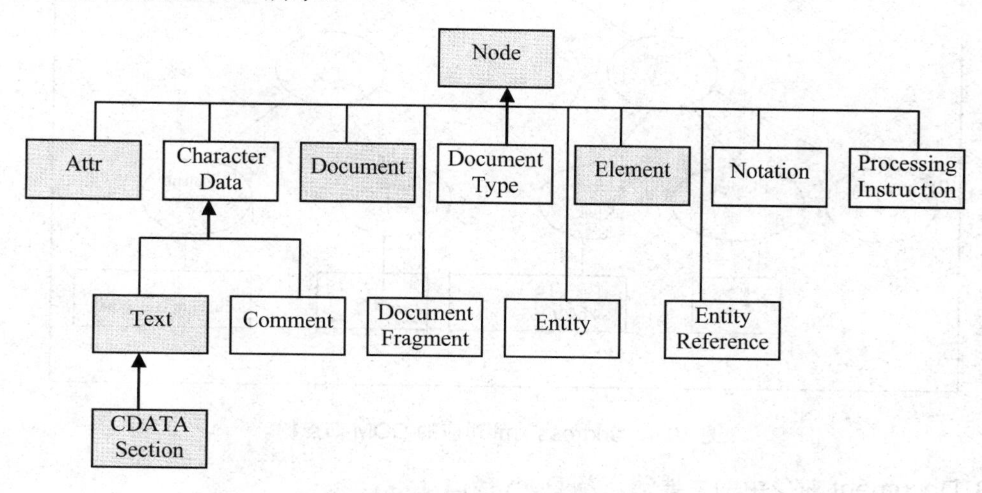

图 10.3　Node 接口及其子接口

Element 类、Text 类和 CDATASection 类都是实现了 Node 接口的类，这些类的对象分别称为 Document 对象中的 Element 节点、Text 节点和 CDATASection 节点。Document 对象调用 getDocumentElement()方法返回 Document 对象中的根节点，根节点的类型为 Element 类型，是 Element 类的一个实例。一个 Element 节点中还可以包含有 Element 节点、Text 节点和 CDATASection 节点。这些节点就形成了一个 Document 对象节点树。下面给出一个 XML 文档及与其对应的 Document 对象。

【例 10.3】描述地址簿的 XML 文件及其对应的 Document 对象(其代码参见 address.xml)。

address.xml 文件中的代码如下：

```
<?xml version="1.0" encoding="UTF-8" ?>
<地址簿>
    <朋友 ID="1001" SEX="男">
        <姓名>张三</姓名>
        <email>zhs@163.com</email>
        <![CDATA[年龄>18 岁
        ]]>
    </朋友>
    <朋友 ID="1002" SEX="女">
        <姓名>李梅</姓名>
        <email>limei@163.com</email>
    </朋友>
</地址簿>
```

用 DOM 分析器得到的 address.xml 文档的 Document 对象如图 10.4 所示。

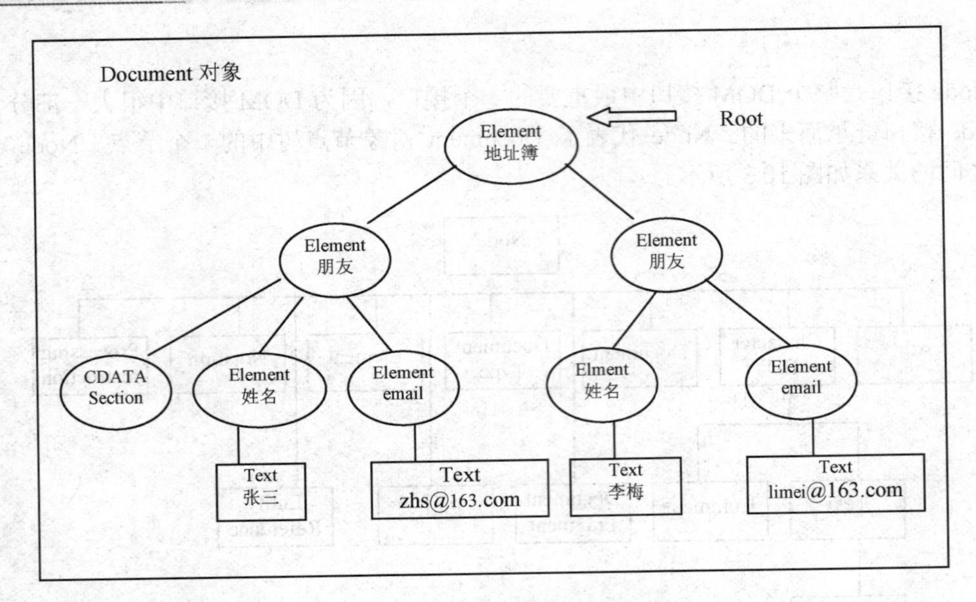

图 10.4　address.xml 对应的 DOM 节点树

3. Document 对象中的主要节点及 Dom 接口

1) Node 节点(对象)

Node 接口是整个 DOM 接口中最重要的一个接口，因为 DOM 接口中很大一部分都是从 Node 接口继承而来的。Node 代表了 Document 对象树中的一个节点。Node 接口提供了访问 Document 对象树中元素内容与属性的途径，并对 Document 对象树中的元素遍历提供了支持，如图 10.5 所示。

Node 节点常用方法如下。

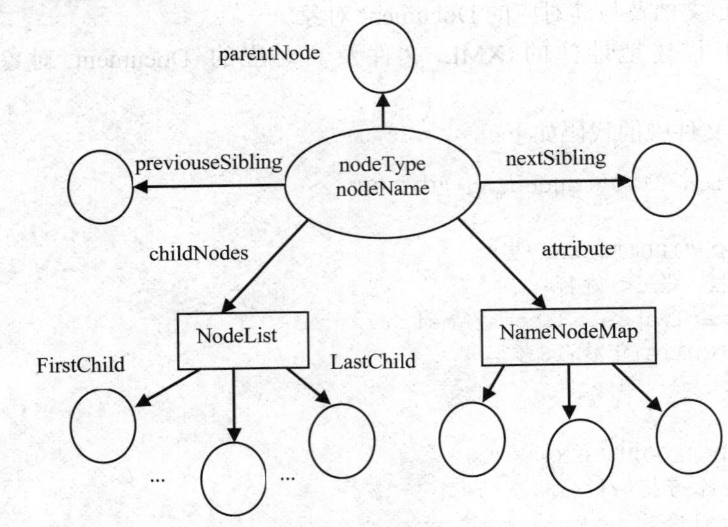

图 10.5　Node 接口

- NodeList getChildNodes()方法返回一个 NodeList 对象，该对象由当前节点的子节点组成。

- NodeList getElementByTagName(String name)方法返回一个 NodeList 对象，该对象由当前节点的 Element 类型子孙节点组成。name 为子孙节点的名字。
- String getTextContent()方法返回当前节点所有 Text 子孙节点中的文本内容。
- boolean hasAttribute(String name)方法判断当前节点是否有名字为 name 的属性。
- Node getFirstChild()方法获得该节点的第一个子节点。
- Node getLastChild()方法获得该节点的最后一个子节点。
- Node getNextSibling()方法获取该节点的下一个兄弟节点。
- Node getPreviousSibling()方法获取该节点的上一个兄弟节点。
- Node getParentNode()方法获取该节点的父节点。
- NamedNodeMap getAttributes()方法获取该节点所有属性的 Attr 节点的映射表。
- String getNodeName()方法返回节点的名字。当该节点是 Attr 节点时，该名字就是属性名。
- String getNodeValue()方法返回节点的值。当该节点是 Attr 节点时，返回属性值。

2) Element 节点(对象)

Element 接口是 org.w3c.dom.Node 接口的子接口，它继承了 Node 接口的方法，获取和该节点有关信息的常用方法如下。

- String getTagName()方法可以返回元素的标签名。例如：root.getName()返回字符串“地址簿”。
- String getAttribute(String name)返回该节点中参数 name 指定的属性值。

3) CharacterData 节点

org.w3c.dom.CharacterData 是 Node 的子接口，CharacterData 节点常用的方法如下。

String getData()方法获取存储在该节点的文本。

4) Text 节点

Text 接口是 CharacterData 接口的子接口，如图 10.3 所示，Text 接口的常用方法如下。

getWholeText()方法获取节点中的文本，包括空白字符。

5) CDATAsection 节点

CDATASection 接口是 Text 接口的子接口，其接口的常用方法如下。

getWholeText 方法获取节点中的文本，即 CDATA 段中的文本，包括其中的空白字符。

6) org.w3c.dom.NodeList 接口

NodeList 接口是一个集合类型，组成这个集合的元素为 Node 对象，其常用方法如下。

- int getLength()方法返回列表中的节点数。
- Node item(int index)方法返回给定节点索引号的节点。索引范围在 0 到 getLength()-1 之间。

7) org.w3c.dom.NamedNodeMap 接口

NamedNodeMap 接口也是一个集合类型，组成这个集合的元素为专门描述属性的 Node 对象。调用元素的 getAttribute()方法就会得到一个 NamedNodeMap 对象。注意 NamedNodeMap 不是继承自 NodeList。NamedNodeMap 常用方法如下。

- int getlength()方法返回该节点映射表中的节点数。

- Node item(int index)方法返回给定索引号的节点。索引范围在 0 到 getLength()-1 之间。

4. 分析文档内容

当用户得到了 Document 对象之后就可以调用它的 getDocumentElement()方法来分析文档的内容了。Document 对象的 getDocumentElement()方法返回根元素。其代码如下：

```
Element root=doc.getDocumentElement();
```

例如用户要处理 address.xml 文件，调用该函数后返回根元素(地址簿元素)。获得根元素之后就可以处理根元素包含的所有子节点了，这些子节点可以是子元素、文本、注释或其他节点。getTagName()可以返回元素的标签名。在前面这个例子中返回字符串“地址簿”。如果要得到这个元素的子元素，包括子元素、文本、注释或其他节点，可以使用 getChildNodes()方法。这个方法返回一个类型为 NodeList 的集合。对于 NodeList 集合，getLength()方法得到它的项的总数，item 方法将得到指定索引号的项。用户可以采用下面的方法获得子元素：

```
NodeList children=root.getChildNodes();
for(int i=0;i<children.getlength();i++)
{
   Node child=children.item(i);
   if(child instanceof Element)
   {
      Element childElement=(Element)child;
      ...
   }
}
```

上面的示例代码在读取 address.xml 文件时会得到两个子元素，它们的名字都是“朋友”。如果用户分析“朋友”元素的子元素“姓名”和“email”元素时，想检索到它们所包含的文本字符串，则需要获得包含这些文本字符串的 Text 类型的子节点，如图 10.4 所示。这些 Text 类型的子节点是“姓名”和“email”元素唯一的子节点，所以可以用 getFirstChild()方法得到这个节点，进而调用 getData()方法得到存储在 Text 节点中的字符串，trim()函数能够将字符串中的空白字符删除掉，示例代码如下：

```
for(int j=1;j<childNodes.getLength();j++)
{
    childNode=childNodes.item(j);
    if(childNode instanceof Element){
        element=(Element)childNode;
        //得到 text 节点
        Text textNode=(Text)element.getFirstChild();
        String str=textNode.getData().trim();
        out.print("<br>-"+element.getTagName()+":"+
                  str);
    }
}
```

如果要枚举节点的属性，可以调用元素的 getAttributes()方法。该方法返回一个 NamedNodeMap 对象，其中包含了描述节点属性的节点对象。可以用遍历 NodeList 的方法遍历 NamedNodeMap。调用 getNodeName()方法可以得到属性节点的名称，调用 getNodeValue()可以得到属性值。示例代码如下。

```
NamedNodeMap namemap=element.getAttributes();
//遍历所有的属性
for(int k=0;k<namemap.getLength();k++)
{    //输出属性名称
     out.print(namemap.item(k).getNodeName()+"=\"");
    //输出属性值
     out.print(namemap.item(k).getNodeValue()+"" ");
}
```

10.2.3　使用 DOM 解析器读取 XML 文件示例

从 XML 文件中解析出所需要的数据是 DOM 编程的基本技术，下面是一个范例。本章的 Web 服务目录为 ch10，创建 Web 虚拟服务目录方法请参见第 1 章的 1.3 节。

【例 10.4】 使用 JSP 读取 address.xml 文件，并显示该文件的内容。页面显示效果如图 10.6 所示。将 address.xml 文件存储在 E:\programJsp\ch10 文件夹中，文件内容参见例 10.3。

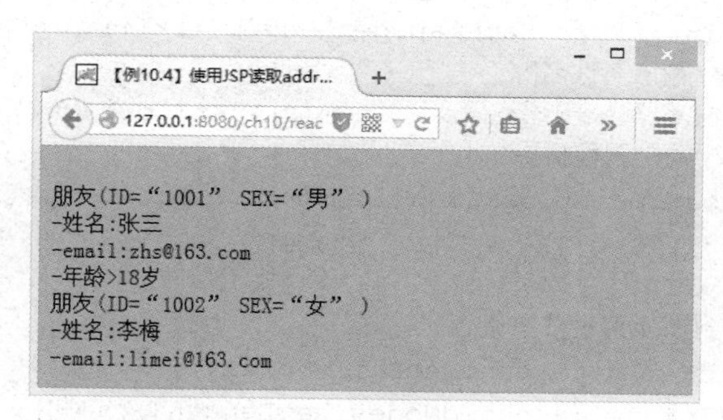

图 10.6　readAddrXml.jsp 页面效果

readAddrXml.jsp 文件中的代码如下：

```
<%@ page contentType="text/html;charset=GB2312" %>
<%@ page import="javax.xml.parsers.DocumentBuilder" %>
<%@ page import="javax.xml.parsers.DocumentBuilderFactory" %>
<%@ page import="org.w3c.dom.*" %>
<%@ page import="java.io.*" %>
<html>
<body bgcolor="cyan">
<%
try{
    //得到 Document 对象
    DocumentBuilderFactory factory=DocumentBuilderFactory.newInstance();
```

```
DocumentBuilder builder=factory.newDocumentBuilder();
Document doc=builder.parse(new
    File("e:\programJsp\ch10\address.xml"));
//得到根元素
Element root=doc.getDocumentElement();
//得到根元素的子节点的集合
NodeList persons=root.getChildNodes();
Node person;//根元素的子节点
NodeList childNodes;//子节点的子节点集合
Node childNode;//子节点的子节点
Element element;//元素节点
//遍历根节点的所有子节点
for(int i=0;i<persons.getLength();i++)
{
    person=persons.item(i);
    //如果是元素节点
    if(person instanceof Element)
    {
        element=(Element)person;//得到元素节点
        out.print("<br>"+element.getTagName()+"(");
        //得到该元素的属性集合
        NamedNodeMap namemap=element.getAttributes();
        //遍历所有的属性
        for(int k=0;k<namemap.getLength();k++)
        {   //输出属性名称
            out.print(namemap.item(k).getNodeName()+"=\"");
            //输出属性值
            out.print(namemap.item(k).getNodeValue()+"\" ");
        }
        out.print(")");
        //得到该元素子节点集合
        childNodes=person.getChildNodes();
        for(int j=1;j<childNodes.getLength();j++)
        {
            childNode=childNodes.item(j);
            if(childNode instanceof Element){
                element=(Element)childNode;
                //得到 text 节点
                Text textNode=(Text)element.getFirstChild();
                String str=textNode.getData().trim();
                out.print("<br>-"+element.getTagName()+":"+
                        str);
            }
            //如果是 CDATASection 节点输出节点文本
            if(childNode.getNodeType()==Node.CDATA_SECTION_NODE){
                CDATASection cdatas=(CDATASection)childNode;
                String str=cdatas.getData().trim();
```

```
                        out.print("<br>-"+str);
                    }
                }
            }
        }
    }
    catch(Exception e)
    {
        out.println(e);
    }
    %>
    </body>
    </html>
```

提示：使用 DOM 接口，应用程序可以随时访问文档中任何一部分数据，NodeList 集合中的节点都是 live 的，对各节点的操作都会反映到 XML 文件中，因此 DOM 接口访问 XML 文件又称为随机访问模式。

10.3 SAX 解析器

DOM 解析器读入的是一个完整的 XML 文档，并将其转换成一个树形结构的文档对象。对于大多数应用来讲 DOM 分析器很适用。但是，当 XML 文档很大时，DOM 方式会占据较多的内存，如果操作的只是部分节点，将文档的所有内容读入内存可能就显得低效了，这种情况就可以使用 SAX 解析器。

10.3.1 什么是 SAX 解析器

SAX(Simple API for XML)的中文意思是 XML 简单应用程序接口。相对于 DOM 接口来讲，SAX 接口规范是 XML 分析器提供的更底层的接口。它能够给应用程序提供更大的灵活性。SAX 解析器是一种事件驱动机制的解析器。它的基本原理是由接口的用户提供符合定义的事件处理器，SAX 解析器在分析 XML 时遇到特定的事件，就去调用事件处理器中特定的处理函数。SAX 解析器的大体框架如图 10.7 所示，最上方的 SAXParserFactory 用来生成一个解析器实例(SAXParser 实例)。XML 文档从左侧箭头读入，当解析器对文档进行分析时，根据文档内容产生相应的事件，并报告这个事件给相应的事件处理器，也就是触发在 DocumentHandler、ErrorHandler、DTDHandler 及 EntityResolver 接口中定义的方法。

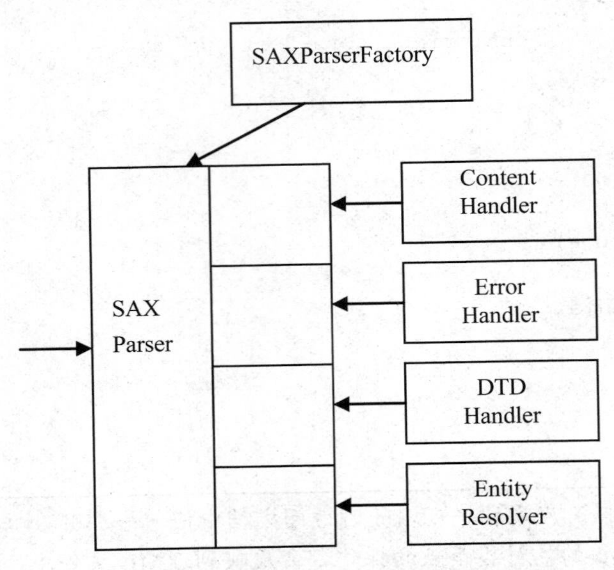

图 10.7　SAX 接口工作示意图

10.3.2　SAX 的常用接口

Sun 公司的 JDK1.5 以后版本提供的 Java API for XML Processing 中，都包含 SAX 解析器及 Simple API for XML 接口。SAX 的主要接口和类说明如下。

1. SAXParserFaetory 类

javax.xml.parsers 包中的 SAXParserFactory 类调用 newInstance()方法可以得到一个 SAXParserFactory 对象，该对象的作用是调用 newSAXParser()方法，按照系统属性中的定义创建一个 SAX 分析器的实例。

2. SAXParser 接口

org.xml.sax.Parser 接口定义了类似于 setDocumentHandler 的方法来创建事件处理函数。另外，该接口中还定义了 parser()方法来对 XML 文档进行实际的分析工作，并把解析事件报告给指定的处理器。parser 函数的几种形式：

```
void parse(File f,DefaultHandler handler)
void parse(String url,DefaultHandler handler)
void parse(InputStream in,DefaultHandler)
```

3. ContentHandler 接口

org.xml.sax 包中的 ContentHandler 接口，当分析器遇到 XML 文档中的起始或结束标记时，就会激活该接口中的 startDocument、endDocument 方法。当分析器遇到元素的起始或结束标志时，激活该接口中的 startElement、endElement 等方法，其语法格式如下：

```
void startElement(String uri,String lname,String qname,Attributes attr)
void endElement(String uri,String lname,String qname)
```

其中的参数说明如下。

- 参数 uri 为名字空间的 URI。
- 参数 lname 不带别名前缀的本地名。
- 参数 qname 为元素名。

另外，charaeters 方法也是在 ContentHandler 接口中实现的。当分析器遇到元素内部的文本内容就会激活 charactcrs 方法。characters 方法的语法格式如下：

```
void characters(char[] data,int start ,int length)
```

其中的参数说明如下。

- 参数 data 为字符数组。
- 参数 start 表示字符数组的子串中第一个字符的索引。
- 参数 length 表示的字符串的长度。

一个典型的 SAX 应用程序至少要提供一个 ContentHandler 方法。

4. ErrorHandler 接口

org.xml.sax 包中的 ErrorHandler 接口，当分析器在分析过程中遇到不同的错误时，ErrorHandler 接口中的 error，fatalError 或者 warning 方法就会被激活。

5. DTDHandler 接口

org.xml.sax 包中的 DTDHandler 接口，当处理 DTD 中的定义时，就会调用该接口中的方法。

6. EntityResolver 接口

org.xml.sax 包中的 EntityResolver 接口，当分析器要识别由 URI 定义的数据时，就会调用该接口中的 resolveEniity 方法。

7. DefaultHandler 类

DefaultHandler 类是 org.xml.sax.helpers 包中的类，该类或其子类的对象称为 XML 解析器的事件处理器。DefaultHandler 类实现了 ContentHandler、DTDHandler、EntityResolver 和 ErrorHandler 四个接口，并为它们定义了事件处理器根据相应事件调用的方法。例如当解析器发现了文档中一个元素开始标记时，就将所发现的数据封装成一个为“标记开始事件”，并报告该事件给处理器，事件处理器就会知道所发生的事件，然后调用 startElement 方法对发现的数据做出处理。

8. Attributes

Attributes 是 org.xml.sax 包中的属性集合，集合中的元素是文档中元素的属性。它提供了遍历集合的方法，下面是其常用方法。

- int getLength()返回存储在该属性集合中的属性数量。
- String getLocalName(int index)返回给定索引的属性的本地名，不支持本地名时返回空串。
- String getURI(int index)返回给定索引的属性的名字空间 URI，不支持名字空间时返回空字符串。

- String getVaule(int index)根据给定索引返回属性值。

10.3.3 使用 SAX 解析器读取文档内容

用户要想使用 SAX 解析器读取 XML 文档内容，首先要在程序中得到一个解释器的实例，也就是 SAXParser 类的一个实例，该实例由 SAXParserFactory 类负责创建，示例代码如下：

```
SAXParserFactory factory=SAXParserFactory.newInstance();
SAXParser saxParser=factory.newSAXParser();
```

有了 SAXParser 对象，用户就可以处理文档了：

```
saxParser.parse(source,handler);
```

这里的 source 可以是一个文件、一个 URL 字符串或者一个输入流，handler 是一个事件处理器。当用户得到一个 SAX 解析器后，还需要一个处理器为不同的解析器事件定义事件处理动作，这个事件处理器就是 DefaultHandler 类的一个子类。在这个 DefaultHandler 类的子类中，用户可以根据需要重写子类的 startElement、endElement、characters、startDocument 和 endDocument 事件处理方法。处理器必须覆盖这些方法，让它们执行用户在解析文件时想要执行的动作。

下面是一个简单的 XML 文件中的片段，解析器会产生以下事件，回调处理器中的方法。

XML 文档中的一个片段：

```
<font>
    <name>TimesRoma</name>
    <size units="pt">36</size>
</font>
```

解析器确保产生以下回调。

(1) startElement，元素名 font。
(2) startElement，元素名 name。
(3) characters，内容 TimesRoma。
(4) endElement，元素名 name。
(5) startElement，元素名 size，属性：untis="pt"。
(6) characters，内容 36。
(7) endElement，元素名 size。
(8) endElement，元素名 font。

下面的示例代码，处理器覆盖了 ContentHandler 接口中的 startElement 方法来查看带有 href 属性的 a 元素。

```
class MyHandler extends DefaultHandler{
   StringBuffer strbuffer;
```

```
    MyHandler(StringBuffer mess){
        strbuffer=mess;
    }
    public void startElement(String namespaceURL,
              String lname,String qname,Attributes attrs){
        if("a".equalsIgnoreCase(lname)&&attrs!=null){
            for(int i=0;i<attrs.getLength();i++){
                String aname=attrs.getLocalName(i);
                if("href".equalsIgnoreCase(aname))
                    strbuffer.append("<br>"+attrs.getValue(i));
            }
        }
    }
}
```

下面通过一个完整的例子，演示 SAX 解析器读取 XML 文档内容的步骤和方法。

【例 10.5】编写一个 JSP 页面，该页面能够显示 W3C 网站主页面(http://www.w3.org)中所有的<a href=" ...">超级链接。页面显示效果如图 10.8 所示。

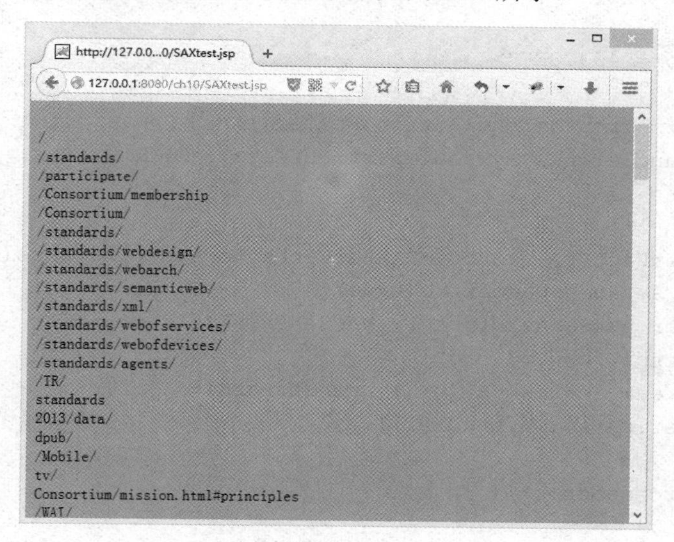

图 10.8　SAXtest.jsp 的页面效果

该例子实现了“网络爬行者”一些功能，即一个沿着链接到达越来越多网页的程序。W3C 的网页是使用 XHTML 语言编写的，同时也是一个结构良好的 XML。该示例程序是一个使用 SAX 解析器的典型例子。SAXtest.jsp 页面显示效果如图 10.9 所示。

SAXtest.jsp 文件中的代码如下：

```
<%@ page contentType="text/html;charset=GB2312" %>
<%@ page import="java.net.*" %>
<%@ page import="org.xml.sax.*" %>
<%@ page import="org.xml.sax.helpers.*" %>
<%@ page import="javax.xml.parsers.*" %>
<%@ page import="java.io.*" %>
```

```
<%!
    class MyHandler extends DefaultHandler{
        StringBuffer strbuffer;
        MyHandler(StringBuffer mess){
            strbuffer=mess;
        }
        public void startElement(String namespaceURL,
                    String lname,String qname,Attributes attrs){
                if("a".equalsIgnoreCase(lname)&&attrs!=null){
                    for(int i=0;i<attrs.getLength();i++){
                        String aname=attrs.getLocalName(i);
                        if("href".equalsIgnoreCase(aname))
                            strbuffer.append("<br>"+attrs.getValue(i));
                    }
                }
        }
    }
%>
<html>
<body bgcolor="#33CCFF">
<%
StringBuffer strbuffer=new StringBuffer();
MyHandler handler=new MyHandler(strbuffer);
try{
    String url;
    SAXParserFactory factory=SAXParserFactory.newInstance();
    factory.setNamespaceAware(true);
    SAXParser saxParser=factory.newSAXParser();
    url=" http://www.w3.org";
    InputStream in=new URL(url).openStream();
    saxParser.parse(in,handler);
    }
    catch(Exception e){}
%>
<%=strbuffer %>
</body>
</html>
```

提示： DOM 解析器读入 XML 文件在内存中建立一棵“树”，XML 文档中的标记、标记的内容都会和内存“树”中的节点对应。通过 DOM 接口，应用程序可以随时访问 XML 文档中的任意部分，是一种随机访问模式。SAX 解析器根据从文件中解析出的数据产生相应的事件，并报告这个事件给事件处理器，事件处理器就会处理所发现的事件。SAX 解析器是顺序访问的，占用内存较少，效率较高。DOM 解析器和 SAX 解析器各有优势，它们适合不同的场合并将长期共存，用户需要根据具体情况选择解析器。

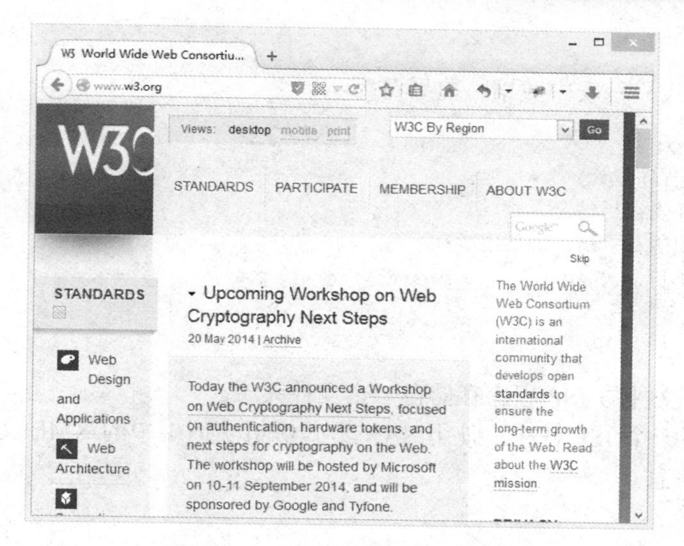

图 10.9 W3C 的主页内容

10.4 上 机 实 训

实训目的

- 理解 XML 的概念和特性。
- 掌握编写和测试 XML 文件的方法和注意事项(字符集类型)。
- 掌握 DOM 解析器读取 XML 文件的编程方法。
- 掌握 SAX 解析器读取 XML 文件的编程方法。

实训内容

实训 1 编写一个 XML 文件(book.xml)，用来表示下面表格中的数据，要求表 10.1 中的记录信息都以属性方法存储。

表 10.1 图书信息

书 名	作 者	出版社	价 格
JSP 程序设计	杨学全	清华大学出版社	35
JSP 应用开发技术	贾素玲	清华大学出版社	27
XML 教程	陈美霖	清华大学出版社	36

要求：

(1) 根标记为<booksinfo>，每本书的标记为<book>。

(2) 每本书的书名、作者、出版社、价格均作为属性处理。

实训 2 编写一个 JSP 程序，使用 DOM 方式解析实训 1 中的 XML 文件，在浏览器中显示所有图书的信息。

要求：

(1) 按缩格形式显示表中数据，前导符为--。

```
booksinfo:
    --JSP 程序设计
        --作者：杨学全
        --出版社：清华大学出版社
        --单价：35 元
 ...
booksinfo
```

(2) book.xml 文件与 JSP 程序存储在一个文件夹中。

实训 3　编写一个 JSP 程序，使用 SAX 方式解析实训 1 中的 XML 文件，在浏览器中显示所有图书的信息。

要求同实训 2。

实训总结

通过本章的上机实训，学员应该能够理解 XML 的概念和特点；掌握 XML 文件的编写、保存和测试方法；掌握 DOM 解析 XML 文件的编程技术；掌握 SAX 解析 XML 文件的编程技术。

10.5 本 章 习 题

思考题

(1) 试写出 HTML 常用的表格元素(标记)及其意义。

(2) 为 XML 文档设置属性时应注意哪些方面？

(3) XML 注释的作用是什么？如何为一个 XML 文档添加注释？

(4) XML 有哪些优点和用途？

(5) DOM 是如何实现对 XML 文档读取的？

(6) SAX 是如何实现对 XML 文档读取的？

拓展实践题

完善例 10.5，将链接的显示名称和链接地址 href 一起显示出来。

例如：

```
<a href="/Consortium/membership">about W3C Membership</a>
```

显示为：

```
Consolidation/membership  about W3C Membership
```

第 11 章　网上报名系统开发案例

学习目的与要求：

本章通过一个网上报名系统开发案例，示范了 Web 应用项目的需求分析、系统设计、数据库设计与实现等内容，演示了三层架构设计与实现，详细介绍了网上报名系统各模块的模型、视图和控制器的实现代码。通过本章的学习，读者要理解并掌握 MVC 模式，掌握实体层、业务逻辑层和数据库访问层的实现技术，掌握 JSP 在 Web 应用开发中的实用编程技术。本章案例项目开发平台：开发工具为 Eclipse，版本为 eclipse-jee-kepler-SR2，JSP 引擎为 Tomcat 7.0，JDK 版本为 jdk1.8.0_05，数据库为 MYSQL 5.6。

11.1　网上报名系统设计

通过互联网完成考试报名，可以方便考生报考，有效地提高报名信息的准确性和实效性。“职称计算机考试网上报名系统”项目实例是根据某省职称计算机报名考试系统需求而开发的项目实例。为了便于学习，本教材对其进行了适当精练。该实例采用 MVC 模式示范 JSP+JavaBean+Servlet 开发技术，展示数据增、删、改、查等基本操作的编程技术，为后期基于框架技术 Struts2+Hibernate+Spring 开发奠定理论和技术基础。

11.1.1　需求分析

在信息高速发展的当今年代，以建立高效的电子政务为核心的行政管理信息化势不可当。作为考试管理部门行政信息化的一个成功应用——基于互联网的网上报名系统为考生报名提供了方便。例如，全国执业医师资格考试、职称计算机考试、高校自主招生考试、全国研究生考试等已实现网上报名。

网上报名系统能够方便考生的报考，节省考生报名开支，提高考试管理部门的工作效率。职称计算机网上报名系统，是收集、管理考生信息的一个平台，主要包括考生报名、登录、信息修改、上传照片、浏览信息、注销等功能。网上报名系统完成的主要任务如下。

(1) 考生报名：考生填写报名表，包括考生身份证号、考生姓名、工作单位、单位地址、联系电话、报考类别、E-mail 地址、备注等信息。系统提供对填写信息的有效性检查。

(2) 考生登录：完成已报名考生身份的验证任务。通过验证的考生才能有权限使用上传照片、浏览信息、修改密码、修改信息、注销考试功能。

(3) 上传照片：登录后上传考生照片文件，实现照片与考生身份证号的一一对应。

(4) 浏览信息：考生登录后，可浏览自己的详细信息和其他考生的报考信息，但不能查看其他考生的密码。

(5) 修改密码：考生成功登录后可以修改自己的密码。

(6) 修改信息：考生登录后，考生可对除身份证号、密码以外的个人信息进行修改。

(7) 注销考试：成功登录的考生可以注销考试，也就是取消这次考试报名。

根据以上任务，绘制网上考试报名系统的 Use Case 图，如图 11.1 所示。

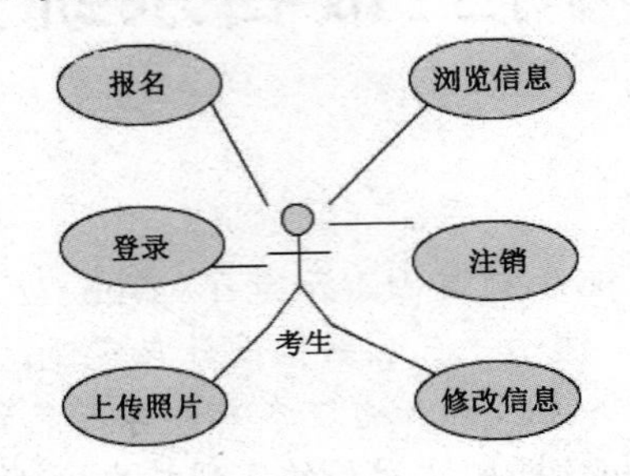

图 11.1 网上报名系统 Use Case 图

11.1.2 总体设计

根据网上考试报名系统的需求分析，可以将网上报名系统划分为 9 个模块，考生报名、考生登录、上传照片、浏览信息、修改密码、修改信息、注销考试、退出登录、返回主页等模块。网上报名系统 UML 活动图如图 11.2 所示。

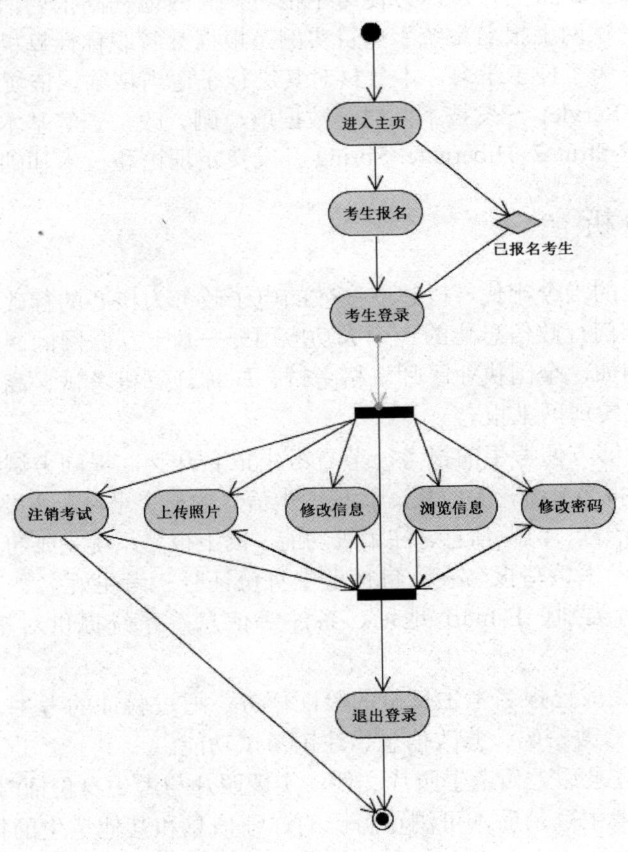

图 11.2 网上报名系统 UML 活动图

网上报名系统上传照片、修改信息、浏览信息、修改密码模块是平行的，考生登录以后可以进行这些模块的操作。注销报考后，自动退出报名系统，失去对平行模块的操作权限。

11.1.3 功能设计与系统组成

1. 模块功能设计

(1) 考生报名模块：考生填写报名表，包括考生身份证号、考生姓名、工作单位、单位地址、联系电话、报考类别、E-mail 地址、备注等信息。系统对身份证号、密码、电子邮件地址进行有效性检查，对没有通过有效性检查的数据，提示用户更正。对考生性别和报考级别限定考生输入，性别限定为“男”或“女”，报考级别限定为“高级”或“中级”。

(2) 考生登录模块：输入考生身份证号、密码登录系统。身份证号和密码正确，则进入系统，可进行浏览信息、上传照片、修改报考信息等操作。如果用户输入的身份证号或密码错误，系统将显示错误信息。

(3) 上传照片模块：成功登录的考生可以使用该模块上传考生照片。如果上传成功，系统将显示上传文件名称、服务器端的文件保存名称，并在页面上显示上传照片。如果失败，则显示错误信息。如果考生未登录，系统将跳转到考生登录页面。

(4) 浏览信息模块：成功登录的考生可以使用该模块浏览自己的报名信息，也可以按身份证号浏览其他考生的详细信息，还可以分页浏览所有考生的部分信息(其他考生的密码信息不显示)。如果用户未登录直接进入该页面，系统则跳转到考生登录页面。

(5) 修改密码模块：成功登录的用户，可以使用该模块修改自己的密码。修改密码必修输入原密码，原密码与新密码不能相同，新密码不能为空。如果用户未登录直接进入该模块，系统将跳转到登录页面。

(6) 修改信息模块：成功登录的考生可以使用该模块修改报名信息，例如考生姓名、性别、工作单位等信息，但不能修改考生身份证号。系统对电子邮件等信息进行有效性检查。如果用户未登录直接使用该模块，系统将跳转到登录页面。

(7) 注销考试模块：成功登录的考生可以使用该模块注销报名信息。执行此操作需要用户再次输入登录密码，系统验证密码通过后将彻底删除考生报名信息，并显示操作结果。

(8) 退出登录模块：成功登录的用户可以使用该模块退出登录。

(9) 返回首页模块：使用该模块，用户可以随时返回主页面。

2. 系统组成

系统组成情况如表 11.1 所示，包括页面、对应的 JSP 文件及页面调用的 Servlet 类和 JavaBean 类的清单。用户在开发自己的 Web 应用系统时，要注意模块名称要符合命名规范，可以使用英文，也可以使用汉语拼音，不管用英文还是汉语拼音都要尽量简明知意，一般名称包含类型和功能的含义。

表 11.1　系统组成情况

页面名称	页面 JSP 文件	JavaBean	Servlet
系统主页	index.jsp	无	无
考生报名	register.jsp showRegisterMess.jsp	registerBean	HandleRegister
登录页面	login.jsp showLoginMess.jsp	loginBean	HandleLogin
上传照片	uploadpic.jsp showUploadMess.jsp	uploadFileBean	HandleUploadFile
浏览信息(分页)	chooseshowtype.jsp showExamineeAll.jsp	showExamineeBydPage	HandleShowExamineeByPage
浏览信息(单个)	showExamineeByID.jsp	showExamineeByIDBean	HandleShowExamineeByID
修改密码	modifypassword.jsp showModifyPassword.jsp	passwordBean	HandleModifyPwd
修改信息	modifymessage.jsp showModifyMessage.jsp	modifyMessageBean	handleGetExamineeMessage HandleModifyMessage
注销考试	unregister.jsp showUnRegister.jsp	unregisterBean	HandleUnRegister
退出登录	无	loginBean	HandleExit

11.2　数据库设计及实现

11.2.1　数据库设计

系统使用 MySQL 5.6 作为后台网络数据库，建立一个数据库 dataExam，dataExam 中有 examinee 表。报名信息存入 examinee 表中，examinee 表有如下字段。

- id：存储考生身份证号，主键，非空。
- name：存储考生姓名，姓名可以是汉字或字母。
- sex：存储考生性别，存储“男”或“女”。
- company：考生工作单位。
- address：单位地址。
- password：登录密码，密码可使用英文字母或数字。
- pic：存储照片文件名。
- phone：考生的联系电话。
- eamil：考生邮箱地址。
- examtype：考生考试类别。
- memo：考生备注信息。

examinee 表的详细信息如表 11.2 所示。

表 11.2　examinee 表结构

字段名	字段含义	字段类型	主键/空否	示例数据
id	身份证号	char(20)	主键	130603198902021218
name	考生姓名	char(20)	允空	杨里海
sex	考生性别	char(4)	允空	男
company	工作单位	varchar(50)	允空	□□大学
address	单位地址	varchar(50)	允空	北京市海淀区
phone	联系电话	varchar(50)	允空	010-88765678
email	邮件地址	varchar(50)	允空	yanglh@163.com
password	用户密码	char(20)	允空	yang
memo	备注	varchar(100)	允空	高校教学人员
examtype	报考类别	varchar(50)	允空	高级
pic	照片	varchar(50)	允空	130603198902021218.JPG

11.2.2　数据库实现

1. 创建 examinee 表的 SQL 语句

根据前面网上报名系统数据库设计，读者可以使用 MYSQL 的图形界面创建库和表，或使用命令行工具编写 SQL 脚本来创建数据库及表。下面给出在命令行工具中创建数据库和表的步骤及 SQL 语句。

(1) 创建 dataExam 数据库：

```
mysql->create database dataExam;
```

(2) 切换到 dataExam 数据库：

```
mysql->use dataExam
```

(3) 创建 examinee 表：

```
CREATE TABLE 'examinee' (
   'id' char(20) NOT NULL,
   'name' char(20) default NULL,
  'sex' char(4) default NULL,
  'company' varchar(50) default NULL,
  'address' varchar(50) default NULL,
  'phone' varchar(50) default NULL,
  'email' varchar(50) default NULL,
  'password' char(20) default NULL,
  'memo' varchar(100) default NULL,
  'examtype' varchar(50) default NULL,
  'pic' varchar(50) default NULL,
```

```
  PRIMARY KEY  ('id')
) ENGINE=InnoDB DEFAULT CHARSET=utf8;
```

2. 插入示例数据

插入示例数据的 SQL 语句如下：

```
insert into examinee ( id, name, sex, company, address, phone, email,
password, memo, examtype, pic) value ("130603198902021218","杨里海","男",
"□□大学","北京市海淀区","010-88765678"," yanglh@163.com "," yang","高校教
学人员","高级","130603198902021218.JPG");
```

11.3 网上报名系统配置

系统在开发之前必须设计好系统的目录结构、页面组成和页面布局、JavaBean 与 Servlet 和配置文件，下面介绍网上报名系统的配置。

11.3.1 系统文件目录结构

本教材案例是在 Eclipse 开发工具下完成的。新建工程的物理路径为 E:\programjsp\ch11\。

系统使用的目录结构如图 11.3 所示。

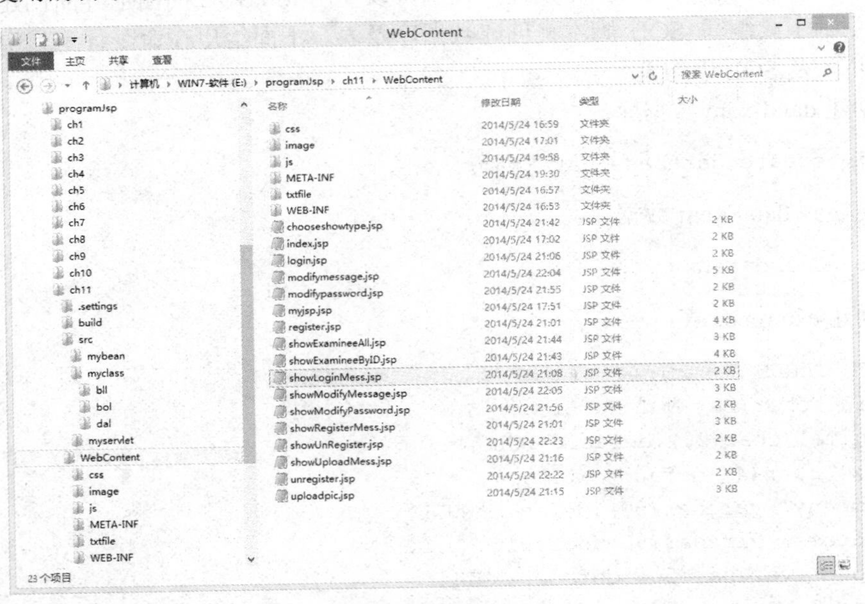

图 11.3 系统文件目录结构

提示：使用 Eclipse 工具进行开发，在 src 目录中创建包时，生成的 class 文件会首先放到 build 文件夹下，在对网站进行发布时，Eclipse 会自动将包中的 class 文件复制到 WebContent\WEB-INF\classes 相应的文件夹中。

11.3.2 主页面管理

本系统的 JSP 页面全部保存在 Eclipse 所建工程的 WebContent 目录中。所有页面的最上面是一个标题和广告，左侧是一列导航链接，下面是一个页脚。导航链接由考生报名、考生登录、上传照片、浏览信息、修改密码、修改信息、注销考试、退出登录、返回主页组成。所有页面布局采用分层技术，使用<div></div>标签，将整个页面分为标题区(header)、左侧导航链接(sidebar1)、主内容区(mainContent)、页脚区(clearfloat)，为了便于维护页面，页面通过使用 JSP 的<%@ include file="……" %>标记将标题文件 head.txt、左侧导航链接文件 left.txt、页脚文件 footer.txt 嵌入其中。为了统一页面风格，页面使用了 css 样式表 mystylesheet.css，页面通过使用<link href="css/mystylesheet.css" rel="stylesheet" type="text/css" />应用定义的样式。

head.txt、left.txt、footer.txt 存储在 WebContent\txtfile 文件夹中。mystylesheet.css 文件存储在 WebContent\css 文件夹中。

1. head.txt 文件

head.txt 文件的内容如下：

```
<%@ page contentType="text/html;charset=GB2312" %>
<h1 align="center">XX 省职称计算机考试网上报名系统</h1>
```

2. left.txt 文件

left.txt 文件的内容如下：

```
<%@ page contentType="text/html;charset=GB2312" %>
<p> </p>
<h3>报名系统功能菜单</h3>
<table width="150" border="1" bordercolor="#EBEBEB">
<tr><th height="36" scope="row"><a href="register.jsp">考试报名
</a></th></tr>
<tr><th height="36" scope="row"><a href="login.jsp">考生登录
</a></th></tr>
<tr><th height="36" scope="row"><a href="uploadpic.jsp">上传照片
</a></th></tr>
<tr><th height="36" scope="row"><a href="chooseshowtype.jsp">浏览信息
</a></th></tr>
<tr><th height="36" scope="row"><a href="modifypassword.jsp">修改密码
</a></th></tr>
<tr><th height="36" scope="row"><a href="helpGetMessage">修改信息
</a></th></tr>
<tr><th height="36" scope="row"><a href="unregister.jsp">注销考试
</a></th></tr>
<tr><th height="36" scope="row"><a href="helplogout">退出登录
</a></th></tr>
<tr>
<th height="36" scope="row"><a href="index.jsp" target="_self">返回首页
```

```
</a></th>
</tr>
</table>
<p> </p>
```

3. footer.txt 文件

footer.txt 文件内容如下：

```
<%@ page contentType="text/html;charset=GB2312" %>
<div id="footer">
<p align="center">版权所有《JSP 编程技术》编写组，2014 年 5 月</p>
</div>
```

4. mystylesheet.css 文件

mystylesheet.css 文件的内容如下：

```
@charset "gb2312";
body  {
    font: 100% 宋体, Arial,仿宋, 幼圆;
    background: #666666;
    margin: 0;
    padding: 0;
    text-align: center;
    color: #000000;
}
.twoColHybLtHdr #container {
    width: 80%;
    background: #FFFFFF;
    margin: 0 auto;
    border: 1px solid #000000;
    text-align: left;
}
.twoColHybLtHdr #header {
    background: #DDDDDD;
    padding: 0 10px;
}
.twoColHybLtHdr #header h1 {
    margin: 0;
    padding: 10px 0;
}
.twoColHybLtHdr #sidebar1 {
    float: left;
    width: 12em;
    background: #EBEBEB;
    padding: 15px 0;
}
.twoColHybLtHdr #sidebar1 p {
    margin-left: 10px;
```

```
        margin-right: 10px;
    }
    .twoColHybLtHdr #sidebar1 h3{
        margin-right: 10px;
    }
    .twoColHybLtHdr #mainContent {
        margin: 0 20px 0 13em;
    }
    .twoColHybLtHdr #footer {
        padding: 0 10px;
        background:#DDDDDD;
    }
    .twoColHybLtHdr #footer p {
        margin: 0;
        padding: 10px 0;
    }
    .fltrt {
        float: right;
        margin-left: 8px;
    }
    .fltlft {
        float: left;
        margin-right: 8px;
    }
    .clearfloat {
        clear:both;
        height:0;
        font-size: 1px;
        line-height: 0px;
    }
```

5. index.jsp 页面

主页面 index.jsp 由标题、左侧导航链接、欢迎语、欢迎图片 welcome.jpg、页脚组成，welcome.jpg 存储在 WebContent\image 文件夹中。index.jsp 文件内容如下：

```
<%@ page language="java" import="java.util.*" pageEncoding="GBK"%>
<%
String path = request.getContextPath();
String basePath = request.getScheme() + "://" + request.getServerName()+
":"+
request.getServerPort()+path+"/";
%>
<!DOCTYPE HTML PUBLIC "-//W3C//DTD HTML 4.01 Transitional//EN">
<html>
  <head>
    <base href="<%=basePath%>">
    <title>职称计算机考试报名</title>
```

```
    <meta http-equiv="pragma" content="no-cache">
    <meta http-equiv="cache-control" content="no-cache">
    <meta http-equiv="expires" content="0">  <!-- 下面 keywords 为搜索引擎而
准备 -->
    <meta http-equiv="keywords" content="网上报名,职称考试,计算机">
    <meta http-equiv="description" content="网上报名首页">
    <link href="css/mystylesheet.css" rel="stylesheet" type="text/css" />
  </head>
  <body class="twoColHybLtHdr"><!--为 css 样式类型选择器指定类名-->
    <div id="container">
      <div id="header">
         <%@ include file="txtfile/header.txt" %>
      </div>
      <div id="sidebar1">
         <%@ include file="txtfile/left.txt" %>
      </div>
      <div id="mainContent">
          <center>
              <p><font size=4 color=red>欢迎您来报名! </font> </p><br>
              <img src="image/welcome.JPG" width=450 height=350></img>
          </center>
      </div>
      <br class="clearfloat" />
      <%@ include file="txtfile/footer.txt" %>
    </div>
  </body>
</html>
```

启动 Eclipse 的运行工具，用户可以在浏览器中看到主页运行效果，如图 11.4 所示。

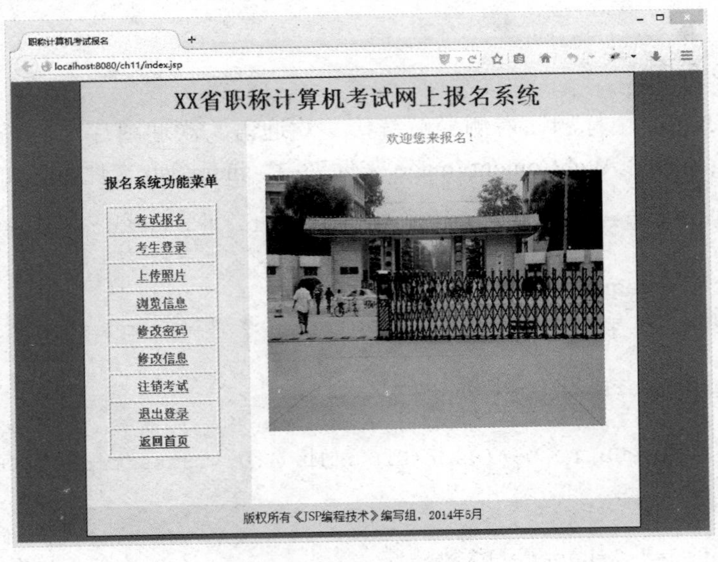

图 11.4　主页运行效果

11.3.3　JavaBean 和 Servlet 管理

本系统 JavaBean 类的包名均为 mybean.data，Servlet 类的包名均为 myservlet.controls。使用 Eclipse 开发环境，在 src 目录创建 mybean.data 包和 myservlet.controls 包，开发工具会自动在 build 目录中创建相应的目录，并在发布时，将 build 目录中的文件复制到 WEB-INF\classes 相应的文件夹中。在 src 目录的包中编写的类，Eclipse 会自动编译。本教材案例是在 Eclipse 开发工具下完成的。

如果使用记事本等代码编辑工具编写类，需要手动编译源文件并复制 calss 类到相应目录中。由于在 Servlet 类中使用了 JavaBean 和其他基础类，为了能顺利地编译 Servlet 类，可将 JavaBean 类保存在 D:\myservlet\controls\mybean\data 目录中，Servlet 类保存在 D:\myservlet\controls 目录中，先进入 D:\myservlet\controls\mybean\data 目录中编译 JavaBean 类，再在 D:\myservlet\controls 目录中编译 Servlet 类，然后将编译通过的 JavaBean 和 Servlet 类的字节码分别复制到 WebContent\WEB-INF\classes\mybean\data 和 WebContent\WEB-INF\classess\myservlet\controls 目录中。

注意： Servlet、JavaBean 以及其他的 Java 类在重新编译后，需要复制新的字节码文件到 E:\programjsp\ch11\WebContent\WEB-INF\classes 文件夹下对应的子目录中，并重新启动 Tomcat 服务器，如果使用 Eclipse 开发工具，系统会自动复制字节码文件并重新启动 Tomcat 服务器。

11.3.4　配置文件管理

本系统的 Servlet 类的包名均为 myservlet.controls，为了让 Tomcat 正确使用各模块的 Servlet，需要配置项目的 WebContent\WEB-INF 目录中的 web.xml 文件。用户在开发该系统时，可先写出 web.xml 文件的框架，随着系统各模块的实现，逐步加入模块所用 Servlet 对象配置片段。web.xml 文件的框架内容如下：

```
<?xml version="1.0" encoding="UTF-8"?>
<web-app version="2.4"
    xmlns="http://java.sun.com/xml/ns/j2ee"
    xmlns:xsi="http://www.w3.org/2001/XMLSchema-instance"
    xsi:schemaLocation="http://java.sun.com/xml/ns/j2ee
    http://java.sun.com/xml/ns/j2ee/web-app_2_4.xsd">
   <welcome-file-list>
     <welcome-file>index.jsp</welcome-file>
   </welcome-file-list>
</web-app>
```

当 Web 服务目录需要提供更多的 Servlet 对象时，需要在 web.xml 文件中根标记 <web-app ....>和</web-app>之间，增加<servlet>和<servlet-maping>元素。例如报名模块的 Servlet 对象配置片段如下：

```
<servlet><!--考生报名模块 Servlet 配置-->
    <servlet-name>register</servlet-name>
```

```
    <servlet-class>myservlet.controls.HandleRegister</servlet-class>
</servlet>
<servlet-mapping>
    <servlet-name>register</servlet-name>
    <url-pattern>/helpRegister</url-pattern>
</servlet-mapping>
```

11.4 三层架构设计与实现

网上报名系统采用 MVC 模式下的三层架构设计，实体层、业务逻辑层、数据库访问层包名为 myclass.bol、myclass.bll 和 myclass.dal 包。视图(JSP 页面)向控制器 Servlet 发送请求；Servlet 根据不同请求调用不同的业务逻辑组件，业务逻辑组件通过数据库访问层访问数据库，将返回数据封装成为实体或实体集并保存到 Bean 中；数据处理完毕后，Servlet 通知视图更新显示；视图从 JavaBean 中获得更新数据并显示。MVC 模式下三层之间的关系如图 11.5 所示。

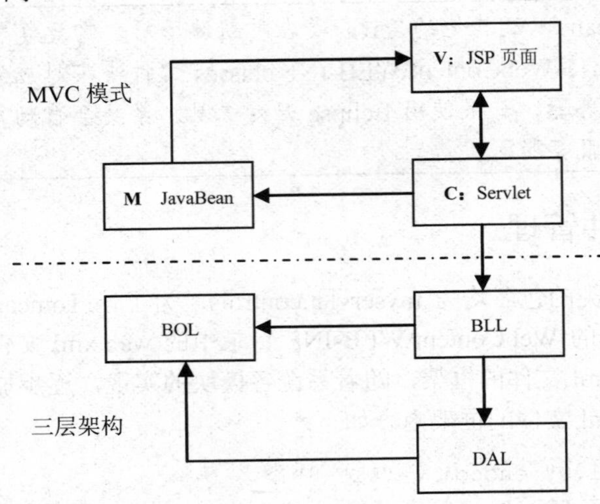

图 11.5 各层之间的关系

数据库访问层 myclass.dal 包中有 DBAccess、examineeDal 和 uploadFileDal 类，实体层 myclass.bol 包中有 examinee 类，业务逻辑层 myclass.bll 包中有 uploadFileBll 和 examineeBll 类。如果用户不使用 Eclipse 开发工具，参照 JavaBean 和 Servlet 类，在 D:\programjsp\ch11\WebContent 目录中的\WEB-INF\classes 子目录中创建各层包所对应目录，并将编译通过后的类字节码复制到其中。

11.4.1 实体层

本系统采用面向对象的系统分析与设计方法，将考生定义为一个类，这样在 examineeDal 类中就可以以对象的方式操作考生信息，因此首先要在 myclass.bol 包中创建 examinee 类。其属性和主要方法列表如表 11.3 所示。

表 11.3 examineeBol 类的属性和方法

属 性 名	类 型	属性说明	set 方法	get 方法
examID	String	身份证号	setExamID	getExamID
examName	String	考生姓名	setExamName	getExamName
sex	String	考生性别	setSex	getSex
company	String	单位地址	setCompany	getCompany
phone	String	联系电话	setPhone	getPhone
email	String	邮件地址	setEmail	getEmail
password	String	密码	setPassword	getPassword
pic	String	照片	setPic	getPic
examType	String	报考类别	setExamType	getExamType
address	String	单位地址	setAddress	getAddress

examinee 类的源代码如下：

```
package myclass.bol;
public class examinee {
    private String examID = null;// 身份证号
    private String examName = null;// 姓名
    private String sex = null;// 性别
    private String company = null;// 工作单位
    private String address = null;// 单位地址
    private String phone = null;// 联系电话
    private String email = null;// email 地址
    private String password = null;// 密码
    private String pic = null;// 照片文件名
    private String examType = null;// 考试类型
    private String memo = null;// 备注
    /**无参构造函数*/
    public examinee() {
    }
   /**有参构造函数*/
    public examinee(String examID, String examName, String sex, String
company,
            String address, String phone, String email, String password,
            String examType, String memo, String pic) {
        this.examID = examID;
        this.examName = examName;
        this.sex = sex;
        this.company = company;
        this.phone = phone;
        this.email = email;
        this.password = password;
        this.examType = examType;
        this.memo = memo;
```

```
        this.pic = pic;
        this.address=address;
    }
    public String getAddress() {
        return address;
    }
    public String getCompany() {
        return company;
    }
    public String getEmail() {
        return email;
    }
    public String getExamID() {
        return examID;
    }
    public String getExamName() {
        return examName;
    }
    public String getExamType() {
        return examType;
    }
    public String getMemo() {
        return memo;
    }
    public String getPassword() {
        return password;
    }
    public String getPhone() {
        return phone;
    }
    public String getPic() {
        return pic;
    }
    public String getSex() {
        return sex;
    }
    public void setAddress(String address) {
        this.address = address;
    }
    public void setCompany(String company) {
        this.company = company;
    }
    public void setEmail(String email) {
        this.email = email;
    }
    public void setExamID(String examID) {
        this.examID = examID;
```

```
    }
    public void setExamName(String examName) {
        this.examName = examName;
    }
    public void setExamType(String examType) {
        this.examType = examType;
    }
    public void setMemo(String memo) {
        this.memo = memo;
    }
    public void setPassword(String password) {
        this.password = password;
    }
    public void setPhone(String phone) {
        this.phone = phone;
    }
    public void setPic(String pic) {
        this.pic = pic;
    }
    public void setSex(String sex) {
        this.sex = sex;
    }
}
```

提示：如果用户使用 Eclipse 工具开发本系统，对于 set 和 get 方法可以在代码编辑器里右键单击，在弹出的快捷菜单中选择 Source→Generate→Getters and Setters 命令，自动生成 examinee 类的 set 和 get 方法。如果使用记事本等代码编写工具，则需要用户手工录入。从这一点也体现了开发工具的高效性。

11.4.2 数据访问层

数据访问层实现对数据的保存和读取操作等功能。与数据库有关的代码，如打开和关闭数据库连接，SELECT、INSERT、UPDATE、DELETE 语句和调用存储过程等代码都应该放到这一层，业务逻辑层和实体层不应包括对数据库的直接操作代码。本系统与文件操作有关的代码也放到了这一层。

1. DBAccess 类

本系统使用数据库连接池来实现对数据库的访问，需要在工程的 META-INF 文件夹下新建 content.xml 文件，并写入以下内容：

```
<?xml version="1.0" encoding="UTF-8"?>
<Context>
<WatchedResource>WEB-INF/web.xml</WatchedResource>    //监听资源
<Resource
name="jdbc/dataExam"
```

```
auth="Container"
type="javax.sql.DataSource"
driverClassName="org.gjt.mm.mysql.Driver"
url="jdbc:mysql://localhost:3306/dataExam"
username="root"
password="admin"
maxActive="5000"
maxIdle="10"
maxWait="-1" />
</Context>
```

DBAccess 类通过连接池对数据库进行访问，它使用纯 Java 驱动程序，包装了 JDBC 对 MYSQL 数据库的访问，增强了代码的重用性和可读性，如果进行代码移植，例如数据库发生变化，只需修改 content.xml 配置文件即可。该类的属性和方法如表 11.4、表 11.5 所示。

表 11.4　DBAccess 类属性列表

属 性 名	类 型	说 明
conn	Connection	数据库连接对象
stmt	Statement	SQL 语句对象
rs	ResultSet	结果集对象
prpSql	PerepareStatement	预处理 SQL 语句对象

表 11.5　DBAccess 类方法列表

方 法 名	返回值类型	说 明
DBAccess	void	构造函数
getConn	Connection	返回数据库连接对象
getConnection	void	连接数据库，为对象 conn 属性赋值
query(String)	ResultSet	执行查询 SQL 语句，返回查询结果集
query(PreparedStatement)	ResultSet	执行查询预处理 SQL 语句，返回查询结果集
insert(String[] sqls)	boolean	执行插入多条记录的 SQL 语句
executeSql(String strSql)	int	执行不返回结果集的查询语句
executeSql(PreparedStatement)	int	执行不返回结果集的查询预处理语句
executeSql(String[] sqls)	boolean	执行多条不返回结果集的查询语句
closeConnection	void	关闭结果集、查询语句、连接对象等
getSysDate()	String	获得 Web 服务器系统时间

DBAccess 类的源代码如下：

```
package myclass.dal;
import java.sql.SQLException;
import java.sql.Connection;
import java.sql.ResultSet;
import java.sql.Statement;
```

```
import java.sql.*;
import javax.sql.DataSource;
import javax.naming.*;
public class DBAccess {
    private Connection conn = null;
    private Statement stmt = null;
    private ResultSet rs = null;
    private PreparedStatement prpSql=null;
    public DBAccess() {}
    /** 返回一个数据库连接 * @return Connection */
    public Connection getConn() {
        if (conn == null) {
            getConnection();
        }
        return conn;
    }
    /** 函数功能:获得连接对象,Statement 对象 */
    public void getConnection() {
        try {
                javax.naming.Context ctx = new javax.naming.InitialContext();
                javax.sql.DataSource ds = (javax.sql.DataSource) ctx.lookup(
                        "java:comp/env/jdbc/dataExam");
                this.conn = ds.getConnection();
                this.stmt = this.conn.createStatement();
                //System.out.println("数据库连接建立(连接池)!");
        }
        catch (NamingException ex1) {
            //System.out.println("请检查数据库连接池配置是否正确!");
            ex1.printStackTrace();
        }
        catch (SQLException ex2) {
            System.out.println("请检查数据库是否启动!");
            ex2.printStackTrace();
        }
    }
    /** 返回结果集* @param strSql 查询语句 * @return 成功返回结果集，失败返回 null */
    public ResultSet query(String strSql) {
        //System.out.println("sql:" + strSql);
        ResultSet rs = null;
        try {
            rs = stmt.executeQuery(strSql);
            return rs;
        } catch (SQLException ex) {
            ex.printStackTrace();
            return rs;
        }
    }
```

```
    /** 返回结果集 * @param prpSql 查询预处理 * @return 成功返回结果集,失败返回null */
    public ResultSet query(PreparedStatement prpSql) {
        //System.out.println(prpSql.toString());
        this.prpSql=prpSql;
        ResultSet rs = null;
        try {
            rs = this.prpSql.executeQuery();
            return rs;
        } catch (SQLException ex) {
            ex.printStackTrace();
            return rs;
        }
    }
    /** 插入多条数据 * @param sqls insert 语句数组 * @return 成功返回 true,失败
返回 false */
    public boolean insert(String[] sqls) {
        boolean breturn = false;
        try {
            conn.setAutoCommit(false);
            for (int i = 0; i < sqls.length; i++) {
                if (sqls[i] != null) {
                    stmt.addBatch(sqls[i]);
                }
            }
            stmt.executeBatch();
            conn.commit();
            conn.setAutoCommit(true);
            breturn = true;
        } catch (SQLException ex) {
        }
        return breturn;
    }
    /** 返回影响行数
     * @param strSql 查询语句 * @return 成功返回影响行数,失败返回0 -1 数据库访问错误 */
    public int executeSql(String strSql) {
        int result = 0;
        try {
            stmt = conn.createStatement();
            result = stmt.executeUpdate(strSql);
        } catch (SQLException ex) {
            System.out.println("产生异常,: at DBAccess.executeSql()");
            ex.printStackTrace();
            result= -1;
        }
        return result;
    }
    /** 返回影响行数 * @param strSql 查询预处理 * @return 影响行数,失败返回0 */
```

```
    public int executeSql(PreparedStatement prpSql){
        int result = 0;
        try {
            this.prpSql=prpSql;
            result = this.prpSql.executeUpdate();
        } catch (SQLException ex) {
            System.out.println("产生异常,: at DBAccess.executeSql()");
            ex.printStackTrace();
            result=-1;
        }
        return result;

    }
    /** 返回结果true或false * @param sqls 查询语句数组 * @return 成功true,
失败false */
    public boolean executeSql(String[] sqls) {
        boolean breturn = false;
        try {
            conn.setAutoCommit(false);
            stmt = conn.createStatement();
            for (int i = 0; i < sqls.length; i++) {
                if (sqls[i] != null) {
                    //System.out.println("sqls[0]="+sqls[i]);
                    stmt.addBatch(sqls[i]);
                }
            }
            stmt.executeBatch();
            conn.commit();
            conn.setAutoCommit(true);
            breturn = true;
        } catch (SQLException ex) {
            System.out.println("产生异常: at DBAccess.executeSql()");
            ex.printStackTrace();
        }
        return breturn;
    }
    /** 关闭连接 */
    public void closeConnection() {
        try {
            if (rs != null) {
                rs.close();
                rs = null;
            }
            if (stmt != null) {
                stmt.close();
                stmt = null;
            }
```

```
            if (conn != null) {
                conn.close();
                conn = null;
            }
            if(prpSql!=null){
              prpSql.close();
              prpSql=null;
            }

            //System.out.println("数据库连接关闭! ");
        } catch (SQLException ex) {
            System.out.println("产生异常: at DBAccess.closeConnection()");
            ex.printStackTrace();
        }
    }
    /** 获取系统时间 * 静态方法,直接用类名调用 */
    public static String getSysDate() {
        DBAccess dba = new DBAccess();
        String sql = "select sysdate() sysdate;";
        try {
            dba.getConnection();
            ResultSet rs = dba.query(sql);
            String currentDate = null;
            if (rs.next()) {
                currentDate = rs.getString("sysdate");
            }
            return currentDate;
        } catch (SQLException ex) {
            System.out.println("产生异常: at DBAccess.getSysDate()");
            ex.printStackTrace();
            return null;
        } finally {
            dba.closeConnection();
        }
    }
}
```

技巧： DBAccess 类的方法中使用了//System.out.println("数据库连接建立(连接池)! ");等输出语句，这样可以从 Console 窗口中查看类的执行情况，也是程序调试的一个实用技巧。

2. examineeDal 类

examineeDal 类使用前面的 DBAccess 类连接数据库，实现对 examinee 表的增、删、改、查等操作方法，为业务逻辑层 examineeBll 类的方法(如考生报名、考生信息修改、考生登录等方法)提供数据访问基础。返回的查询结果集被封装为 examinee 对象或其对象集合，该类为系统的关键类，其方法列表如表 11.6 所示。

表 11.6　examineeDal 类主要方法列表

方 法 名	返回值类型	说　明
CreateExaminee(…)	int	在考生表 examinee 中插入一条记录
getExamineeByID(String)	examinee 对象	按身份证号得到一个考生对象
getExamineeAll()	ArrayList<examinee>	得到所有考生对象集列表
getExamineeByIdPwd( )	examinee 对象	按身份证号和密码得到一个考生对象
setExaminePwd( )	int	按身份证号，设置登录密码
updateExamineeByID(…)	int	按身份证号，修改考生信息
setExamineePic( )	int	按身份证号，设置考生照片
deleteExmineeByID( )	int	按身份证号，删除考生信息

(1) examineeDal 类的源代码如下。

```
package myclass.dal;
import java.sql.ResultSet;
import java.sql.PreparedStatement;
import java.util.ArrayList;
import myclass.dal.DBAccess;
import myclass.bol.examinee;

public class examineeDal {
    /**在考生表中插入一条记录 * @return 成功 1,失败 0 */
    public int CreateExaminee(String examID,String examName,String sex,
            String company,String address,String phone,String email,
            String password,String examType,String memo,String pic){
        int result=0;
        String createSql="insert into examinee(id,name,sex,company," +
                "address,phone,email,password,examType,memo,pic)"+
               "values('"+examID+"','"+examName+"','"+sex+"','"+
               company+"','"+address+"','"+phone+"','"+email+"','"+
               password+"','"+examType+"','"+memo+"','"+pic+"')";
        DBAccess dba=new DBAccess();
       try{
           if(dba.getConn()!=null)
           {
             result=dba.executeSql(createSql);
           }
        }
        catch(Exception ne)
        {
           System.out.println("examineedal 出现如下错误：<br>");
           System.out.println(ne);
        }
        finally{
```

```
            dba.closeConnection();
        }
        return result;
    }

    /**按考生身份证号获得一个考生对象(考生信息)* @return 成功返回 examinee 对象,
失败返回 null * @param examID */
    public examinee getExamineeByID(String examID){
        DBAccess dba=new DBAccess();
        examinee exam=null;
        try{
            if(dba.getConn()!=null&&examID!=null)
            {  //System.out.println("数据库连接已建立");
                String str="select * from examinee where id='"+examID+"'";
                ResultSet rst=dba.query(str);//从一个已存在的表中读取数据
            if(rst!=null&&rst.next()){//将数据赋值给实体
                exam=new examinee(rst.getString(1),rst.getString(2),
                    rst.getString(3),
                    rst.getString(4),rst.getString(5),rst.getString(6),
                    rst.getString(7),rst.getString(8),rst.getString(10),
                    rst.getString(9),rst.getString(11));
                //System.out.println(rst.getString(5));
                }
            }
        }
        catch(Exception ne)
        {
         System.out.println("examineedal:getExamineeByID 发生错误");
        }
        finally{
         dba.closeConnection();
        }
        return exam;
    }

    /**功能:返回 examinee 对象数组 *@return 成功 examinee 对象数组,否则 null * /
    public ArrayList<examinee> getExamineeAll(){
         DBAccess dba=new DBAccess();
        ArrayList<examinee> examList=new ArrayList<examinee>();
        try{
            if(dba.getConn()!=null)
            {
                String str="select * from examinee";
                ResultSet rst=dba.query(str);//从一个已存在的表中读取数据
            while (rst!=null&&rst.next())//循环赋值
            {
                if(rst.getString("id")!=null){
```

```
                examinee exam = new examinee(rst.getString(1),rst.getString(2),
                    rst.getString(3),
                    rst.getString(4),rst.getString(5),rst.getString(6),
                    rst.getString(7),rst.getString(8),rst.getString(10),
                    rst.getString(9),rst.getString(11));
                  examList.add(exam);
                }
              }
            }
        }
        catch(Exception ne)
        {
         System.out.println("examineedal:getExamineeall 发生错误");
        }
        finally{
         dba.closeConnection();
        }
        return examList;
    }
    /** 根据身份证和密码返回考生对象 * @return 成功 examinee 对象，失败返回 null */
    public examinee getExamineeByIdPwd(String examID,String password){
        DBAccess dba=new DBAccess();
       examinee exam=null;
       try{
           if(dba.getConn()!=null&&examID!=null&&password!=null)
           {  System.out.println("数据库连接已建立");
              String str="select * from examinee where id=? and password=?";
              PreparedStatement prpSql;
              prpSql=dba.getConn().prepareStatement(str);
              prpSql.setString(1, examID);
              prpSql.setString(2, password);
              ResultSet rst=dba.query(prpSql);//从一个已存在的表中读取数据
            if(rst!=null&&rst.next()){
                exam=new examinee(rst.getString(1),rst.getString(2),
                    rst.getString(3),
                    rst.getString(4),rst.getString(5),rst.getString(6),
                    rst.getString(7),rst.getString(8),rst.getString(10),
                    rst.getString(9),rst.getString(11));
             }
             if(prpSql!=null){
               prpSql.close();
               prpSql=null;
             }
          }
       }
       catch(Exception ne)
       {
```

```
        System.out.println("examineedal:getexamineebyIDPwd 发生错误");
        }
        finally{
        dba.closeConnection();
        }
        return exam;
    }
    /** 函数功能修改用户密码* @return 成功返回 1，否则返回 0 */
    public int setExaminePwd(String examID,String newPassword){
        DBAccess dba=new DBAccess();
        int result=0;
        try{
            if(dba.getConn()!=null){
                PreparedStatement prpSql;
                String strSql="Update examinee set password=? where id=?";
                prpSql=dba.getConn().prepareStatement(strSql);
                prpSql.setString(1, newPassword);
                prpSql.setString(2, examID);
                result=dba.executeSql(prpSql);
                if(prpSql!=null){
                    prpSql.close();
                    prpSql=null;
                }
            }
        }
        catch(Exception ne)
        {
            System.out.println("exmaineedal:setPassword 发生错误");
            result=-1;
        }
        finally
        {
            dba.closeConnection();
        }
        return result;
    }
    /** 修改考生信息* @return 影响行数，失败 0 */
    public int updateExamineeByID(String examID,String examName,String sex,
            String company,String address,String phone,String email,
            String examType,String memo){
        DBAccess dba=new DBAccess();
        int result=0;
        try{
            if(dba.getConn()!=null){
                PreparedStatement prpSql;
                String strSql="update examinee set name=?,sex=?,company=?,"+
                        "address=?,phone=?,email=?,examType=?,memo=? where id=?";
```

```
            prpSql=dba.getConn().prepareStatement(strSql);
            prpSql.setString(1, examName);
            prpSql.setString(2, sex);
            prpSql.setString(3, company);
            prpSql.setString(4, address);
            prpSql.setString(5, phone);
            prpSql.setString(6, email);
            prpSql.setString(7, examType);
            prpSql.setString(8, memo);
            prpSql.setString(9, examID);
            result=prpSql.executeUpdate();
           if(prpSql!=null){
               prpSql.close();
               prpSql=null;
           }
        }
    }
    catch(Exception ne){
        ne.printStackTrace();
        System.out.println(ne.toString());
        return -1;
    }
    finally{
        dba.closeConnection();
    }
    return result ;
}

/** 设置图片文件名 * @param examID * @param picStr * @return 成功1,失败0 */
public int setExamineePic(String examID,String picStr){
    DBAccess dba=new DBAccess();
    int result=0;
    try{
      if(dba.getConn()!=null){
          String strSql="update examinee set pic='"+picStr+
            "' where id='"+examID+"'";
          result=dba.executeSql(strSql);
      }
    }
    catch(Exception ne)
    {
        System.out.println("发生异常"+ne);
        ne.printStackTrace();
    }
    finally{
        dba.closeConnection();
    }
```

```
        return result;
    }

    /** 按身份证号删除考生 * @param examID * @return 1 成功，0 失败，出现异常 */
    public int deleteExmineeByID(String examID){
        DBAccess dba=new DBAccess();
        int result=0;
        try{
            if(dba.getConn()!=null){
                PreparedStatement prpSql;
                String strSql="delete from examinee where id=?";
                prpSql=dba.getConn().prepareStatement(strSql);
                prpSql.setString(1, examID);
                result=prpSql.executeUpdate();
                if(prpSql!=null){
                    prpSql.close();
                    prpSql=null;
                }
            }
        }
        catch(Exception ne){
            ne.printStackTrace();
            System.out.println(ne.toString());
            return -1;
        }
        finally{
            dba.closeConnection();
        }
        return result ;
    }
}
```

技巧： DBAccess 类和 examineeDal 类是本系统两个重要的类，它们是系统对数据操作的基础，必须首先测试通过，确保逻辑上没有问题。用户在编写这两个类时，可以先写一个简单的 JSP 测试页面，通过这个页面来分层测试这两个类，这样可以有效地减小用户的开发难度。用户要特别注意随时将最新类的字节码复制到 WebRoot-WEB-INF\classes 的相应目录中。

(2) 测试页面。

测试 DBAccess 类和 examineeDal 类的 myjsp.jsp 页面代码如下：

```
<%@ page contentType="text/html; charset=gb2312" %>
<%@ page import="javax.naming.Context" %>
<%@ page import="javax.sql.DataSource"%>
<%@ page import="javax.naming.InitialContext"%>
<%@ page import="java.sql.*"%>
```

```
<%@ page import="myclass.dal.*" %>
<%@ page import="myclass.bol.*" %>
<%@ page import="java.util.ArrayList;" %>
<html>
<body>
1234
<%
   DBAccess dba=new DBAccess();
   examineeDal examd=new examineeDal();
   ArrayList<examinee> list=new ArrayList<examinee>();
   try{ //测试数据添加方法
      int  rsult=
         examd.CreateExaminee("123456789123456789","杨里海","男","河北农大",
            "河北保定", "838838383","kdiek@die.kdi","yang","高级","计算机类","");
      out.println(rsult);
      dba.getConnection();
      if(dba.getConn()!=null)
      {
        out.println("已经获得 DataSource!");
        out.println("<br>");
        String str="select * from examinee";
        ResultSet rst=dba.query(str);//从一个已存在的表中读取数据
        out.println("以下是从数据库中读取出来的数据:<br>");
        while(rst.next()){//测试数据查询
           out.println("<br>");
           out.println(rst.getString(1));}
      }
      out.println("<br>系统时间: "+DBAccess.getSysDate());
      examinee exambean=null;
      exambean=examd.getExamineeByID("123456789123456789");
      out.println("<br><br>"+exambean.getExamID());
      list=examd.getExamineeAll();
      for(examinee e:list){
       out.println("<br>这是通过 ArrayList 得到的值: "+e.getExamID());
      }
   }
   catch(Exception ne)
   {
      out.println("出现如下错误: <br>");
      out.println(ne);
   }
   finally{
      dba.closeConnection();
      out.println("<br>已经关闭 DataSource!");
   }
%>
</body>
```

```
</html>
```

3. uploadFileDal 类

uploadFileDal 类是上传文件类，它封装了上传文件的属性和上传方法。该类通过输入流，首先将上传文件保存到服务器的临时文件中(文件名为用户的 sessionID)，然后从临时文件中获得上传文件名、上传文件内容，并将文件内容以新名保存到服务器指定目录中。uploadFileDal 类的属性列表和主要方法列表如表 11.7、表 11.8 所示。

表 11.7 uploadFileDal 类的属性列表

属 性 名	类 型	说 明
driverPath	String	Web 服务路径
tempFileName	String	临时文件名
saveFileName	String	文件最后保存名称
fileSource	PerepareStatement	文件输入流
uploadFileName	String	上传文件名称
backMessage	String	返回消息
flag	String	成功失败标志

表 11.8 uploadFileDal 类的主要方法列表

方 法 名	返回值类型	说 明
uploadFileDal(String,String, String,InputStream)	void	构造函数
uploadFileMethod()	boolean	上传文件的方法

uploadFileDal 类的源代码如下：

```
package myclass.dal;
import java.io.*;
public class uploadFileDal {
    private String driverPath=null;
    private String tempFileName=null;
    private String saveFileName=null;
    private InputStream fileSource=null;
    private String uploadFileName=null;
    private String backMessage=null;
    boolean flag=false;
    public boolean isFlag() {
        return flag;
    }
    public void setFlag(boolean flag) {
        this.flag = flag;
    }
    public String getTempFileName() {
        return tempFileName;
    }
```

```
public void setTempFileName(String tempFileName) {
    this.tempFileName = tempFileName;
}
public String getDriverPath() {
    return driverPath;
}
public void setDriverPath(String driverPath) {
    this.driverPath = driverPath;
}
public InputStream getFileSource() {
    return fileSource;
}
public void setFileSource(InputStream fileSource) {
    this.fileSource = fileSource;
}
public String getSaveFileName() {
    return saveFileName;
}
public void setSaveFileName(String saveFileName) {
    this.saveFileName = saveFileName;
}
public String getUploadFileName() {
    return uploadFileName;
}
public void setUploadFileName(String uploadFileName) {
    this.uploadFileName = uploadFileName;
}
public String getBackMessage() {
    return backMessage;
}
public void setBackMessage(String backMessage) {
    this.backMessage = backMessage;
}
public uploadFileDal(String driverPath, String tempFileName,
        String saveFileName, InputStream fileSource) {
    super();
    this.driverPath = driverPath;
    this.tempFileName = tempFileName;
    this.saveFileName = saveFileName;
    this.fileSource = fileSource;
}
public boolean uploadFileMethod(){
    try{//上传文件保存到临时文件中
        File f1=new File(driverPath,tempFileName);
        FileOutputStream fos=new FileOutputStream(f1);
        byte b[]=new byte[10000];
        int n;
```

```
            while((n=fileSource.read(b))!=-1){
                fos.write(b, 0, n);
            }
            fos.close();
            fileSource.close();
            //读取临时文件中第二行的内容
            RandomAccessFile random=new RandomAccessFile(f1,"r");
            int second=1;
            String secondLine=null;
            while(second<=2){
                secondLine=random.readLine();
                second++;
            }
            //得到上传文件的文件名
            int position=secondLine.lastIndexOf('\\');
            uploadFileName=secondLine.substring(position+1,
secondLine.length()-1);
            //转换编码，识别汉字文件名
            byte cc[]=uploadFileName.getBytes("iso-8859-1");
            uploadFileName=new String(cc);
            //得到上传文件的扩展名
            int extposition=uploadFileName.lastIndexOf('.');
            String extName=uploadFileName.substring(extposition+1,
uploadFileName.length());
            //获取上传临时文件第四行回车符的位置
            random.seek(0);
            long forthEndPosition=0;
            int forth=1;
            while((n=random.readByte())!=-1&&forth<=4){
                if(n=='\n'){
                    forthEndPosition=random.getFilePointer();
                    forth++;
                }
            }
            //删除重名的文件
            saveFileName=saveFileName.concat("."+extName);
            File dir=new File(driverPath);
            dir.mkdir();
            File file[]=dir.listFiles();
            for(int k=0;k<file.length;k++){
                if(file[k].getName().equals(saveFileName)){
                    file[k].delete();
                }
            }
            //按新文件名保存文件
            File savingFile=new File(driverPath,saveFileName);
            RandomAccessFile random2=new RandomAccessFile(savingFile,"rw");
```

```
            //在临时文件中获得上传文件结束位置
            random.seek(random.length());
            long endPosition=random.getFilePointer();
            long mark=endPosition;
            int j=1;
            while((mark>=0)&&(j<=6)){
                mark--;
                random.seek(mark);
                n=random.readByte();
                if(n=='\n'){
                    endPosition=random.getFilePointer();
                    j++;
                }
            }
            random.seek(forthEndPosition);
            long startPoint=random.getFilePointer();
            while(startPoint<endPosition-1){
                n=random.readByte();
                random2.write(n);
                startPoint=random.getFilePointer();
            }
            random2.close();
            random.close();
            f1.delete();
            flag=true;
            backMessage="成功上传！";
        }
        catch(Exception exp)
        {
            System.out.println("uploadFileMethod"+exp.toString());
            backMessage=""+exp;
            flag=false;
        }
        return flag;
    }
}
```

技巧： uploadFileDal 类是本系统实现文件上传的基础类，用户在编写这个类时，可以先写一个简单的 JSP 测试页面，通过这个页面来分层测试这个类，这样可以有效地减小用户的开发难度。如果用户没有使用 Eclipse 工具进行开发时，需要注意随时将最新类的字节码复制到 WebContent\WEB-INF\classes 的相应目录中。

11.4.3　业务逻辑层

业务逻辑层主要完成系统中的业务逻辑，如字段值的校验、是否为空及一些异常处理等。业务逻辑层的方法主要供控制器(Servlet 类)调用，同时调用数据访问层与数据交互，起到一个承上启下的作用。本系统主要有 examineeBll 和 uploadFileBll 类，这两个类的包名为 myclass.bll。

1. examineeBll 类

examineeBll 类主要处理与考生有关的业务逻辑，例如考生报名、信息修改、注销考试等业务。通过 examineeBll 类，可将业务处理的数据保存到数据库表中或从表中提取数据，其基本方法如表 11.9 所示。

表 11.9　examineeBll 类的方法

方 法 名	返回值类型	说　明
CreateExaminee( )	int	构造函数
examineeLogin()	boolean	考生登录
setExamineePic()	int	按身份证号设置考生照片
getExamineeByID()	examinee	按身份证号获得考生信息
getExamineeAll()	ArrayList<examinee>	获得全部考生的信息
setExamineePwd()	int	设置考生登录密码
updateExaminByID( )	int	保存考生修改后的信息
deleteExmineeByID( )	int	删除考生信息

examineeBll 类的源代码如下：

```
package myclass.bll;
import java.sql.*;
import java.io.Serializable;
import java.util.ArrayList;
import myclass.dal.*;
import myclass.bol.*;
public class examineeBll {
    /** 添加考生数据的方法 * @return 0 身份证号或密码不符合要求 1 报名成功 -1 或数
据库访问发生错误 -2 身份证号已存在 */
    public int CreateExaminee(String examID,String examName,String sex,
            String company,String address,String phone,String email,
            String password,String examType,String memo,String pic){
        examineeDal examDal=new examineeDal();
        int result=0;
        //下面是数据校验
        if((examID.length()==15||examID.length()==18)&&
                password.length()>0){
            boolean isLD=true;
            for(int i=0;i<examID.length();i++){
                char c=examID.charAt(i);
                if(!((c<='z'&&c>='a')||(c<='Z')&&(c>='A')||
                        (c<='9')&&(c>='0'))){
                    isLD=false;
```

```
                }
            }
            if(isLD==true && examDal.getExamineeByID(examID)!=null){
                return result=-2;
            }
            if(isLD==true){
                result=examDal.CreateExaminee(examID, examName, sex, company,
                    address, phone, email, password, examType, memo, pic);
            }
        }
        else{
            System.out.println("身份证号或密码不符合要求！");
            System.out.println("函数返回值："+result);
        }
        return result;
    }
    /** 考生登录方法 * @param examID  身份证号 * @param password  密码
     * @return 0 身份证号或密码不符合要求 -1 身份证号或密码不正确 1 登录成功  */
    public int examineeLogin(String examID,String password){
        examineeDal examDal=new examineeDal();
        int result=0;

    if((examID.length()==15||examID.length()==18)&&password.length()>0){
            examinee exam=null;
            exam=examDal.getExamineeByIdPwd(examID, password);
            if(exam!=null){
                result=1;
            }
            else{
                result=-1;
            }
        }
        return result;
    }

    /** 存储照片文件名* @param examID * @param picStr * @return 1 成功 0 失败 */
    public int setExamineePic(String examID,String picStr){
        examineeDal examDal=new examineeDal();
        //int result;
        //if (){
            //业务逻辑
        //}
        return examDal.setExamineePic(examID, picStr);
    }
    /** 返回examinee 对象 * @param examID* @return  成功examinee 对象,否则null*/
    public examinee getExamineeByID(String examID){
        examinee exam  =null;
```

```
        examineeDal examdal=new examineeDal();
        exam=examdal.getExamineeByID(examID);
        return exam;

    }
    /** 返回 examinee 对象集 * @return 成功返回对象集,否则返回 null  */
    public ArrayList<examinee> getExamineeAll(){
    examineeDal examdal=new examineeDal();
    return examdal.getExamineeAll();
    }

    /** 设置用户密码 @return  0 密码空或新旧相同 1 成功设置密码 -1 旧密码不正确  */
    public int setExamineePwd(String examID,String newPassword,
          String oldPassword){
    int result=0;
    examineeDal examdal=new examineeDal();
    if(newPassword =="" || newPassword.equals(oldPassword)){
        return result;
    }
    if(examdal.getExamineeByIdPwd(examID, oldPassword)!=null){
        result=examdal.setExaminePwd(examID, newPassword);
    }
    else {
        result=-1;
    }
    return result;
    }
    /** 更新考生信息 * @return -1 考生姓名错误，0 修改失败，1 修改成功 */
    public int updateExaminByID(String examID,String examName,String sex,
            String company,String address,String phone,String email,
            String examType,String memo){
    examineeDal examdal=new examineeDal();
    int result=0;
    if(examName.length()>0){
        result=examdal.updateExamineeByID(examID, examName, sex,
            company, address, phone, email, examType, memo);
    }
    else{
        return result=-1;
    }
    return result;
    }
    /** 删除考生数据  * @return -2 考生号空，1 删除成功 -1 出现异常 0 删除失败  */
    public int deleteExmineeByID(String examID){
        examineeDal examdal=new examineeDal();
        int result=0;
    if(examID==null){
```

```
            result=-2;
            }
        if(examID.length()>0){
            result=examdal.deleteExmineeByID(examID);
        }
        return result;
        }

    }
```

注意：examineeBll 类的包名为 myclass.bll，在编写该类时要将所需 examineeDal 类引入，在不使用 Eclipse 进行开发的情况下需在 Web 服务目录的 classes 子目录中要创建对应文件目录，并将编译测试通过的字节码复制到 Web 服务目录中。

2. uploadFileBll 类

uploadFileBll 类主要处理与文件上传相关的业务逻辑，使用 uploadFileDal 类完成数据上传。可见业务逻辑层不包含任何关于数据库连接、执行等操作，采用分层思想使各层职责分明。uploadFileBll 类的基本属性和方法如表 11.10 所示。

表 11.10　uploadFileBll 类的主要属性和方法

方法名或属性	返回值类型	说　明
backMessag 属性	String	上传文件返回信息
flag 属性	boolean	成功与否
uploadFileName 属性	String	上传文件名
savedFileName 属性	String	上传文件保存名称
uploadFileMethod( )方法	void	上传文件方法

uploadFileBll 类的源代码如下：

```
package myclass.bll;
import java.io.*;
import myclass.dal.uploadFileDal;

public class uploadFileBll {
    private String backMessage=null;
    private boolean flag=false;
    private  String uploadFileName=null;
    private String savedFileName=null;
     public String getSavedFileName() {
         return savedFileName;
     }
     public void setSavedFileName(String savedFileName) {
         this.savedFileName = savedFileName;
     }
     public String getUploadFileName() {
```

```
        return uploadFileName;
    }
    public void setUploadFileName(String uploadFileName) {
        this.uploadFileName = uploadFileName;
    }
    public String getBackMessage() {
        return backMessage;
    }
    public void setBackMessage(String backMessage) {
        this.backMessage = backMessage;
    }
    public boolean isFlag() {
        return flag;
    }
    public void setFlag(boolean flag) {
        this.flag = flag;
    }
    public boolean uploadFileMethod(String driverPath, String tempFileName,
            String saveFileName, InputStream fileSource){
    //业务逻辑
//      String strchecked=uploadFileName.substring(uploadFileName.indexOf(".")+1,
//              uploadFileName.length());
//      //检查上传文件扩展名是否符合要求
////    String strext[]={"jpg","bmp","gif"};
//      flag=false;
//      for(String str:strext){
//          if(str.equalsIgnoreCase(strchecked)){
//              flag=true;
//              break;
//          }
//      }
//      //不符合要求则返回 false
//      if(flag==false){
//          backMessage="上传文件扩展名不符合要求！";
//          return flag;}
    //上传文件
    try{
        uploadFileDal upFile=new uploadFileDal(driverPath,tempFileName,
            saveFileName,fileSource);
        flag=upFile.uploadFileMethod();
        backMessage=upFile.getBackMessage();
        this.setUploadFileName(upFile.getUploadFileName());
        this.setSavedFileName(upFile.getSaveFileName());
    }
    catch(Exception exp){
        flag=false;
        backMessage="上传文件失败";
        this.setUploadFileName("");
        this.setSavedFileName("");
```

```
        }
        return  flag;
    }

}
```

提示：三层架构的数据访问层类、业务逻辑层类、实体层类的编写很重要，正确地编译和测试这些类是顺利完成系统开发的关键。用户可以采用一些测试技巧，比如在测试点设置断点、增加输出语句、编写简单的 JSP 网页辅助测试等。

11.5　考生报名模块

当考生报名时需要考生输入身份证号、性别、密码、工作单位、单位地址、邮箱地址、联系电话、报考类别等信息，如果身份证号、密码、邮箱地址不符合要求，则不允许考生报名。考生报名的信息通过有效性检查后会自动存储到数据库 dataExam 的 examinee 表中。有效性检查在客户端采用 JS 代码检查，在服务器端由业务逻辑层负责检查。

该模块的模型 JavaBean 描述考生的信息。该模块视图部分有两个 JSP 页面构成，一个页面负责提交考生的信息到控制器，另一个页面负责显示报考是否成功的信息。该模块的控制器 Servlet 负责调用业务逻辑层 examineeBll 类将报考信息写入数据库 examinee 表中，并负责更新视图显示。

11.5.1　模型(JavaBean)

下面的模型(JavaBean)用来存描述报名信息。

registerBean.java 文件代码如下：

```
package mybean.data;
public class registerBean {
    private String examID = null;// 身份证号
    private String examName = null;// 姓名
    private String sex = null;// 性别
    private String company = null;// 工作单位
    private String address = null;// 单位地址
    private String phone = null;// 联系电话
    private String email = null;// email 地址
    private String password = null;// 密码
    private String pic = null;// 照片文件名
    private String examType = null;// 考试类型
    private String memo = null;// 备注
    private String backMessage;
    public String getExamID() {
        return examID;
    }
    public void setExamID(String examID) {
```

```
        this.examID = examID;
    }
    public String getExamName() {
        return examName;
    }
    public void setExamName(String examName) {
        this.examName = examName;
    }
    public String getSex() {
        return sex;
    }
    public void setSex(String sex) {
        this.sex = sex;
    }
    public String getCompany() {
        return company;
    }
    public void setCompany(String company) {
        this.company = company;
    }
    public String getAddress() {
        return address;
    }
    public void setAddress(String address) {
        this.address = address;
    }
    public String getPhone() {
        return phone;
    }
    public void setPhone(String phone) {
        this.phone = phone;
    }
    public String getEmail() {
        return email;
    }
    public void setEmail(String email) {
        this.email = email;
    }
    public String getPassword() {
        return password;
    }
    public void setPassword(String password) {
        this.password = password;
    }
    public String getPic() {
        return pic;
    }
    public void setPic(String pic) {
        this.pic = pic;
```

```
    }
    public String getExamType() {
        return examType;
    }
    public void setExamType(String examType) {
        this.examType = examType;
    }
    public String getMemo() {
        return memo;
    }
    public void setMemo(String memo) {
        this.memo = memo;
    }
    public String getBackMessage() {
        return backMessage;
    }
    public void setBackMessage(String backMessage) {
        this.backMessage = backMessage;
    }
}
```

11.5.2　视图

本模块视图有两个 JSP 页面：register.jsp 和 showRegisterMess.jsp 页面。register.jsp 页面负责输入考生信息，通过<script src="js/check.js" language="JavaScript" type= "text/javascript"></script>引入 JS 文件，JS 文件负责客户端对考生输入信息的验证。showRegisterMess.jsp 负责显示考生报名反馈信息，比如报名成功与否、报名身份证号、姓名等信息，以便用户核对报名信息。

register.jsp 页面效果如图 11.6 所示。

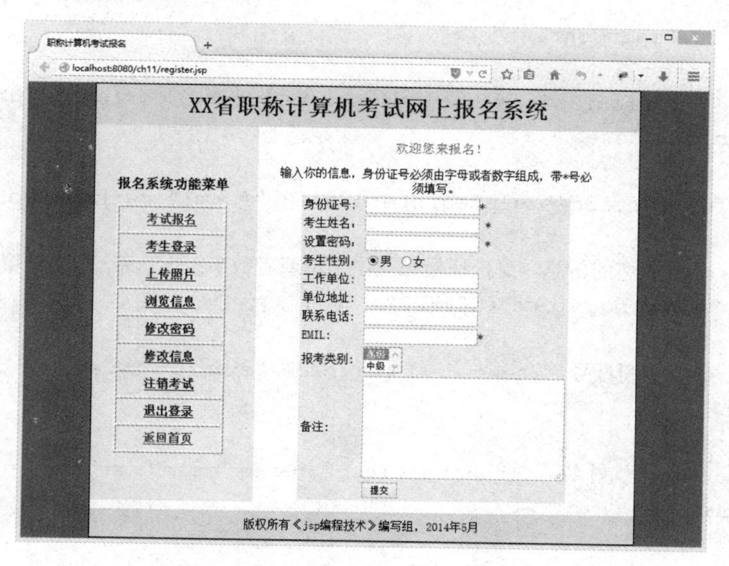

图 11.6　考生报名页面

register.jsp 页面代码：

```
<%@ page language="java" import="java.util.*" pageEncoding="GBK"%>
<% String path = request.getContextPath();
String basePath = request.getScheme() +"://"+ request.getServerName()
+":"+ request.getServerPort() + path + "/";
%>
<!DOCTYPE HTML PUBLIC "-//W3C//DTD HTML 4.01 Transitional//EN">
<html>
  <head>
    <base href="<%=basePath%>">
    <title>职称计算机考试报名</title>
     <meta http-equiv="pragma" content="no-cache">
     <meta http-equiv="cache-control" content="no-cache">
     <meta http-equiv="expires" content="0">
     <meta http-equiv="keywords" content="网上报名,职称考试,计算机">
     <meta http-equiv="description" content="网上报名首页">
     <link href="css/mystylesheet.css" rel="stylesheet" type="text/css" />
    <!-- 引入 js 文件，完成客户端验证-->
    <script src="js/check.js" language="JavaScript" type="text/javascript">
</script>
  </head>
  <body class="twoColHybLtHdr">
    <div id="container">
      <div id="header">
         <%@ include file="txtfile/header.txt" %>
      </div>
      <div id="sidebar1">
         <%@ include file="txtfile/left.txt" %>
      </div>
      <div id="mainContent"  >
          <center >
             <p><font size=4 color=red>欢迎您来报名！</font></p></center>
          <blockquote>
          <center>
          <form action="helpRegister" name="form1" method="post"
onsubmit="return check()" >
           <p>输入您的信息，身份证号必须由字母或者数字组成，带*号必须填写。
           <table bgcolor="#CCFFCC">
            <tr>
            <td>身份证号:</td><td><Input name="examID" type=text id="examID" >
*</td></tr>
            <tr>
            <td>考生姓名：</td><td><Input name="examName" type=text
id="examName" > *
             </td></tr>
            <tr>
```

```
                <td>设置密码：</td><td><Input name="password" type=password
id="password"> *
                </td> </tr>
              <tr>
                <td>考生性别：</td><td><Input name="sex" type=radio id="sex"
value="男" checked
                 = "default">男 <input type=radio name="sex" value="女"
id="sex" >女</td></tr>
              <tr>
              <td>工作单位：</td><td><Input name="company" type=text
id="company"></td></tr>
              <tr>
              <td>单位地址：</td><td><input name="address" type=text
id="address"></td></tr>
              <tr>
              <td>联系电话:</td><td><input name="phone" type=text
id="phone"></td></tr>
              <tr>
              <td>EMIL:</td><td><input name="email" type=text id="email">*</td></tr>
              <tr>
              <td>报考类别:</td><td><Select name="examType" size=2>
                         <option Selected value="高级">高级
                         <option value="中级">中级
                         </Select></td></tr>
              <tr><td>备注:</td><td><TextArea name="memo" Rows="6"
Cols="30"></TextArea>
              </td></tr>
              <tr><td></td><td><input type="submit" value="提交"
name="submit"  ></td></tr>

              </table>
            </form>
            </center>
          </blockquote>
      </div>
      <br class="clearfloat" />
      <%@ include file="txtfile/footer.txt" %>
    </div>
  </body>
</html>
```

check.js 文件(JavaScript 程序)存储在 E:\programjsp\ch11\ch11\webroot\js 文件夹中，文件内容如下：

```
    function check(){
          var txtid=document.form1.examID.value;
          if(txtid.search("^[A-Za-z0-9]+$")!=0 ){
```

```
        alert("请输入正确的身份证号");
        form1.examID.focus();
        return false;
        }
      if(form1.password.value==""){
        alert("密码不能为空！");
        form1.password.focus();
        return false;
        }
      var txt=document.form1.email.value;
      if(txt.search("^[\\w-]+(\\.[\\w-]+)*@[\\w-]+(\\.[\\w-]+)+$")!=0){
        alert("请输入正确的电子邮件地址");
        document.form1.email.select();
        return false;
        }
}
```

提示： JavaScript 程序可以直接写入 JSP 页面中，也可以单独保存在 JS 文件中，单独保存有利于 JSP 页面与程序逻辑的分离，也便于 JavaScript 程序的维护。

showRegisterMess.jsp 页面效果如图 11.7 所示。

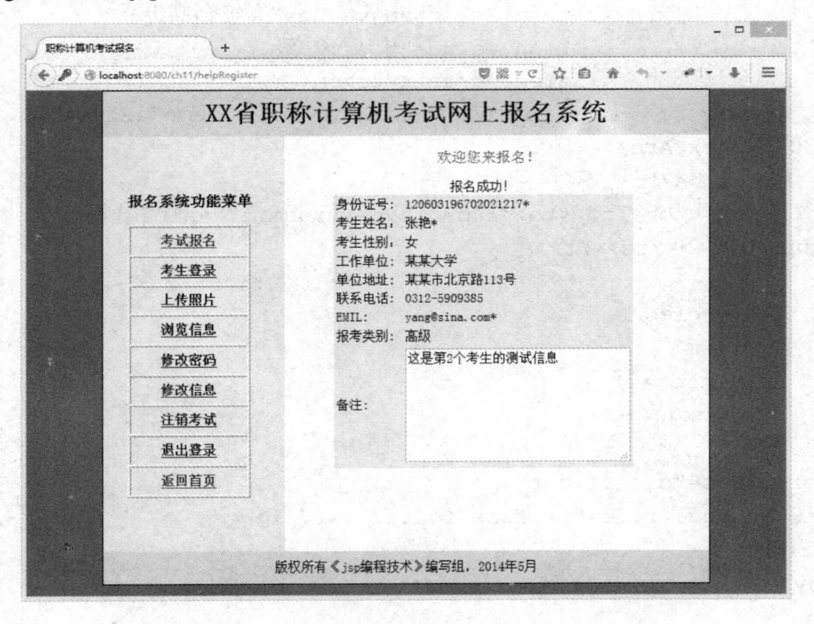

图 11.7　显示考生报名信息

页面代码如下：

```
<%@ page language="java" import="java.util.*" pageEncoding="GBK"%>
<%@ page import="mybean.data.*" %>
<jsp:useBean id="register" type="mybean.data.registerBean"
scope="request"/>
```

```
<%
String path = request.getContextPath(); String basePath =
request.getScheme()+ "://"+
request.getServerName() +":"+request.getServerPort()+path+"/";
%>
<!DOCTYPE HTML PUBLIC "-//W3C//DTD HTML 4.01 Transitional//EN">
<html>
  <head>
    <base href="<%=basePath%>">

    <title>职称计算机考试报名</title>
     <meta http-equiv="pragma" content="no-cache">
     <meta http-equiv="cache-control" content="no-cache">
     <meta http-equiv="expires" content="0">
     <meta http-equiv="keywords" content="网上报名,职称考试,计算机">
     <meta http-equiv="description" content="网上报名首页">
     <link href="css/mystylesheet.css" rel="stylesheet" type="text/css" />
  </head>
  <body class="twoColHybLtHdr">
    <div id="container">
      <div id="header">
         <%@ include file="txtfile/header.txt" %>
      </div>
      <div id="sidebar1">
         <%@ include file="txtfile/left.txt" %>
      </div>
      <div id="mainContent"  >
          <center >
             <p><font size=4 color=red>欢迎您来报名! </font></p></center>
          <blockquote>
            <center>
            <p><jsp:getProperty name="register" property="backMessage"/>
            <table bgcolor="#CCFFCC">
            <tr>
            <td>身份证号:</td><td><jsp:getProperty name="register"
            property="examID "/>*
            </td></tr>
            <tr>
            <td>考生姓名: </td><td><jsp:getProperty name="register"
            property="examName" />*
            </td></tr>
            <tr>
            <td>考生性别: </td><td><jsp:getProperty name="register"
            property="sex" /></td></tr>
            <tr>
            <td>工作单位: </td><td><jsp:getProperty name="register"
            property="company" />
```

```
        </td></tr>
        <tr>
        <td>单位地址: </td><td><jsp:getProperty name="register"
        property="address"/>
        </td></tr>
        <tr>
        <td>联系电话:</td><td><jsp:getProperty name="register"
        property="phone" />
        </td></tr>
        <tr>
        <td>EMIL:</td><td><jsp:getProperty name="register"
        property="email" />*</td></tr>
        <tr>
        <td>报考类别:</td><td><jsp:getProperty name="register"
        property="examType" />
        </td></tr>
        <tr><td>备注:</td><td><TextArea name="memo" Rows="6"
        Cols="30" ><jsp:getProperty  name="register" property
        ="memo"/> </TextArea>
        </td></tr>
        <tr><td></td><td></td></tr>
        </table>
        </center>
       </blockquote>
    </div>
    <br class="clearfloat" />
    <%@ include file="txtfile/footer.txt" %>
</div>
</body>
</html>
```

11.5.3 控制器(Servlet)

该模块控制器 Servlet 类包名是 myservlet.controls，类名是 HandleRegister.java，对象名字为 register，url 映射名字为 helpRegister，url 映射名与 register.jsp 页面 Form 表单 action 属性值相对应。控制器 register 负责调用 examineeBll 类的对象，将考生报名信息写入 examinee 表，并将页面跳转到 showRegisterMess.jsp 页面，通过 registerBean 对象将报名信息显示给用户。

ch11 服务目录的 web.xml 配置文件框架已在 11.3.4 节中经给出。为了使用该模块的控制器 HandleRegister，应在 web.xml 文件中适当位置增加如下配置片段：

```
<servlet>      <!--考生报名模块 Servlet 配置-->
    <servlet-name>register</servlet-name>
    <servlet-class>myservlet.controls.HandleRegister</servlet-class>
</servlet>
```

```
<servlet-mapping>
    <servlet-name>register</servlet-name>
    <url-pattern>/helpRegister</url-pattern>
</servlet-mapping>
```

HandleRegister.java 代码如下：

```
package myservlet.controls;
import mybean.data.*;//引入javabean包
import java.io.*;
import myclass.bll.*;//引入业务逻辑层包
import javax.servlet.*;
import javax.servlet.http.*;
public class HandleRegister extends HttpServlet{
    public void init(ServletConfig config) throws
    ServletException{
        super.init(config);
    }
    public String handleString(String s){
        try{ byte bb[]=s.getBytes("iso-8859-1");
            s=new String(bb);
        }
        catch(Exception ee){System.out.println(ee.toString());}
        return s;
    }
    public void doPost(HttpServletRequest request ,
            HttpServletResponse response) throws ServletException,
            IOException {
        registerBean reg=new registerBean();
        request.setAttribute("register", reg);
        String examID=request.getParameter("examID").trim(),
        examName=request.getParameter("examName").trim(),
        sex=request.getParameter("sex").trim(),
        company=request.getParameter("company").trim(),
        address=request.getParameter("address").trim(),
        phone=request.getParameter("phone").trim(),
        email=request.getParameter("email").trim(),
        examType=request.getParameter("examType").trim(),
        memo=request.getParameter("memo").trim(),
        password=request.getParameter("password"),
        pic="",
        backMessage="";
        int result=0;
        try{//调用业务逻辑组件
            examineeBll exambll=new examineeBll();
            //0 身份证为空 -1 身份证号或密码不符合要求 1 报名成功
            result=exambll.CreateExaminee(examID, examName, sex, company,
            address, phone, email, password, examType, memo, pic);
```

```
            if (result==0){
            backMessage="身份证号或密码不符合要求！请重新报考！";
            reg.setBackMessage(backMessage);
            }
            if(result==-1){
                backMessage="数据库访问发生错误!";
                reg.setBackMessage(backMessage);
            }
            if(result==-2){
                backMessage="身份证号码已存在!";
                reg.setBackMessage(backMessage);
            }
            if(result==1){//给 bean 赋值
                backMessage="报名成功！";
                reg.setAddress(handleString(address));
                reg.setBackMessage(backMessage);
                reg.setCompany(handleString(company));
                reg.setEmail(email);
                reg.setExamName(handleString(examName));
                reg.setExamID(examID);
                reg.setExamType(handleString(examType));
                reg.setMemo(handleString(memo));
                reg.setPassword(password);
                reg.setPhone(phone);
                reg.setPic(pic);
                reg.setSex(handleString(sex));
            }
        }
        catch(Exception ex){
            backMessage="发生错误！"+ex.toString();
            reg.setBackMessage(backMessage);
        }
        RequestDispatcher dispatcher=
            request.getRequestDispatcher("showRegisterMess.jsp");  //页面跳转
        dispatcher.forward(request, response);
    }
    public void doGet(HttpServletRequest request,
            HttpServletResponse response) throws ServletException,IOException
        {
        doPost(request,response);
        }
}
```

技巧：考生报名模块是本系统的一个基础模块，这个模块的编写完成是整个系统完成的关键，这标志着用户对三层架构和 MVC 模式已经理解和驾驭。在各层类的编码、编译、测试过程中，要注意观察变量的值和方法执行的路线。要注意浏览器中给出的异常错误提示，例如"java.lang.NullPointerException"错误，往往是request.getParameter("...")得到了一个空指针错误。用户可核对属“...”是否正确，尝试解决错误。

11.6 考生登录模块

考生可在该模块输入自己的身份证号和密码登录，系统对身份证号和密码进行验证，通过验证的用户将进入系统，如果输入的身份证号或密码错误，将提示用户身份证号或密码错误。

该模块的模型 JavaBean 描述考生的登录信息，视图分两部分，login.jsp 页面负责提交考生的登录信息到控制器，showLoginMess.jsp 页面负责显示考生登录是否成功的信息。该模块的控制器 Servlet 负责调用 examineeBll 对象验证考生登录信息是否正确，并更新视图。

11.6.1 模型

下面的 JavaBean 用来描述考生登录信息。

loginBean.java 文件代码如下：

```
package mybean.data;
public class loginBean {
    String loginName,password,backMessage;
    boolean success=false;

    public String getLoginName() {
        return loginName;
    }
    public void setLoginName(String loginName) {
        this.loginName = loginName;
    }
    public String getPassword() {
        return password;
    }
    public void setPassword(String password) {
        this.password = password;
    }
    public String getBackMessage() {
        return backMessage;
    }
    public void setBackMessage(String backMessage) {
        this.backMessage = backMessage;
    }
    public boolean isSuccess() {
        return success;
```

```
    }
    public void setSuccess(boolean success) {
        this.success = success;
    }
}
```

11.6.2 视图

本模块有两个 JSP 页面，login.jsp 负责提供考生登录界面，showLoginMess.jsp 负责显示登录是否成功的信息。

login.jsp 页面显示效果如图 11.8 所示。

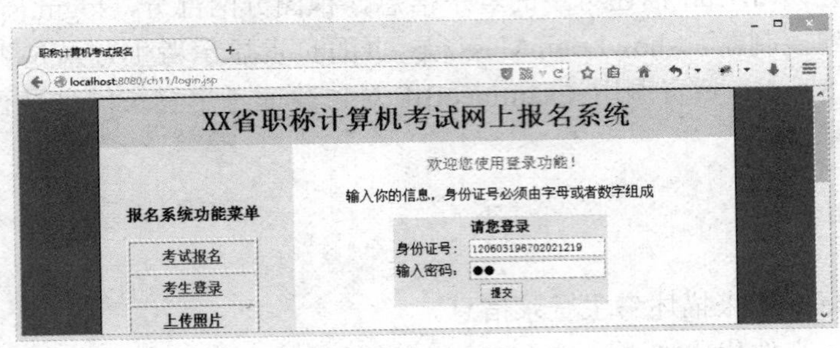

图 11.8 登录视图

页面代码如下：

```
<%@ page language="java" import="java.util.*" pageEncoding="GBK"%>
<%
String path = request.getContextPath();
String basePath =
request.getScheme()+"://"+request.getServerName()+":"+request.getServerP
ort()+
path + "/";
%>
<!DOCTYPE HTML PUBLIC "-//W3C//DTD HTML 4.01 Transitional//EN">
<html>
  <head>
    <base href="<%=basePath%>">

    <title>职称计算机考试报名</title>

    <meta http-equiv="pragma" content="no-cache">
    <meta http-equiv="cache-control" content="no-cache">
    <meta http-equiv="expires" content="0">
    <meta http-equiv="keywords" content="网上报名,职称考试,计算机">
    <meta http-equiv="description" content="网上报名首页">
    <link href="css/mystylesheet.css" rel="stylesheet" type="text/css" />
```

```
 </head>

   <body class="twoColHybLtHdr">
     <div id="container">
       <div id="header">
          <%@ include file="txtfile/header.txt" %>
       </div>
       <div id="sidebar1">
          <%@ include file="txtfile/left.txt" %>
       </div>
       <div id="mainContent"  >
            <center >
               <p><font size=4 color=red>欢迎您使用登录功能！</font></p></center>
            <blockquote>
              <center>
              <form action="helpLogin" name=form method="post" >
              <p>输入您的信息，身份证号必须由字母或者数字组成
              <p>
              <table bgcolor="#CCFFCC">
              <tr><th colspan="2">请您登录</th>
              </tr>
              <tr>
              <td>身份证号:</td><td><Input name="loginName" type=text
id="loginName" ></td>
              </tr>
              <tr>
              <td>输入密码：</td><td><Input name="password" type=password
id="password">
              </td></tr>
              <tr align="center"><td colspan="2"><input type="submit" value="提交"
                        name="submit" ></td></tr>
              </table>
              </form>
              </center>
            </blockquote>
       </div>
       <br class="clearfloat" />
       <%@ include file="txtfile/footer.txt" %>
     </div>
   </body>
</html>
```

showLoginMess.jsp 页面效果如图 11.9 所示。

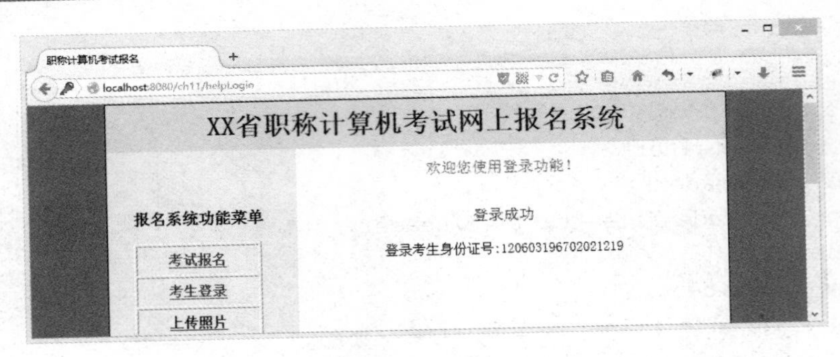

图 11.9　显示登录信息

页面代码如下：

```
<%@ page language="java" import="java.util.*" pageEncoding="GBK"%>
<%@ page import="mybean.data.loginBean" %>
<jsp:useBean id="login" type="mybean.data.loginBean" scope="session"/>

<%
String path = request.getContextPath();
String basePath = request.getScheme()+"://"+request.getServerName()+":"+
request.getServerPort() +
path+"/";
%>

<!DOCTYPE HTML PUBLIC "-//W3C//DTD HTML 4.01 Transitional//EN">
<html>
  <head>
    <base href="<%=basePath%>">

    <title>职称计算机考试报名</title>

     <meta http-equiv="pragma" content="no-cache">
     <meta http-equiv="cache-control" content="no-cache">
     <meta http-equiv="expires" content="0">
     <meta http-equiv="keywords" content="网上报名,职称考试,计算机">
     <meta http-equiv="description" content="网上报名首页">
     <link href="css/mystylesheet.css" rel="stylesheet" type="text/css" />

 </head>

   <body class="twoColHybLtHdr">
     <div id="container">
       <div id="header">
          <%@ include file="txtfile/header.txt" %>
       </div>
       <div id="sidebar1">
          <%@ include file="txtfile/left.txt" %>
       </div>
       <div id="mainContent"  >
```

```
            <center >
            <p><font size=4 color=red>欢迎您使用登录功能！</font></p></center>
            <br><center><font size=4 color=blue>
            <jsp:getProperty name="login" property="backMessage" />
            </font></center>
            <% if(login.isSuccess()==true){%>
              <center><br>登录考生身份证号:<jsp:getProperty name="login"
              property="loginName" /></center>
            <%}
              else {
            %><br><center>登录考生身份证号：<jsp:getProperty name="login"
              property="loginName" /></center>
              <br><center>登录使用的密码为：<jsp:getProperty name="login"
               property="password"/></center>
            <%} %>
        </div>
        <br class="clearfloat" />
        <%@ include file="txtfile/footer.txt" %>
</div>
</body>
</html>
```

11.6.3 控制器

控制器 Servlet 类包名是 myservlet.controls，类名是 HandleLogin，对象名字为 login，URL 映射名字为 helpLogin，URL 映射名与 login.jsp 页面 Form 表单 action 属性值相对应。控制器 login 负责调用 examineeBll 类的对象，查询 examinee 表，验证考生的身份证号和密码是否正确，并将页面跳转到 showLoginMess.jsp 页面，通过 login 对象将报名信息显示给用户。login 对象的生存周期为 session，这样用户转到其他页面时，可通过判断 login 对象来判断用户是否已经登录。

为了使用该模块的控制器 HandleLogin，应在 web.xml 文件中适当位置增加如下配置片段：

```
<servlet>
   <servlet-name>login</servlet-name>
   <servlet-class>myservlet.controls.HandleLogin</servlet-class>
</servlet>
<servlet-mapping>
   <servlet-name>login</servlet-name>
   <url-pattern>/helpLogin</url-pattern>
</servlet-mapping>
```

HandleLogin.java 的源代码如下：

```
package myservlet.controls;
import mybean.data.*;
```

```
import java.io.*;
import myclass.bll.*;
import javax.servlet.*;
import javax.servlet.http.*;
public class HandleLogin extends HttpServlet{
    public void init(ServletConfig config) throws ServletException{
        super.init(config);
    }
    public String handleString(String s){
        try{ byte bb[]=s.getBytes("iso-8859-1");
            s=new String(bb);
        }
        catch(Exception ee){}
        return s;
    }
    public void doPost(HttpServletRequest request,HttpServletResponse response)
        throws ServletException,IOException{
        loginBean log=null;
        String backMessage="";
        HttpSession session=request.getSession(true);
        try{ log=(loginBean)session.getAttribute("login");
            if(log==null)
            { log=new loginBean();
              session.setAttribute("login", log);
            }
        }
        catch(Exception ee){
            log=new loginBean();
            session.setAttribute("login", log);
        }
        String loginName=request.getParameter("loginName").trim(),
            password=request.getParameter("password").trim();
        loginName=handleString(loginName);
        boolean ok=log.isSuccess();
        if(ok==true && loginName.equals(log.getLoginName())){
            backMessage="已经登录了";
            log.setBackMessage(backMessage);
        }
        else{
            examineeBll exambll=new examineeBll();
            int result=exambll.examineeLogin(loginName, password);
            if(result==1){
                backMessage="登录成功";
                log.setBackMessage(backMessage);
                log.setSuccess(true);
                log.setLoginName(loginName);
            }
            else if(result==0){
```

```
                backMessage="您输入的身份证号或密码不符合要求！";
                log.setBackMessage(backMessage);
                log.setSuccess(false);
                log.setLoginName(loginName);
                log.setPassword(password);
            }
            else{
                backMessage="您输入的身份证号不存在或密码不正确!";
                log.setBackMessage(backMessage);
                log.setSuccess(false);
                log.setLoginName(loginName);
                log.setPassword(password);
            }
        }
        RequestDispatcher dispatcher=
            request.getRequestDispatcher("showLoginMess.jsp");
        dispatcher.forward(request, response);
    }
}
```

11.7 上传照片模块

登录的考生可在该模块上传报名照片。如果考生已经上传过照片，新上传的照片将替换原有照片。

该模块的模型 JavaBean 描述上传文件的信息，如上传文件名、保存文件名、返回信息等。视图分为两个 JSP 页面，uploadpic.jsp 页面负责提交图像文件到控制器，showUploadMess.jsp 页面负责显示上传操作返回的信息。该模块的控制器负责调用 uploadFileBll 对象将文件上传到服务器并保存文件名到 examinee 表中，另外还负责视图显示和阻止未登录用户上传照片。

11.7.1 模型

下面 uploadFileBean 描述上传文件的信息。

uploadFileBean.java 文件代码如下：

```
package mybean.data;

public class uploadFileBean {
    String fileName,savedFileName,backMessage="";
    public String getFileName() {
        return fileName;
    }
    public void setFileName(String fileName) {
        this.fileName = fileName;
    }
```

```
    public String getSavedFileName() {
        return savedFileName;
    }
    public void setSavedFileName(String savedFileName) {
        this.savedFileName = savedFileName;
    }
    public String getBackMessage() {
        return backMessage;
    }
    public void setBackMessage(String backMessage) {
        this.backMessage = backMessage;
    }
}
```

11.7.2 视图

本视图有两个页面组成，uploadpic.jsp 和 showUploadMess.jsp。uploadpic.jsp 负责提供上传文件界面，showUploadMess.jsp 负责显示上传文件的反馈信息。

uploadpic.jsp 页面效果如图 11.10 所示。

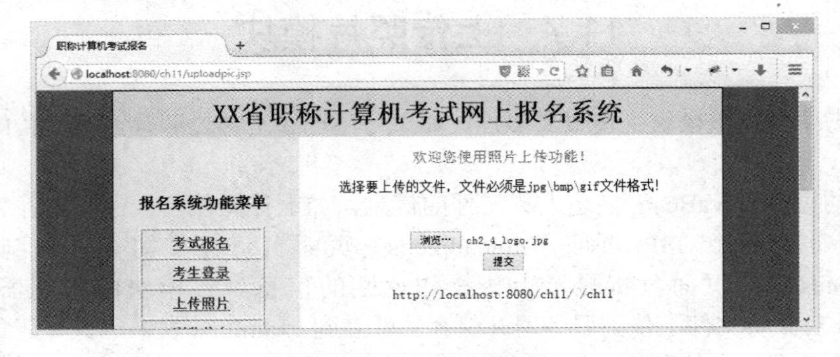

图 11.10　上传文件页面

页面代码如下：

```
<%@ page language="java" import="java.util.*" pageEncoding="GBK"%>
<%
String path = request.getContextPath();
String basePath =
request.getScheme()+"://"+request.getServerName()+":"+request.getServerP
ort()+
path+"/";
%>

<!DOCTYPE HTML PUBLIC "-//W3C//DTD HTML 4.01 Transitional//EN">
<html>
  <head>
    <base href="<%=basePath%>">
```

```
    <title>职称计算机考试报名</title>
    <meta http-equiv="pragma" content="no-cache">
    <meta http-equiv="cache-control" content="no-cache">
    <meta http-equiv="expires" content="0">
    <meta http-equiv="keywords" content="网上报名,职称考试,计算机">
    <meta http-equiv="description" content="网上报名首页">
    <link href="css/mystylesheet.css" rel="stylesheet" type="text/css" />
    <script type="text/javascript">
        function check()
        {
         if(myform.uploadFileName.value==""){
           alert("文件名不正确");
           myform.uploadFileName.focus();
           return false;
           }
        }
    </script>
</head>
  <body class="twoColHybLtHdr">
    <div id="container">
      <div id="header">
        <%@ include file="txtfile/header.txt" %>
      </div>
      <div id="sidebar1">
        <%@ include file="txtfile/left.txt" %>
      </div>
      <div id="mainContent"  >
          <center >
             <p><font size=4 color=red>欢迎您使用照片上传功能! </font></p></center>
          <center>
           <form action="helpUpload" method="post" name="myform" ENCTYPE=
           "multipart/form-data"  onsubmit="return check()" >
           <p>选择要上传的文件，文件必须是 jpg\bmp\gif 文件格式！
           <br>
           <br><br>
           <table>
           <tr><td><input type=file name="uploadFileName" size="40"
id="uploadFileName" >
           </td></tr>
           <tr><td align="center"><input type=submit name="g" value="提交">
           </table>
           </form>
           <p> <% out.println(basePath); out.println(path);%>
         </center>
     </div>
     <br class="clearfloat" />
     <%@ include file="txtfile/footer.txt" %>
```

```
</div>
</body>
</html>
```

showUploadMess.jsp 页面效果如图 11.11 所示。

图 11.11　显示上传操作信息

页面代码如下：

```
<%@ page language="java" import="java.util.*" pageEncoding="GBK"%>
<%@ page import="mybean.data.uploadFileBean" %>
<jsp:useBean id="upFile" type="mybean.data.uploadFileBean"
scope="request"/>
<%
String path = request.getContextPath();
String basePath =
request.getScheme()+"://"+request.getServerName()+":"+request.getServerP
ort()+
path+"/";
%>
<!DOCTYPE HTML PUBLIC "-//W3C//DTD HTML 4.01 Transitional//EN">
<html>
  <head>
    <base href="<%=basePath%>">
    <title>职称计算机考试报名</title>
     <meta http-equiv="pragma" content="no-cache">
     <meta http-equiv="cache-control" content="no-cache">
     <meta http-equiv="expires" content="0">
     <meta http-equiv="keywords" content="网上报名,职称考试,计算机">
     <meta http-equiv="description" content="网上报名首页">
     <link href="css/mystylesheet.css" rel="stylesheet" type="text/css" />
</head>
  <body class="twoColHybLtHdr">
    <div id="container">
      <div id="header">
         <%@ include file="txtfile/header.txt" %>
      </div>
```

```
        <div id="sidebar1">
           <%@ include file="txtfile/left.txt" %>
        </div>
        <div id="mainContent" >
            <center >
             <p><font size=4 color=red>欢迎您使用照片上传功能!
</font></p></center>
             <center><p><font size=2 color=blue>
             <br><jsp:getProperty name="upFile" property="backMessage" />
               上传的文件名：<jsp:getProperty name="upFile"
property="fileName"/>
             <br>保存后的文件名：<jsp:getProperty name="upFile"
property="savedFileName"/>
             <br></font>
             <img src=image/<jsp:getProperty name="upFile"
property="savedFileName"/>
             width=150 height=120>图像效果
          </center>
       </div>
       <br class="clearfloat" />
       <%@ include file="txtfile/footer.txt" %>
</div>
</body>
</html>
```

11.7.3　控制器

该模块控制器 Servlet 对象的名字是 uploadFile，类名是 HandleUploadFile，URL 映射名是 helpUpload。uploadFile 控制器负责检查考生是否登录。如果没有登录则将页面跳转到 login.jsp 登录页面。如果已登录，则调用 uploadFileBll 对象上传文件，将文件保存到 Web 服务目录中的 image 文件夹中。保存图像文件的名字为考生的身份证号+上传文件的扩展名。文件上传成功后，uploadFile 将文件名保存到 examinee 表的 pic 字段，并更新成视图显示。

为了使用该模块的控制器 HandleUploadFile，应在 web.xml 文件中适当位置增加如下配置片段：

```
<servlet>
   <servlet-name>uploadFile</servlet-name>
   <servlet-class>myservlet.controls.HandleUploadFile</servlet-class>
</servlet>
<servlet-mapping>
   <servlet-name>uploadFile</servlet-name>
   <url-pattern>/helpUpload</url-pattern>
</servlet-mapping>
```

HandleUploadFile.java 源代码如下：

```
package myservlet.controls;
import java.io.*;
import mybean.data.loginBean;
import mybean.data.uploadFileBean;
import myclass.bll.*;
import javax.servlet.*;
import javax.servlet.http.*;

public class HandleUploadFile extends HttpServlet{
    public void init(ServletConfig config)throws ServletException{
        super.init(config);
    }
    public String handleString(String s){
        try{byte bb[]=s.getBytes("iso-8859-1");
            s=new String(bb);
        }
        catch(Exception ee){}
        return s;
    }
    public void doPost(HttpServletRequest request ,HttpServletResponse
            response) throws ServletException,IOException{
     HttpSession session=request.getSession(true);
     loginBean log=(loginBean)session.getAttribute("login");
     if(log==null){
          response.sendRedirect("login.jsp");
     }
     else{
       String loginName=log.getLoginName();
          uploadFileMethod(request,response,loginName);
     }
    }
    public void  uploadFileMethod(HttpServletRequest
request,HttpServletResponse
            response,String loginName) throws
ServletException,IOException{
        uploadFileBean upFile=new uploadFileBean();
        String backMessage="";
        try{
            HttpSession session=request.getSession(true);
            request.setAttribute("upFile", upFile);
            String tempFileName=(String)session.getId();
            InputStream in=request.getInputStream();
            String saveFileName=loginName,
            //上传文件保存的物理路径
            driverPath=request.getRealPath("image");
```

```
            uploadFileBll upbll=new uploadFileBll();
            boolean flag=upbll.uploadFileMethod(driverPath, tempFileName,
                    saveFileName, in);
            //System.out.println("执行到这里的");
            if(flag){
                examineeBll exambll=new examineeBll();
                int n=exambll.setExamineePic(loginName,
upbll.getSavedFileName());
                if(n==1){
                    backMessage="文件上传成功！";
                    upFile.setBackMessage(backMessage);
                    upFile.setFileName(upbll.getUploadFileName());
                    upFile.setSavedFileName(upbll.getSavedFileName());
                }
                else{
                    backMessage="文件上失败！";
                }
            }
            else{
                backMessage="文件上传失败！";
                upFile.setBackMessage(backMessage);
            }
        }
        catch(Exception exp){
            backMessage=""+exp;
            upFile.setBackMessage(backMessage);
        }
        RequestDispatcher dispatcher=
            request.getRequestDispatcher("showUploadMess.jsp");
        dispatcher.forward(request, response);
    }
    public void doGet(HttpServletRequest request,HttpServletResponse
            response) throws ServletException,IOException{
        doPost(request,response);
    }
}
```

11.8 浏览信息模块

该模块负责查找考生并显示考生信息，包括考生姓名、性别、工作单位、单位地址、联系电话、备注、考试类别、照片等信息。同时提供了分页显示所有考生姓名、性别、工作单位、报考类别等信息的功能。

该模块有两个 JavaBean，分别描述考生信息和分页信息。视图部分由三个 JSP 页面构成，一个页面选择浏览方式，另两个页面分别显示单个考生信息和全部考生信息。控制器 Servlet 有两个，通过 examineeBll 对象从 examinee 表中获得考生数据，分别控制分页显示和显示特定考生信息。控制器可以阻止未登录考生浏览信息。

11.8.1 模型

该模块模型有两个 JavaBean。showExamineeByIDBean 描述考生信息，showExamineeByPage 模型负责描述分页信息。

showExamineeByIDBean.java 代码如下：

```
package mybean.data;
public class showExamineeByIDBean {
    private String examID = null;// 身份证号
    private String examName = null;// 姓名
    private String sex = null;// 性别
    private String company = null;// 工作单位
    private String address = null;// 单位地址
    private String phone = null;// 联系电话
    private String email = null;// email 地址
    private String examType = null;// 考试类型
    private String memo = null;// 备注
    private String backMessage;//返回信息
    private String pic=null;//照片信息

    public String getExamID() {
        return examID;
    }
    public void setExamID(String examID) {
        this.examID = examID;
    }
    public String getExamName() {
        return examName;
    }
    public void setExamName(String examName) {
        this.examName = examName;
    }
    public String getSex() {
        return sex;
    }
    public void setSex(String sex) {
        this.sex = sex;
    }
    public String getCompany() {
        return company;
    }
    public void setCompany(String company) {
        this.company = company;
    }
    public String getAddress() {
```

```
        return address;
    }
    public void setAddress(String address) {
        this.address = address;
    }
    public String getPhone() {
        return phone;
    }
    public void setPhone(String phone) {
        this.phone = phone;
    }
    public String getEmail() {
        return email;
    }
    public void setEmail(String email) {
        this.email = email;
    }
    public String getExamType() {
        return examType;
    }
    public void setExamType(String examType) {
        this.examType = examType;
    }
    public String getMemo() {
        return memo;
    }
    public void setMemo(String memo) {
        this.memo = memo;
    }
    public String getBackMessage() {
        return backMessage;
    }
    public void setBackMessage(String backMessage) {
        this.backMessage = backMessage;
    }
    public String getPic() {
        return pic;
    }
    public void setPic(String pic) {
        this.pic = pic;
    }
}
```

showExamineeByPage.java 代码如下：

```
package mybean.data;
import myclass.bol.*;
import java.util.ArrayList;
public class showExamineeByPage {
    ArrayList<examinee> list=null;
```

```
    String backMessage="";
    int pageSize=10;
    int pageAllCount=0;
    int showPage=1;
    StringBuffer presentPageResult;
    public ArrayList<examinee> getList() {
        return list;
    }
    public void setList(ArrayList<examinee> list) {
        this.list = list;
    }
    public int getPageSize() {
        return pageSize;
    }
    public void setPageSize(int pageSize) {
        this.pageSize = pageSize;
    }
    public int getPageAllCount() {
        return pageAllCount;
    }
    public void setPageAllCount(int pageAllCount) {
        this.pageAllCount = pageAllCount;
    }
    public int getShowPage() {
        return showPage;
    }
    public void setShowPage(int showPage) {
        this.showPage = showPage;
    }
    public StringBuffer getPresentPageResult() {
        return presentPageResult;
    }
    public void setPresentPageResult(StringBuffer presentPageResult) {
        this.presentPageResult = presentPageResult;
    }
    public String getBackMessage() {
        return backMessage;
    }
    public void setBackMessage(String backMessage) {
        this.backMessage = backMessage;
    }
}
```


11.8.2 视图

该模块有视图由三个页面组成，chooseshowtype.jsp 负责选择显示考生信息的方式，并将其交给不同的控制器。showExamineeByID.jsp 页面负责特定考生信息的显示，

showExamineeByAll.jsp 负责分页显示全体考生的四项信息。

chooseshowtype.jsp 页面效果如图 11.12 所示。

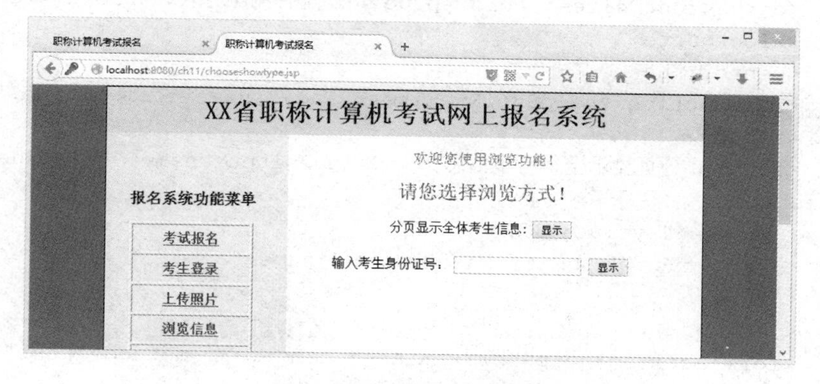

图 11.12　选择浏览方式

页面代码如下：

```
<%@ page language="java" import="java.util.*" pageEncoding="GBK"%>
<%
String path = request.getContextPath();
String basePath =
request.getScheme()+"://"+request.getServerName()+":"+request.getServerP
ort()+
path+"/";
%>

<!DOCTYPE HTML PUBLIC "-//W3C//DTD HTML 4.01 Transitional//EN">
<html>
  <head>
    <base href="<%=basePath%>">
    <title>职称计算机考试报名</title>
     <meta http-equiv="pragma" content="no-cache">
     <meta http-equiv="cache-control" content="no-cache">
     <meta http-equiv="expires" content="0">
     <meta http-equiv="keywords" content="网上报名,职称考试,计算机">
     <meta http-equiv="description" content="网上报名首页">
     <link href="css/mystylesheet.css" rel="stylesheet" type="text/css" />
</head>
  <body class="twoColHybLtHdr">
    <div id="container">
      <div id="header">
         <%@ include file="txtfile/header.txt" %>
      </div>
      <div id="sidebar1">
         <%@ include file="txtfile/left.txt" %>
      </div>
      <div id="mainContent"  >
```

```
        <center >
        <p><font size=4 color=red>欢迎您使用浏览功能！</font></p>
        <p><p><font size=5 color=blue>请您选择浏览方式！</font></center>
        <blockquote>
        <center><p><p>
        <form action="helpShowExamineeByPage" name=form method="post" >
        <table >
        <tr><td>分页显示全体考生信息:</td><td><input name="showPage"
type="hidden"
            size=6" value="1" >
        <input type="submit" value="显示" name="submit" ></td></tr>
        </table>
        </form>
        <form action="helpShowExaminee" name=form method="post">
         <table>
         <tr><td>输入考生身份证号: </td><td><Input name="examID" type=text
id="examID">
         <input type="submit" value="显示" name="submit" ></td></tr>
         </table>
        </form>
        </center>
        </blockquote>
     </div>
     <br class="clearfloat" />
     <%@ include file="txtfile/footer.txt" %>
</div>
</body>
</html>
```

showExamineeByID.jsp 页面显示效果如图 11.13 所示。

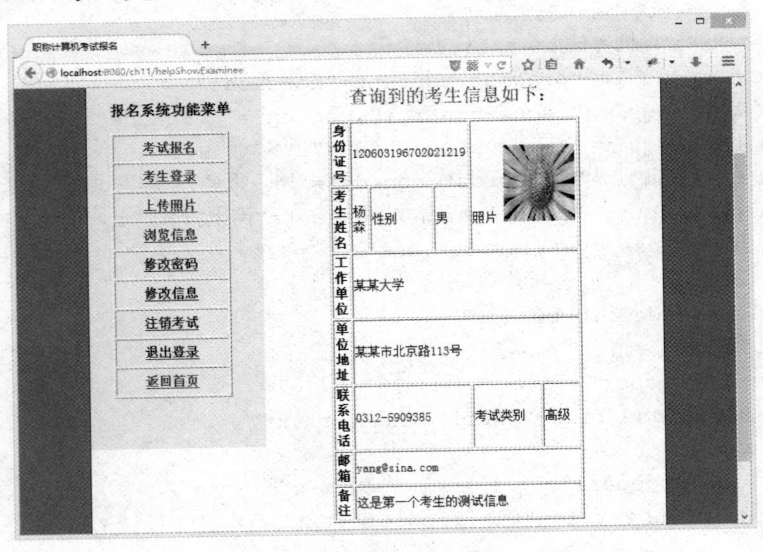

图 11.13　显示单个考生的信息页面

showExamineeByID.jsp 页面代码如下：

```
<%@ page language="java" import="java.util.*" pageEncoding="GBK"%>
<%@ page import="mybean.data.showExamineeByIDBean" %>
<jsp:useBean id="examineeinfo" type="mybean.data.showExamineeByIDBean"
scope="request"/>
<%
String path = request.getContextPath();
String basePath = request.getScheme() + ":// "+ request.getServerName()
+ ":" + request.getServerPort()
    +path+"/";
%>
<!DOCTYPE HTML PUBLIC "-//W3C//DTD HTML 4.01 Transitional//EN">
<html>
  <head>
    <base href="<%=basePath%>">
    <title>职称计算机考试报名</title>
     <meta http-equiv="pragma" content="no-cache">
     <meta http-equiv="cache-control" content="no-cache">
     <meta http-equiv="expires" content="0">
     <meta http-equiv="keywords" content="网上报名,职称考试,计算机">
     <meta http-equiv="description" content="网上报名首页">
     <link href="css/mystylesheet.css" rel="stylesheet" type="text/css" />
</head>
  <body class="twoColHybLtHdr">
    <div id="container">
      <div id="header">
         <%@ include file="txtfile/header.txt" %>
      </div>
      <div id="sidebar1">
         <%@ include file="txtfile/left.txt" %>
      </div>
      <div id="mainContent"  >
          <center >
          <p><font size=4 color=red>欢迎您使用浏览功能! </font></p></center>
          <p>
          <center><p><font size=5 color=blue>
             <jsp:getProperty name="examineeinfo"
property="backMessage"/></font></p>
           <table width="318" border="1">
             <tr>
               <th width="19" scope="row">身份证号</th>
               <td colspan="3"><jsp:getProperty name="examineeinfo"
                    property="examID" /></td>
               <td colspan="2" rowspan="2">照片
                    <img src=image/<jsp:getProperty name="examineeinfo"
property="pic"/>
```

```
                         width=90 height=100></td>
                    </tr>
                    <tr>
                      <th width="19" height="15" scope="row">考生姓名</th>
                      <td width="18" ><jsp:getProperty name="examineeinfo"
                          property="examName"/></td>
                      <td >性别</td>
                      <td><jsp:getProperty name="examineeinfo"
property="sex"/></td>
                    </tr>
                    <tr>
                      <th scope="row">工作单位</th>
                      <td colspan="5"><jsp:getProperty name="examineeinfo"
                           property="company"/></td>
                    </tr>
                    <tr>
                      <th scope="row">单位地址</th>
                      <td colspan="5"><jsp:getProperty name="examineeinfo"
property="address"/></td>
                    </tr>
                    <tr>
                      <th scope="row">联系电话</th>
                      <td colspan="3"><jsp:getProperty name="examineeinfo"
property="phone"/></td>
                      <td>考试类别</td>
                      <td><jsp:getProperty name="examineeinfo"
property="examType"/></td>
                    </tr>
                    <tr>
                      <th scope="row">邮箱</th>
                      <td colspan="5"><jsp:getProperty name="examineeinfo"
property="email"/></td>
                    </tr>
                    <tr>
                      <th scope="row">备注</th>
                      <td colspan="5"><jsp:getProperty name="examineeinfo"
property="memo"/></td>
                    </tr>
                  </table>
                  <p> </p>
                  </center>
          </div>
          <br class="clearfloat" />
          <%@ include file="txtfile/footer.txt" %>
</div>
</body>
</html>
```

showExamineeAll.jsp 页面显示效果如图 11.14 所示。

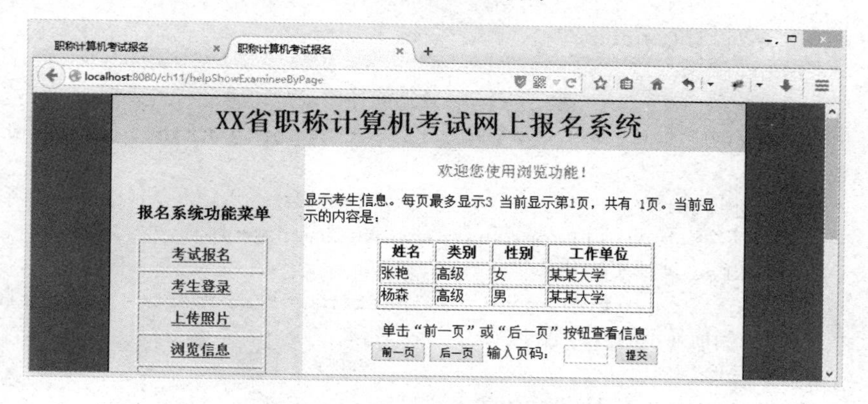

图 11.14　分页显示页面效果

页面代码如下：

```
<%@ page language="java" import="java.util.*" pageEncoding="GBK"%>
<%@ page import="mybean.data.showExamineeByPage" %>
<jsp:useBean id="examineebypage" type="mybean.data.showExamineeByPage"
scope="session"/>
<%
String path = request.getContextPath();
String basePath = request.getScheme()+"://"+request.getServerName()+":"
+request.getServerPort()+
path+"/";
%>

<!DOCTYPE HTML PUBLIC "-//W3C//DTD HTML 4.01 Transitional//EN">
<html>
  <head>
    <base href="<%=basePath%>">

    <title>职称计算机考试报名</title>
    <meta http-equiv="pragma" content="no-cache">
    <meta http-equiv="cache-control" content="no-cache">
    <meta http-equiv="expires" content="0">
    <meta http-equiv="keywords" content="网上报名,职称考试,计算机">
    <meta http-equiv="description" content="网上报名首页">
    <link href="css/mystylesheet.css" rel="stylesheet" type="text/css" />
</head>
  <body class="twoColHybLtHdr">
    <div id="container">
      <div id="header">
        <%@ include file="txtfile/header.txt" %>
      </div>
      <div id="sidebar1">
```

```
        <%@ include file="txtfile/left.txt" %>
      </div>
      <div id="mainContent" >
         <center >
         <p><font size=4 color=red>欢迎您使用浏览功能！</font></p></center>
         <p>显示考生信息。每页最多显示<jsp:getProperty name="examineebypage"
property=
            "pageSize"/>
         当前显示第<jsp:getProperty name="examineebypage"
property="showPage"/>页，共有
         <jsp:getProperty name="examineebypage" property="pageAllCount"
         />页。当前显示的内容是:
         <center>
         <table width="318" border="1">
          <tr><th>姓名</th><th>类别</th><th>性别</th><th>工作单位</th></tr>
          <jsp:getProperty name="examineebypage"
property="presentPageResult"/>
         </table>
         <p>单击"前一页"或"后一页"按钮查看信息
         <table>
         <tr><td>
         <form action="helpShowExamineeByPage" method="post" >
           <input type=hidden name="showPage"
             value="<%=examineebypage.getShowPage()-1 %>">
           <input type="submit" name="g" value="前一页">
         </form></td>
         <td><form action="helpShowExamineeByPage" method="post">
           <input type=hidden name="showPage" value=
             "<%=examineebypage.getShowPage()+1 %>">
           <input type="submit" name="g" value="后一页">
         </form></td>
         <td><form action="helpShowExamineeByPage" method="post">输入页码:
            <input type=text name="showPage" size=5>
            <input type=submit name="g" value="提交">
         </form>
         </table>
         </center>
      </div>
      <br class="clearfloat" />
      <%@ include file="txtfile/footer.txt" %>
 </div>
 </body>
 </html>
```

11.8.3 控制器

该模块有两个控制器，一个是控制器 showExaminee，类名为 HandleShowExamineeByID，

URL 映射名称为 helpShowExaminee，chooseshowtype.jsp 页面将用户的单个查询提交给控制器 showExaminee，该控制器负责调用 examineeBll 对象查询特定考生信息，并将转发给 showExamineeByID.jsp 显示查询信息。另一个是控制器 showExamineeAll，URL 映射名为 helpShowExamineeByPage，类名为 HandleShowExamineeByPage，chooseshowtype.jsp 页面将用户分页查询提交给控制器 showExaminee，该控制器负责调用 examineeBll 对象获得全部考生信息，控制 showExamineeAll.jsp 页面显示查询信息。该模块控制器能够阻止未登录用户的访问。

为了使用该模块的控制器 HandleShowExamineeByID、HandleshowExamineeByPage，应在 web.xml 文件中适当位置增加如下配置片段：

```
<servlet>
   <servlet-name>showExaminee</servlet-name>
   <servlet-class>myservlet.controls.HandleShowExamineeByID</servlet-
class>
</servlet>
<servlet-mapping>
   <servlet-name>showExaminee</servlet-name>
   <url-pattern>/helpShowExaminee</url-pattern>
</servlet-mapping>
<servlet>
   <servlet-name>showExamineeAll</servlet-name>
   <servlet-class>myservlet.controls.HandleShowExamineeByPage</servlet-
class>
</servlet>
<servlet-mapping>
   <servlet-name>showExamineeAll</servlet-name>
   <url-pattern>/helpShowExamineeByPage</url-pattern>
</servlet-mapping>
```

HandleShowExamineeByID.java 源代码如下：

```
package myservlet.controls;
import java.io.*;
import mybean.data.loginBean;
import mybean.data.showExamineeByIDBean;
import myclass.bol.*;
import myclass.bll.*;
import javax.servlet.*;
import javax.servlet.http.*;
public class HandleShowExamineeByID extends HttpServlet{
    public void init(ServletConfig config)throws ServletException{
        super.init(config);
    }
    public String handleString(String s){
        try{byte bb[]=s.getBytes("iso-8859-1");
            s=new String(bb);
        }
```

```
        catch(Exception ee){}
        return s;
    }
    public void doPost(HttpServletRequest request ,HttpServletResponse
            response) throws ServletException,IOException{
     HttpSession session=request.getSession(true);
     loginBean log=(loginBean)session.getAttribute("login");
     if(log==null){
          response.sendRedirect("login.jsp");
     }
     else{
          showExamineeByID(request,response);
     }
    }
    public void showExamineeByID(HttpServletRequest request,HttpServletResponse
            response) throws ServletException,IOException{
        showExamineeByIDBean examineeinfo=new  showExamineeByIDBean();
        request.setAttribute("examineeinfo", examineeinfo);
        String examID=request.getParameter("examID");
        examineeBll exambll=new examineeBll();
        examinee exam=new examinee();
        exam=exambll.getExamineeByID(examID);
        if(exam!=null){
            examineeinfo.setAddress(handleString(exam.getAddress()));
            examineeinfo.setBackMessage("查询到的考生信息如下：");
            examineeinfo.setCompany(handleString(exam.getCompany()));
            examineeinfo.setEmail(handleString(exam.getEmail()));
            examineeinfo.setExamID(exam.getExamID());
            examineeinfo.setExamName(handleString(exam.getExamName()));
            examineeinfo.setExamType(handleString(exam.getExamType()));
            examineeinfo.setMemo(handleString(exam.getMemo()));
            examineeinfo.setPhone(exam.getPhone());
            examineeinfo.setPic(exam.getPic());
            examineeinfo.setSex(handleString(exam.getSex()));
        }
        else{
            examineeinfo.setBackMessage("未查询到该考生信息！");
        }
        RequestDispatcher dispatcher=
            request.getRequestDispatcher("showExamineeByID.jsp");
        dispatcher.forward(request, response);
    }
}
```

HandleShowExamineeByPage.java 源代码如下：

```
package myservlet.controls;
import java.io.*;
```

```
import mybean.data.loginBean;
import mybean.data.showExamineeByPage;
import myclass.bol.*;
import myclass.bll.*;
import javax.servlet.*;
import javax.servlet.http.*;
public class HandleShowExamineeByPage extends HttpServlet{
    public void init(ServletConfig config)throws ServletException{
        super.init(config);
    }
    public String handleString(String s){
        try{byte bb[]=s.getBytes("iso-8859-1");
            s=new String(bb);
        }
        catch(Exception ee){}
        return s;
    }
    public void doPost(HttpServletRequest request ,HttpServletResponse
            response) throws ServletException,IOException{
     HttpSession session=request.getSession(true);
     loginBean log=(loginBean)session.getAttribute("login");
     if(log==null){
            response.sendRedirect("login.jsp");
     }
     else{
            showExamineeByPage(request,response);
     }
    }
    public void showExamineeByPage(HttpServletRequest
request,HttpServletResponse
            response) throws ServletException,IOException{
        HttpSession session=request.getSession(true);
        StringBuffer presentPageResult=new StringBuffer();
        showExamineeByPage showBean=null;
        //如果 showBean 不存在则创建
        try{ showBean=(showExamineeByPage)session.getAttribute("examineebypage");
            if(showBean==null){
             showBean=new showExamineeByPage();
             session.setAttribute("examineebypage", showBean);
            }
        }
        catch(Exception exp){
            showBean=new showExamineeByPage();
            session.setAttribute("examineebypage", showBean);
        }
        //设置每页显示记录数
        showBean.setPageSize(3);
```

```
        //检查显示页数文本框中输入数据合法性
        String strShowPage=request.getParameter("showPage");
        for(int i=0;i<strShowPage.length();i++){
            char c=strShowPage.charAt(i);
            if(!(c>='0' && c<='9'))
            {
                strShowPage="1";
                break;
            }
        }
        if(strShowPage==""){
            strShowPage="1";
        }
        int showPage=Integer.parseInt(strShowPage);
        int pageSize=showBean.getPageSize();
        //设置 showBean 的属性值
        try{
            examineeBll exambll=new examineeBll();
            showBean.setList(exambll.getExamineeAll());
            int m=showBean.getList().size();
            int n=pageSize;
            int pageAllCount=((m%n)==0)?(m/n):(m/n+1);
            showBean.setPageAllCount(pageAllCount);
            if(showPage>showBean.getPageAllCount()){
                showPage=1;
            }
            if(showPage<=0){
                showPage=showBean.getPageAllCount();
            }
            showBean.setShowPage(showPage);
            presentPageResult=show(showPage,pageSize,showBean);
            showBean.setPresentPageResult(presentPageResult);
        }
        catch(Exception exp){
            System.out.println(exp.toString());
        }
        RequestDispatcher dispatcher=
            request.getRequestDispatcher("showExamineeAll.jsp");
        dispatcher.forward(request, response);
    }
    /**
     * 得到带 html 标记的字符串
     * @param page
     * @param pageSize
     * @param showBean
     * @return
     */
```

```
    public StringBuffer show(int page,int pageSize,showExamineeByPage
showBean){
        StringBuffer str=new StringBuffer();
        try {
          for(int i=(page-1)*pageSize;i<=page*pageSize-1;i++){
           str.append("<tr>");

    str.append("<td>"+handleString(showBean.getList().get(i).getExamName
())+"</td>");

    str.append("<td>"+handleString(showBean.getList().get(i).getExamType
())+"</td>");

    str.append("<td>"+handleString(showBean.getList().get(i).getSex())+"
</td>");

    str.append("<td>"+handleString(showBean.getList().get(i).getCompany(
))+"</td>");
           str.append("</tr>");
           //System.out.println("执行到这里 i="+i);
          }
           //System.out.println("执行到这里 page="+page);
        }
       catch(Exception exp){//System.out.println("执行到这里 page="+page);
            return str;
        }
        return str;
    }
    public void doGet(HttpServletRequest request,HttpServletResponse
            response) throws ServletException,IOException{
        doPost(request,response);
    }
}
```

11.9　修改密码模块

考生可在该模块修改自己的登录密码。该模块的 JavaBean 负责描述密码的有关信息。视图有两个页面，modifypassword.jsp 页面和 showModifyPassword.jsp 页面。控制器 HandleModifPwd 负责修改密码和页面跳转。

11.9.1　模型

模型实例 passwordBean 负责描述修改密码的信息。

passwordBean.java 文件代码如下：

```
package mybean.data;
public class passwordBean {
    String oldPassword,newPassword,backMessage="";
    public String getOldPassword() {
        return oldPassword;
    }
    public void setOldPassword(String oldPassword) {
        this.oldPassword = oldPassword;
    }
    public String getNewPassword() {
        return newPassword;
    }
    public void setNewPassword(String newPassword) {
        this.newPassword = newPassword;
    }
    public String getBackMessage() {
        return backMessage;
    }
    public void setBackMessage(String backMessage) {
        this.backMessage = backMessage;
    }
}
```

11.9.2 视图

本模块视图的 modifypassworfd.jsp 页面负责提供密码修改界面，showModifyPassword.jsp 页面负责显示修改密码的反馈信息。

modifypassword.jsp 页面的显示效果如图 11.15 所示。

图 11.15　密码修改页面效果

页面代码如下：

```
<%@ page language="java" import="java.util.*" pageEncoding="GBK"%>
<%
String path = request.getContextPath();
String basePath =
request.getScheme()+"://"+request.getServerName()+":"+request.getServerP
```

```
ort()+
path+"/";
%>

<!DOCTYPE HTML PUBLIC "-//W3C//DTD HTML 4.01 Transitional//EN">
<html>
  <head>
    <base href="<%=basePath%>">
    <title>职称计算机考试报名</title>
     <meta http-equiv="pragma" content="no-cache">
     <meta http-equiv="cache-control" content="no-cache">
     <meta http-equiv="expires" content="0">
     <meta http-equiv="keywords" content="网上报名,职称考试,计算机">
     <meta http-equiv="description" content="网上报名首页">
     <link href="css/mystylesheet.css" rel="stylesheet" type="text/css" />
</head>

  <body class="twoColHybLtHdr">
    <div id="container">
      <div id="header">
         <%@ include file="txtfile/header.txt" %>
      </div>
      <div id="sidebar1">
         <%@ include file="txtfile/left.txt" %>
      </div>
      <div id="mainContent"  >
          <center >
              <p><font size=4 color=red>欢迎您使用密码修改功能!
</font></p></center>
          <blockquote>
            <center>
              <form action="helpModifyPassword" name=form method="post" >
              <p><p>
               <table bgcolor="#CCFFCC">
               <tr><th colspan="2">请您输入密码</th></tr>
               <tr><td>输入旧密码: </td>
               <td><Input name="oldPassword" type=text id="oldPassword"
></td></tr>
               <tr><td>输入新密码: </td>
               <td><Input name="newPassword" type=password
id="newPassword"></td></tr>
               <tr align="center"><td colspan="2"><input type="submit"
               value="提交" name="submit" ></td></tr>
               </table>
              </form>
            </center>
          </blockquote>
```

```
        </div>
        <br class="clearfloat" />
        <%@ include file="txtfile/footer.txt" %>
</div>
</body>
</html>
```

showModifyPassword.jsp 页面效果如图 11.16 所示。

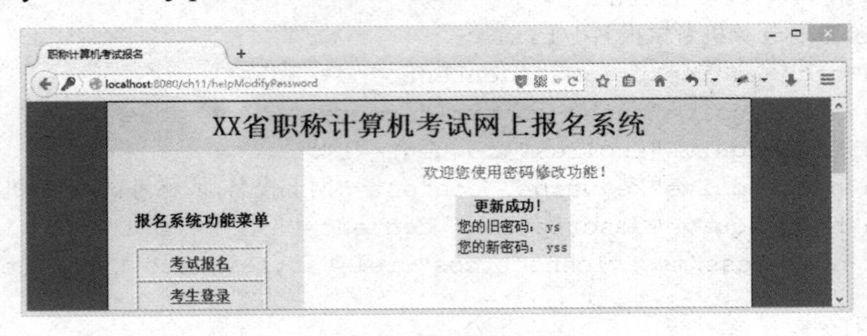

图 11.16　显示密码修改结果

页面代码如下：

```
<%@ page language="java" import="java.util.*" pageEncoding="GBK"%>
<%@ page import="mybean.data.passwordBean" %>
<jsp:useBean id="password" type="mybean.data.passwordBean"
scope="request"/>
<%  String path = request.getContextPath();
    String basePath =
request.getScheme()+"://"+request.getServerName()+":"+
    request.getServerPort()+  path+"/";
%>
<!DOCTYPE HTML PUBLIC "-//W3C//DTD HTML 4.01 Transitional//EN">
<html> <head>
    <base href="<%=basePath%>">
    <title>职称计算机考试报名</title>
     <meta http-equiv="pragma" content="no-cache">
     <meta http-equiv="cache-control" content="no-cache">
     <meta http-equiv="expires" content="0">
     <meta http-equiv="keywords" content="网上报名,职称考试,计算机">
     <meta http-equiv="description" content="网上报名首页">
     <link href="css/mystylesheet.css" rel="stylesheet" type="text/css" />
</head>
  <body class="twoColHybLtHdr">
    <div id="container">
      <div id="header">
         <%@ include file="txtfile/header.txt" %>
      </div>
      <div id="sidebar1">
         <%@ include file="txtfile/left.txt" %>
```

```
        </div>
        <div id="mainContent" >
            <center >
                <p><font size=4 color=red>欢迎您使用密码修改功能！
</font></p></center>
            <blockquote>
             <center><p><p><table bgcolor="#CCFFCC">
             <tr><th colspan="2"><jsp:getProperty name="password"
             property="backMessage" /></th></tr>
             <tr><td>您的旧密码：</td>
             <td><jsp:getProperty name="password" property="oldPassword"
/></td></tr>
             <tr><td>您的新密码：</td>
             <td><jsp:getProperty name="password"
property="newPassword"/></td></tr>
             </table> </center>
            </blockquote>
        </div>
        <br class="clearfloat" />
        <%@ include file="txtfile/footer.txt" %>
</div>
</body>
</html>
```

11.9.3 控制器

控制器 modifypassword，类名为 HandleModifPwd，URL 映射名为 helpModifyPassword。控制器 modifypassword 被 modifypassworfd.jsp 页面调用，负责生成 examineeBll 对象，并调用该对象的方法修改 examinee 表中的 password 字段，然后将请求转发给 showModifyPassword.jsp 页面显示修改信息。另外，控制器能够阻止未登录用户进行修改操作。

使用该模块的控制器 HandleShowModifPwd，应在 web.xml 文件中适当位置增加如下配置片段：

```
<servlet>
   <servlet-name>modifypassword</servlet-name>
   <servlet-class>myservlet.controls.HandleModifPwd</servlet-class>
</servlet>
<servlet-mapping>
   <servlet-name>modifypassword</servlet-name>
   <url-pattern>/helpModifyPassword</url-pattern>
</servlet-mapping>
```

HandleModifPwd.java 源文件如下：

```
package myservlet.controls;
import mybean.data.*;
```

```
import java.io.*;
import myclass.bll.*;
import javax.servlet.*;
import javax.servlet.http.*;
public class HandleModifPwd extends HttpServlet{
    public void init(ServletConfig config) throws
          ServletException{
        super.init(config);
    }
    public void doPost(HttpServletRequest request ,HttpServletResponse
                  response) throws ServletException,IOException{
        HttpSession session=request.getSession(true);
        loginBean log=(loginBean)session.getAttribute("login");
        if(log==null){  response.sendRedirect("login.jsp");
        }
        else{       modifyPassword(request,response);
        }
    }
    public void modifyPassword(HttpServletRequest request,
          HttpServletResponse response) throws ServletException,
                IOException {
        HttpSession session=request.getSession(true);
        loginBean log=(loginBean)session.getAttribute("login");
        String loginName=log.getLoginName();
        passwordBean pwd=new passwordBean();
        request.setAttribute("password", pwd);
        String oldPassword=request.getParameter("oldPassword");
        String newPassword=request.getParameter("newPassword");
        examineeBll exambll=new examineeBll();
        int result=exambll.setExamineePwd(loginName, newPassword,
oldPassword);
        if (result==0){
            pwd.setBackMessage("密码不符合要求，未更新！");
        }
        else if(result==-1){
            pwd.setBackMessage("旧密码不正确，密码更新失败！");
        }
        else {
            pwd.setBackMessage("更新成功！");
            pwd.setNewPassword(newPassword);
            pwd.setOldPassword(oldPassword);
        }
        RequestDispatcher dispatcher=
            request.getRequestDispatcher("showModifyPassword.jsp");
        dispatcher.forward(request, response);
    }
    public void doGet(HttpServletRequest request,
```

```
            HttpServletResponse response)throws
ServletException,IOException{
        doPost(request,response);
    }
}
```

11.10　修改报名信息模块

考生可在该模块修改除密码、照片、身份证号以外的考生报名信息。该模块的 JavaBean 负责描述考生的报考信息。模块的视图有两个，一个视图负责显示当前信息，并提交修改后的信息；另一个负责显示修改结果信息。该模块的控制器有两个，一个负责读取考生当前数据，并调用视图显示。另一个负责调用业务逻辑层类对象保存修改信息，并控制更新视图，显示修改结果。

11.10.1　模型

模型中需要两个 JavaBean，一个是 registerBean，保存读出的修改前信息，第 11 章 11.5 节中已经给出代码；另一个是 modifyMessageBean，用来描述考生的修改信息。

modifyMessageBean.java 文件代码如下：

```
package mybean.data;
public class modifyMessageBean {
    private String examID = null;// 身份证号
    private String examName = null;// 姓名
    private String sex = null;// 性别
    private String company = null;// 工作单位
    private String address = null;// 单位地址
    private String phone = null;// 联系电话
    private String email = null;// email 地址
    private String examType = null;// 考试类型
    private String memo = null;// 备注
    private String backMessage;
    public String getExamID() {
        return examID;
    }
    public void setExamID(String examID) {
        this.examID = examID;
    }
    public String getExamName() {
        return examName;
    }
    public void setExamName(String examName) {
        this.examName = examName;
    }
```

```
    public String getSex() {
        return sex;
    }
    public void setSex(String sex) {
        this.sex = sex;
    }
    public String getCompany() {
        return company;
    }
    public void setCompany(String company) {
        this.company = company;
    }
    public String getAddress() {
        return address;
    }
    public void setAddress(String address) {
        this.address = address;
    }
    public String getPhone() {
        return phone;
    }
    public void setPhone(String phone) {
        this.phone = phone;
    }
    public String getEmail() {
        return email;
    }
    public void setEmail(String email) {
        this.email = email;
    }
    public String getExamType() {
        return examType;
    }
    public void setExamType(String examType) {
        this.examType = examType;
    }
    public String getMemo() {
        return memo;
    }
    public void setMemo(String memo) {
        this.memo = memo;
    }
    public String getBackMessage() {
        return backMessage;
    }
    public void setBackMessage(String backMessage) {
        this.backMessage = backMessage;
    }
}
```

11.10.2　视图

本模块有两个视图，modifymessage.jsp 和 showModifyMessage.jsp 页面。用户单击“修改信息”链接，就会将调用控制器 handleGetExamineeMessage 获得登录考生已有报名信息；获得信息后，该 Servlet 将请求转发到 modifymessage.jsp 页面；modifymessage.jsp 页面显示本次修改前的信息，提供修改界面并负责提交修改信息到 haandleModifyMesage 控制器，showModifyMessage.jsp 页面负责显示修改结果。

modifymessage.jsp 页面显示效果如图 11.17 所示。

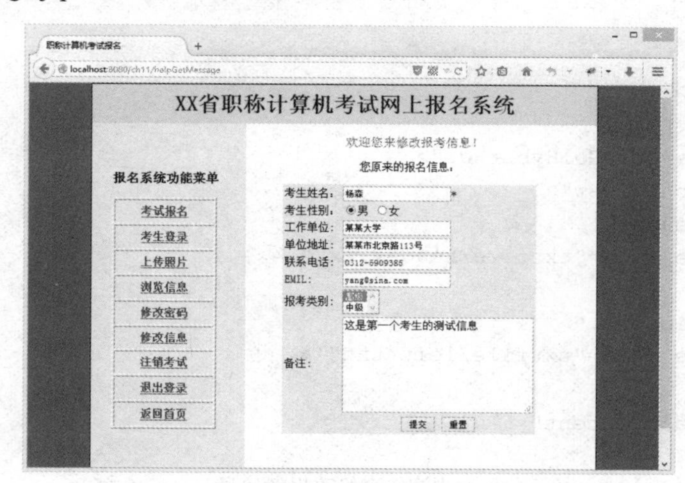

图 11.17　修改信息界面

页面代码如下：

```
<%@ page language="java" import="java.util.*" pageEncoding="GBK"%>
<%@ page import="mybean.data.registerBean" %>
<jsp:useBean id="register" type="mybean.data.registerBean"
scope="request"/>
<%
String path = request.getContextPath();
String basePath =
request.getScheme()+"://"+request.getServerName()+":"+request.getServerP
ort()+
path+"/";
%>
<!DOCTYPE HTML PUBLIC "-//W3C//DTD HTML 4.01 Transitional//EN">
<html>
<head>
    <base href="<%=basePath%>">
    <title>职称计算机考试报名</title>
     <meta http-equiv="pragma" content="no-cache">
     <meta http-equiv="cache-control" content="no-cache">
     <meta http-equiv="expires" content="0">
```

```
    <meta http-equiv="keywords" content="网上报名,职称考试,计算机">
    <meta http-equiv="description" content="网上报名首页">
    <link href="css/mystylesheet.css" rel="stylesheet" type="text/css" />
    <script type="text/javascript">
    function check(){
       var txt=document.form1.email.value;
       if(txt.search("^[\\w-]+(\\.[\\w-]+)*@[\\w-]+(\\.[\\w-]+)+$")!=0){
         alert("电子邮件地址不正确! ");
         document.form1.email.select();
         return false;
         }
       }
    </script>
  </head>
  <body class="twoColHybLtHdr">
  <div id="container">
  <div id="header">
  <%@ include file="txtfile/header.txt" %>
  </div>
  <div id="sidebar1">
  <%@ include file="txtfile/left.txt" %>
  </div>
  <div id="mainContent"  >
  <center >
  <p><font size=4 color=red>欢迎您来修改报考信息!
  </font></p></center>
  <blockquote>
  <center>
  <form action="helpModifyMessage" name=form1 method="post"
  onsubmit="return check()" >
  <p><jsp:getProperty name="register" property="backMessage"/>
  <p>
  <table bgcolor="#CCFFCC">
  <tr>
  <td>考生姓名: </td>
  <td><Input name="examName" type=text id="examName"
value=<jsp:getProperty name="register" property="examName" /> >*</td></tr>
        <tr>
        <td>考生性别: </td><td>
        <% if("男".equals(register.getSex())){
        %><Input name="sex" type=radio id="sex" value="男" checked="default">男
        <input type=radio name="sex" value="女" id="sex" >女
        <%}
        else{%>
        <Input name="sex" type=radio id="sex" value="男" >男
        <input type=radio name="sex" value="女" id="sex" checked="default" >女
        <%}%>
```

```
        </td></tr>
        <tr>
        <td>工作单位: </td><td><Input name="company" type=text id="company"
        value=<jsp:getProperty name="register" property="company" /> ></td></tr>
        <tr>
        <td>单位地址: </td><td><input name="address" type=text id="address"
        value=<jsp:getProperty name="register" property="address"/> ></td></tr>
        <tr>
        <td>联系电话:</td><td><input name="phone" type=text id="phone"
        value=<jsp:getProperty name="register" property="phone"/> ></td></tr>
        <tr>
        <td>EMIL:</td><td><input name="email" type=text id="email"
        value=<jsp:getProperty name="register" property="email"/>></td></tr>
        <tr>
        <td>报考类别:</td><td><Select name="examType" size=2>
        <% if("高级".equals(register.getExamType())){ %>
                <option Selected value="高级">高级
                <option value="中级">中级
        <%}
        else{ %>
                <option value="高级">高级
                <option Selected value="中级">中级
        <%} %>
        </Select></td></tr><tr><td>备注:</td>
        <td><TextArea name="memo" Rows="6" Cols="30"><jsp:getProperty
        name="register" property="memo"/>
        </TextArea>
        </td></tr>
        <tr><td></td><td align="center" >
        <input type="submit" value="提交" name="submit" >
        <input type="reset" value="重置" name="submit" ></td></tr>
        </table>
        </form>
        </center>
        </blockquote>
    </div>
    <br class="clearfloat" />
    <%@ include file="txtfile/footer.txt" %>
</div>
</body>
</html>
```

showModifyMessage.jsp 页面的显示效果如图 11.18 所示。

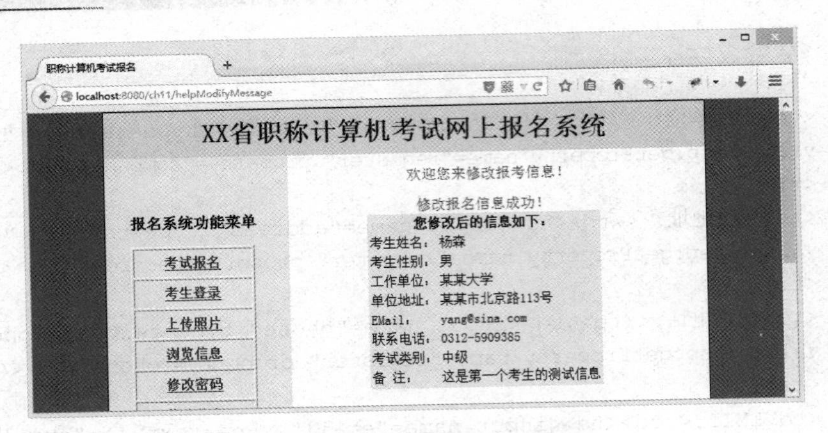

图 11.18 显示修改结果页面

页面代码如下：

```
<%@ page language="java" import="java.util.*" pageEncoding="GBK"%>
<%@ page import="mybean.data.modifyMessageBean" %>
<jsp:useBean id="modify" type="mybean.data.modifyMessageBean"
scope="request"/>
<%
String path = request.getContextPath();
String basePath =
request.getScheme()+"://"+request.getServerName()+":"+request.getServerP
ort()+
path+"/";
%>
<!DOCTYPE HTML PUBLIC "-//W3C//DTD HTML 4.01 Transitional//EN">
<html>
  <head>
    <base href="<%=basePath%>">
    <title>职称计算机考试报名</title>
     <meta http-equiv="pragma" content="no-cache">
     <meta http-equiv="cache-control" content="no-cache">
     <meta http-equiv="expires" content="0">
     <meta http-equiv="keywords" content="网上报名,职称考试,计算机">
     <meta http-equiv="description" content="网上报名首页">
     <link href="css/mystylesheet.css" rel="stylesheet" type="text/css" />
  </head>
  <body class="twoColHybLtHdr">
    <div id="container">
      <div id="header">
        <%@ include file="txtfile/header.txt" %>
      </div>
      <div id="sidebar1">
        <%@ include file="txtfile/left.txt" %>
      </div>
      <div id="mainContent"  >
```

```
        <center >
        <p><font size=4 color=red>欢迎您来修改报考信息！</font></p></center>
        <blockquote> <center>
        <p><font size=4 color=red><jsp:getProperty name="modify"
property="backMessage"/>
        </font><br>
        <table bgcolor="#CCFFCC">
          <tr><th colspan="2">您修改后的信息如下：</th></tr>
          <tr><td>考生姓名：</td>
          <td><jsp:getProperty name="modify"
property="examName"/></td></tr>
          <tr><td>考生性别：</td>
          <td><jsp:getProperty name="modify" property="sex"/></td></tr>
          <tr><td>工作单位：</td>
          <td><jsp:getProperty name="modify"
property="company"/></td></tr>
          <tr><td>单位地址：</td>
          <td><jsp:getProperty name="modify"
property="address"/></td></tr>
          <tr><td>EMail：</td>
          <td><jsp:getProperty name="modify" property="email" /></td></tr>
          <tr><td>联系电话：</td>
          <td><jsp:getProperty name="modify"
property="phone"/></td></tr>
          <tr><td>考试类别：</td>
          <td><jsp:getProperty name="modify"
property="examType"/></td></tr>
          <tr><td>备    注：</td>
          <td><jsp:getProperty name="modify"
property="memo"/></td></tr>
        </table>
        </center>
        </blockquote>
    </div>
    <br class="clearfloat" />
    <%@ include file="txtfile/footer.txt" %>
</div>
</body>
</html>
```

11.10.3　控制器

1. 查询考生报名信息的 Servlet

查询报名信息的 Servlet 对象名为 getmessage，类名为 handleGetExamineeMessage，URL 映射名为 helpGetMessage，参见系统服务目录的 Web.xml 文件。getmessage 对象创建 examineeBll 类的实例，从 examinee 表中取得考生信息，然后调用 modifymessage.jsp 页

面显示考生信息。

使用控制器 HandleShowModifPwd，应在 web.xml 文件中适当位置增加如下配置片段：

```
<servlet>
   <servlet-name>getmessage</servlet-name>
   <servlet-class>myservlet.controls.handleGetExamineeMessage</servlet-
class>
</servlet>
<servlet-mapping>
   <servlet-name>getmessage</servlet-name>
   <url-pattern>/helpGetMessage</url-pattern>
</servlet-mapping>
```

handleGetExammineeMessage.java 源文件如下：

```
package myservlet.controls;
import mybean.data.*;
import java.io.*;
import myclass.bol.*;
import myclass.bll.*;
import javax.servlet.*;
import javax.servlet.http.*;
public class handleGetExamineeMessage extends HttpServlet{
    public void init(ServletConfig config) throws ServletException{
        super.init(config);
    }
    public String handleString(String s){
        try{byte bb[]=s.getBytes("iso-8859-1");
            s=new String(bb);
        }
        catch(Exception ee){}
        return s;
    }
    public void doPost(HttpServletRequest request ,HttpServletResponse
            response) throws ServletException,IOException{
       HttpSession session=request.getSession(true);
       loginBean log=(loginBean)session.getAttribute("login");
       if(log==null){
           response.sendRedirect("login.jsp");
       }
       else{
           getExamMessage(request,response);
       }
    }
    public void getExamMessage(HttpServletRequest request,HttpServletResponse
            response) throws ServletException,IOException{
       HttpSession session=request.getSession(true);
       loginBean log=(loginBean)session.getAttribute("login");
```

```
            String loginName=log.getLoginName();
            registerBean reg =new registerBean();
            request.setAttribute("register", reg);
            examineeBll exambll=new examineeBll();
            examinee exam=new examinee();
            exam=exambll.getExamineeByID(loginName);
            if(exam!=null)
            {
                reg.setAddress(handleString(exam.getAddress()));
                //System.out.println(exam.getAddress());
                reg.setBackMessage("您原来的报名信息：");
                reg.setCompany(handleString(exam.getCompany()));
                reg.setEmail(exam.getEmail());
                reg.setExamID(exam.getExamID());
                reg.setExamName(handleString(exam.getExamName()));
                reg.setExamType(handleString(exam.getExamType()));
                reg.setMemo(handleString(exam.getMemo()));
                reg.setPhone(exam.getPhone());
                reg.setSex(handleString(exam.getSex()));
                reg.setPassword(exam.getPassword());
                reg.setPic(exam.getPic());
            }
            else{
                reg.setBackMessage("获取用户信息发生错误！");
            }
            RequestDispatcher dispatcher
                =request.getRequestDispatcher("modifymessage.jsp");
            dispatcher.forward(request, response);
        }
        public void doGet(HttpServletRequest request,HttpServletResponse
                response) throws ServletException,IOException{
            doPost(request,response);
        }
    }
```

2. 负责修改注册信息的 Servlet

负责显示修改信息的 Servlet 对象名为 modifymessage，类名为 handleModifyMessage，URL 映射名为 helpModifyMessage，参见系统服务目录的 Web.xml 文件。modifymessage 对象创建 examineeBll 类的实例，负责将用户提交的信息写入 examinee 表中，并调用 showModifyMessage.jsp 页面显示修改信息。

使用控制器 handleShowModifPwd，应在 web.xml 文件中的适当位置增加如下配置片段：

```
<servlet>
   <servlet-name>modifymessage</servlet-name>
   <servlet-class>myservlet.controls.handleModifyMessage</servlet-class>
</servlet>
```

```
<servlet-mapping>
   <servlet-name>modifymessage</servlet-name>
   <url-pattern>/helpModifyMessage</url-pattern>
</servlet-mapping>
```

handleModifyMessage.java 源文件如下：

```
package myservlet.controls;
import java.io.*;
import mybean.data.loginBean;
import mybean.data.modifyMessageBean;
import myclass.bol.*;
import myclass.bll.*;
import javax.servlet.*;
import javax.servlet.http.*;
public class handleModifyMessage extends HttpServlet{
    public void init(ServletConfig config)throws ServletException{
        super.init(config);
    }
    public String handleString(String s){
        try{byte bb[]=s.getBytes("iso-8859-1");
            s=new String(bb);
        }
        catch(Exception ee){}
        return s;
    }
    public void doPost(HttpServletRequest request ,HttpServletResponse
            response) throws ServletException,IOException{
      HttpSession session=request.getSession(true);
      loginBean log=(loginBean)session.getAttribute("login");
      if(log==null){
          response.sendRedirect("login.jsp");
      }
      else{
          modifyExamMessage(request,response);
      }
    }
    public void modifyExamMessage(HttpServletRequest
  request,HttpServletResponse response)
        throws ServletException,IOException{
        HttpSession session=request.getSession(true);
        loginBean log=(loginBean)session.getAttribute("login");
        String loginName=log.getLoginName();
        modifyMessageBean modify=new modifyMessageBean();
        request.setAttribute("modify", modify);
        String examName=request.getParameter("examName").trim(),
            examType=request.getParameter("examType").trim(),
            address=request.getParameter("address").trim(),
```

```
            company=request.getParameter("company").trim(),
            email=request.getParameter("email").trim(),
            memo=request.getParameter("memo").trim(),
            phone=request.getParameter("phone").trim(),
            sex=request.getParameter("sex").trim();
        examineeBll exambll=new examineeBll();
        int result=exambll.updateExaminByID(loginName, examName, sex,
                company, address, phone, email, examType, memo);
        if(result==1){
            modify.setAddress(handleString(address));
            modify.setBackMessage("修改报名信息成功！");
            modify.setCompany(handleString(company));
            modify.setEmail(email);
            modify.setExamID(loginName);
            modify.setExamName(handleString(examName));
            modify.setExamType(handleString(examType));
            modify.setMemo(handleString(memo));
            modify.setPhone(phone);
            modify.setSex(handleString(sex));
        }
        else if(result==-1){
            modify.setBackMessage("考生姓名有错误！信息未修改！");
        }
        else {
            modify.setBackMessage("信息修改失败！");
        }
        RequestDispatcher dispatcher=
            request.getRequestDispatcher("showModifyMessage.jsp");
        dispatcher.forward(request, response);
    }
    public void doGet(HttpServletRequest request,HttpServletResponse
            response) throws ServletException,IOException{
        doPost(request,response);
    }
}
```

11.11 注销考试模块

考生可在该模块注销考试报名。注销考试报名，需要用户输入登录密码，确保考生信息安全。该模块 JavaBean 描述考生注销信息。模块的视图部分有两部分组成，第一个 JSP 页面负责确认考生的密码，并提交给控制器；第二个 JSP 页面负责显示注销结果信息。该模块控制器调用 exmaineeBll 类的对象实例，负责验证考生密码，验证通过后删除考生信息，并调用视图显示注销结果。

11.11.1 模型

模型 unregisterBean 描述考生注销信息。

unregisterBean.java 文件代码如下：

```
package mybean.data;
public class unregisterBean {
    private String backMessage=null;
    private String loginName=null;
    private boolean flag=false;

    public String getBackMessage() {
        return backMessage;
    }
    public void setBackMessage(String backMessage) {
        this.backMessage = backMessage;
    }
    public String getLoginName() {
        return loginName;
    }
    public void setLoginName(String loginName) {
        this.loginName = loginName;
    }
    public boolean isFlag() {
        return flag;
    }
    public void setFlag(boolean flag) {
        this.flag = flag;
    }
}
```

11.11.2 视图

本模块视图有两个，unregister.sjp 页面负责输入考生密码，并提交给控制器。showUnRegister.jsp 页面负责显示注销结果信息。

unregister.jsp 页面效果如图 11.19 所示。

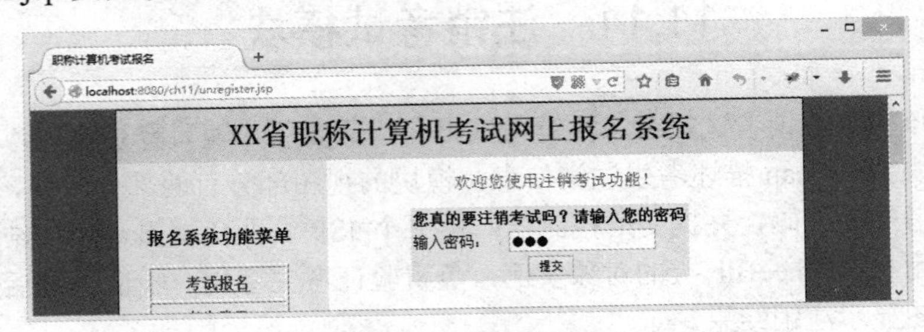

图 11.19 注销考生信息页面

页面代码如下：

```
<%@ page language="java" import="java.util.*" pageEncoding="GBK"%>
<%
String path = request.getContextPath();
String basePath =
request.getScheme()+"://"+request.getServerName()+":"+request.getServerP
ort()+
path+"/";
%>
<!DOCTYPE HTML PUBLIC "-//W3C//DTD HTML 4.01 Transitional//EN">
<html>
  <head>
    <base href="<%=basePath%>">
    <title>职称计算机考试报名</title>
     <meta http-equiv="pragma" content="no-cache">
     <meta http-equiv="cache-control" content="no-cache">
     <meta http-equiv="expires" content="0">
     <meta http-equiv="keywords" content="网上报名,职称考试,计算机">
     <meta http-equiv="description" content="网上报名首页">
     <link href="css/mystylesheet.css" rel="stylesheet" type="text/css"
/>
</head>
  <body class="twoColHybLtHdr">
    <div id="container">
      <div id="header">
         <%@ include file="txtfile/header.txt" %>
      </div>
      <div id="sidebar1">
         <%@ include file="txtfile/left.txt" %>
      </div>
      <div id="mainContent"  >
          <center >
              <p><font size=4 color=red>欢迎您使用注销考试功能!
</font></p></center>
          <blockquote>
            <center>
            <form action="helpUnregister" name=form method="post" >
            <p><p>
            <table bgcolor="#CCFFCC">
            <tr><th colspan="2">您真的要注销考试吗？请输入您的密码</th>
            </tr>
            <tr><td>输入密码：</td>
            <td><Input name="password" type=password
id="password"></td></tr>
            <tr align="center"><td colspan="2">
            <input type="submit" value="提交" name="submit" ></td></tr>
```

```
            </table>
            </form>
            </center>
          </blockquote>
      </div>
      <br class="clearfloat" />
      <%@ include file="txtfile/footer.txt" %>
</div>
</body>
</html>
```

showUnRegister.jsp 页面效果如图 11.20 所示。

图 11.20　考生注销结果信息

页面代码如下:

```
<%@ page language="java" import="java.util.*" pageEncoding="GBK"%>
<%@ page import="mybean.data.unregisterBean" %>
<jsp:useBean id="unregister" type="mybean.data.unregisterBean"
scope="request"/>
<%
String path = request.getContextPath();
String basePath =
request.getScheme()+"://"+request.getServerName()+":"+request.
getServerPort()+
path+"/";
%>
<!DOCTYPE HTML PUBLIC "-//W3C//DTD HTML 4.01 Transitional//EN">
<html>
  <head>
    <base href="<%=basePath%>">
    <title>职称计算机考试报名</title>
     <meta http-equiv="pragma" content="no-cache">
     <meta http-equiv="cache-control" content="no-cache">
     <meta http-equiv="expires" content="0">
     <meta http-equiv="keywords" content="网上报名,职称考试,计算机">
     <meta http-equiv="description" content="网上报名首页">
     <link href="css/mystylesheet.css" rel="stylesheet" type="text/css" />
 </head>
```

```
    <body class="twoColHybLtHdr">
      <div id="container">
        <div id="header">
          <%@ include file="txtfile/header.txt" %>
        </div>
        <div id="sidebar1">
          <%@ include file="txtfile/left.txt" %>
        </div>
        <div id="mainContent"  >
            <center >
             <p><font size=4 color=red>欢迎您使用注销功能！</font></p></center>
            <br><center><font size=4 color=blue>
            <jsp:getProperty name="unregister" property="backMessage" />
            </font></center>
            <% if(unregister.isFlag()){%>
              <center><br>考生:<jsp:getProperty name="unregister"
              property="loginName" />已经删除</center>
            <%}
              else {
            %><br><center>考生: <jsp:getProperty name="unregister"
              property="loginName" />未删除</center>
            <%} %>
        </div>
        <br class="clearfloat" />
        <%@ include file="txtfile/footer.txt" %>
    </div>
    </body>
    </html>
```

11.11.3 控制器

该模块控制器的 Servlet 对象名为 HandleUnRegister，类名为 HandleUnRegister，URL 映射名为 helpUnregister，参见系统服务目录的 Web.xml 文件。HandleUnRegister 对象创建 examineeBll 类的实例，首先验证考生密码，验证密码通过后从 examinee 表中删除考生信息。然后调用 showUnRegister.jsp 页面注销结果信息。

为了使用控制器 HandleUnRegister，应在 web.xml 文件中适当位置增加如下配置片段：

```
<servlet>
  <description>注销考试</description>
  <display-name>helpUnregister</display-name>
  <servlet-name>HandleUnRegister</servlet-name>
  <servlet-class>myservlet.controls.HandleUnRegister</servlet-class>
</servlet>
<servlet-mapping>
  <servlet-name>HandleUnRegister</servlet-name>
  <url-pattern>/helpUnregister</url-pattern>
```

```
</servlet-mapping>
```

HandleUnRegister.java 源文件代码如下：

```
package myservlet.controls;
import java.io.*;
import javax.servlet.RequestDispatcher;
import javax.servlet.ServletConfig;
import javax.servlet.ServletException;
import javax.servlet.http.HttpServlet;
import javax.servlet.http.HttpServletRequest;
import javax.servlet.http.HttpServletResponse;
import javax.servlet.http.HttpSession;
import mybean.data.loginBean;
import mybean.data.unregisterBean;
import myclass.bll.*;
public class HandleUnRegister extends HttpServlet {
    public void doGet(HttpServletRequest request,HttpServletResponse
            response) throws ServletException,IOException{
        doPost(request,response);
    }
    public void doPost(HttpServletRequest request, HttpServletResponse
response)
            throws ServletException, IOException {
        HttpSession session=request.getSession(true);
        loginBean log=(loginBean)session.getAttribute("login");
        if(log==null){
            response.sendRedirect("login.jsp");
        }
        else{
         String loginName=log.getLoginName();
            unRegister(loginName,request,response);
        }
    }
    public void init(ServletConfig config) throws ServletException {
        super.init(config);
    }
    public void unRegister(String loginName,HttpServletRequest
request,HttpServletResponse
            response) throws ServletException, IOException{
        HttpSession session=request.getSession(true);
        String password=request.getParameter("password");
        unregisterBean unreg=new unregisterBean();
        request.setAttribute("unregister",unreg);
        unreg.setLoginName(loginName);
        int result=0;
        String backMessage=null;
        if(password==null){
```

```
            password="";
        }
        examineeBll exambll=new examineeBll();
        try{
            result=exambll.examineeLogin(loginName, password);
            if(result==1){
                if(exambll.deleteExmineeByID(loginName)==1){
                    backMessage="注销考试成功！";
                    unreg.setBackMessage(backMessage);
                    unreg.setFlag(true);
                    session.invalidate();
                }
                else{
                    backMessage="未删除该考生，注销失败";
                    unreg.setBackMessage(backMessage);
                    unreg.setFlag(false);
                }
            }
            else{
                backMessage="输入的密码不正确，注销失败！";
                unreg.setBackMessage(backMessage);
                unreg.setFlag(false);
            }
        }
        catch(Exception exp){
            unreg.setBackMessage("发生异常："+exp);
            unreg.setFlag(false);
        }
        RequestDispatcher dispatcher=
            request.getRequestDispatcher("showUnRegister.jsp");
        dispatcher.forward(request, response);
    }
}
```

11.12 退出登录与返回主页模块

退出登录是该系统中较简单的模块。返回主页就是一个简单的链接。

1. 退出登录

退出登录从功能上讲就是从登录状态中退出来。该模块只有一个名为 logout 的 Servlet 控制器，类名为 HandleExit。logout 负责销毁用户的 session 对象，导致登录失效。

使用控制器 HandleExit，应在 web.xml 文件中适当位置增加如下配置片段：

```
<servlet>
   <servlet-name>logout</servlet-name>
```

```
   <servlet-class>myservlet.controls.HandleExit</servlet-class>
</servlet>
<servlet-mapping>
   <servlet-name>logout</servlet-name>
   <url-pattern>/helplogout</url-pattern>
</servlet-mapping>
```

HandleExit.java 源文件代码如下：

```
package myservlet.controls;
import mybean.data.*;
import java.io.*;
import myclass.bol.*;
import myclass.bll.*;
import javax.servlet.*;
import javax.servlet.http.*;
public class HandleExit extends HttpServlet{
    public void init(ServletConfig config) throws ServletException{
        super.init(config);
    }
    public String handleString(String s){
        try{byte bb[]=s.getBytes("iso-8859-1");
            s=new String(bb);
        }
        catch(Exception ee){}
        return s;
    }
    public void doPost(HttpServletRequest request ,HttpServletResponse
            response) throws ServletException,IOException{
      HttpSession session=request.getSession(true);
      loginBean log=(loginBean)session.getAttribute("login");
      if(log==null){
           response.sendRedirect("login.jsp");
      }
      else{
           logoutExam(request,response);
      }
    }
    public void logoutExam(HttpServletRequest request,HttpServletResponse
            response) throws ServletException,IOException{
        HttpSession session=request.getSession(true);
        session.invalidate();
        response.sendRedirect("index.jsp");
    }
    public void doGet(HttpServletRequest request,HttpServletResponse
            response) throws ServletException,IOException{
```

```
        doPost(request,response);
    }
}
```

2. 返回主页

返回主页就是在导航链接中的一个链接，此链接在 left.txt 文件中是由“<th height="36" scope="row"><a href="index.jsp" target="_self">返回首页</a></th>”实现的。

11.13 本章习题

实训题

(1) 构建开发环境，实现本章项目案例的代码编写、编译、测试。

(2) 在本章项目基础上，开发该项目的系统管理模块。系统管理模块的需求如下。

① 由考务管理人员使用，通过用户名和密码登录系统；

② 系统管理模块实现对考生报名信息删、改、查基本操作；

③ 按考试级别、性别、地区对报考信息进行统计。

附录　Tomcat 7.0 的 server.xml 文件

server.xml 文件内容：

```
<!-- Note:  A "Server" is not itself a "Container", so you may not
     define subcomponents such as "Valves" at this level.
     Documentation at /docs/config/server.html
 -->
<Server port="8005" shutdown="SHUTDOWN">

  <!--APR library loader. Documentation at /docs/apr.html -->
  <Listener className="org.apache.catalina.core.AprLifecycleListener"
SSLEngine="on" />
  <!--Initialize Jasper prior to webapps are loaded. Documentation at
/docs/jasper-howto.html -->
  <Listener className="org.apache.catalina.core.JasperListener" />
  <!-- JMX Support for the Tomcat server. Documentation at /docs/non-
existent.html -->
  <Listener
className="org.apache.catalina.mbeans.ServerLifecycleListener" />
  <Listener
className="org.apache.catalina.mbeans.GlobalResourcesLifecycleListener"
/>

  <!-- Global JNDI resources
       Documentation at /docs/jndi-resources-howto.html
  -->
  <GlobalNamingResources>
    <!-- Editable user database that can also be used by
         UserDatabaseRealm to authenticate users
    -->
    <Resource name="UserDatabase" auth="Container"
              type="org.apache.catalina.UserDatabase"
              description="User database that can be updated and saved"

factory="org.apache.catalina.users.MemoryUserDatabaseFactory"
              pathname="conf/tomcat-users.xml" />
  </GlobalNamingResources>

  <!-- A "Service" is a collection of one or more "Connectors" that
share
       a single "Container" Note:  A "Service" is not itself a
"Container",
```

```
        so you may not define subcomponents such as "Valves" at this level.
        Documentation at /docs/config/service.html
    -->
   <Service name="Catalina">
     <!--The connectors can use a shared executor, you can define one or
 more named thread pools-->
     <!--
     <Executor name="tomcatThreadPool" namePrefix="catalina-exec-"
         maxThreads="150" minSpareThreads="4"/>
     -->


     <!-- A "Connector" represents an endpoint by which requests are
 received
          and responses are returned. Documentation at:
          Java HTTP Connector: /docs/config/http.html (blocking & non-
 blocking)
          Java AJP  Connector: /docs/config/ajp.html
          APR (HTTP/AJP) Connector: /docs/apr.html
          Define a non-SSL HTTP/1.1 Connector on port 8080
     -->
     <Connector port="8080" protocol="HTTP/1.1"
                connectionTimeout="20000"
                redirectPort="8443" />
     <!-- A "Connector" using the shared thread pool-->
     <!--
     <Connector executor="tomcatThreadPool"
                port="8080" protocol="HTTP/1.1"
                connectionTimeout="20000"
                redirectPort="8443" />
     -->
     <!-- Define a SSL HTTP/1.1 Connector on port 8443
          This connector uses the JSSE configuration, when using APR, the
          connector should be using the OpenSSL style configuration
          described in the APR documentation -->
     <!--
     <Connector port="8443" protocol="HTTP/1.1" SSLEnabled="true"
                maxThreads="150" scheme="https" secure="true"
                clientAuth="false" sslProtocol="TLS" />
     -->

     <!-- Define an AJP 1.3 Connector on port 8009 -->
     <Connector port="8009" protocol="AJP/1.3" redirectPort="8443" />


     <!-- An Engine represents the entry point (within Catalina) that
 processes
```

```
         every request.  The Engine implementation for Tomcat stand alone
         analyzes the HTTP headers included with the request, and passes
them
         on to the appropriate Host (virtual host).
         Documentation at /docs/config/engine.html -->

    <!-- You should set jvmRoute to support load-balancing via AJP ie :
    <Engine name="Standalone" defaultHost="localhost" jvmRoute="jvm1">
    -->
    <Engine name="Catalina" defaultHost="localhost">

      <!--For clustering, please take a look at documentation at:
          /docs/cluster-howto.html  (simple how to)
          /docs/config/cluster.html (reference documentation) -->
      <!--
      <Cluster className="org.apache.catalina.ha.tcp.SimpleTcpCluster"/>
      -->

      <!-- The request dumper valve dumps useful debugging information
about
           the request and response data received and sent by Tomcat.
           Documentation at: /docs/config/valve.html -->
      <!--
      <Valve className="org.apache.catalina.valves.RequestDumperValve"/>
      -->

      <!-- This Realm uses the UserDatabase configured in the global JNDI
           resources under the key "UserDatabase".  Any edits
           that are performed against this UserDatabase are immediately
           available for use by the Realm.  -->
      <Realm className="org.apache.catalina.realm.UserDatabaseRealm"
             resourceName="UserDatabase"/>

      <!-- Define the default virtual host
           Note: XML Schema validation will not work with Xerces 2.2.
       -->
      <Host name="localhost"  appBase="webapps"
            unpackWARs="true" autoDeploy="true"
            xmlValidation="false" xmlNamespaceAware="false">

        <!-- SingleSignOn valve, share authentication between web
applications
             Documentation at: /docs/config/valve.html -->
        <!--
        <Valve className="org.apache.catalina.authenticator.SingleSignOn" />
        -->
```

```
        <!-- Access log processes all example.
             Documentation at: /docs/config/valve.html -->
        <!--
        <Valve className="org.apache.catalina.valves.AccessLogValve"
directory="logs"
               prefix="localhost_access_log." suffix=".txt"
pattern="common" resolveHosts="false"/>
        -->
         <Context path="/ch2" docBase="E:\programJsp\ch2" debug="0"
reloadalbe="true"/>
         <Context path="/ch3" docBase="E:\programJsp\ch3" dubug="0"
reloadalbe="true"/>
         <Context path="/ch4" docBase="E:\programJsp\ch4" dubug="0"
reloadalbe="true"/>
        <Context path="/ch5" docBase="E:\programJsp\ch5" debug="0"
reloadable="true"/>
         <Context path="/ch6" docBase="E:\programJsp\ch6" debug="0"
reloadable="true"/>

        <Context path="/ch7" docBase="E:\programJsp\ch7" debug="0"
            reloadable="true">
            <Resource name="jdbc/dataBook" auth="Container"
                type="javax.sql.DataSource"
                driverClassName="org.gjt.mm.mysql.Driver"
                url="jdbc:mysql://localhost:3306/booklib"
                username="root"  password="" maxActive="5000"
                maxIdle="10" maxWait="-1" />
        </Context>

        <Context path="/ch8" docBase="E:\programJsp\ch8" debug="0"
reloadable="true"/>

         <Context path="/ch1" docBase="E:\programJsp\ch1" debug="0"
reloadable="true">
           <Resource name="jdbc/dataExam" auth="Container"
type="javax.sql.DataSource"
            driverClassName="org.gjt.mm.mysql.Driver"
url="jdbc:mysql://localhost:3306/dataExam"
            username="root"  password="" maxActive="5000" maxIdle="10"
maxWait="-1" />
         </Context>
        <Context path="/ch11" docBase="E:\programJsp\ch11\ch11\WebRoot"
debug="0"
                    reloadable="true">
         <Resource name="jdbc/dataExam" auth="Container"
type="javax.sql.DataSource"
          driverClassName="org.gjt.mm.mysql.Driver"
```

```
url="jdbc:mysql://localhost:3306/dataExam"
         username="root"  password="" maxActive="5000" maxIdle="10"
maxWait="-1" />
       </Context>
        <Context path="/ch9" docBase="E:\programJsp\ch9\" debug="0"
reloadable="true">
         <Resource name="jdbc/dataExam" auth="Container"
type="javax.sql.DataSource"
         driverClassName="org.gjt.mm.mysql.Driver"
url="jdbc:mysql://localhost:3306/dataExam"
         username="root"  password="" maxActive="5000" maxIdle="10"
maxWait="-1" />
       </Context>
        <Context path="/ch10" docBase="E:\programJsp\ch10\" debug="0"
reloadable="true"/>
</Host>
</Engine>
</Service>
</Server>
```

参 考 文 献

[1] 李兴华，王月清. Java Web 开发实战经典基础篇[M]. 北京：清华大学出版社，2010.

[2] 耿祥义. JSP 大学实用教程[M]. 2 版. 北京：电子工业出版社，2012.

[3] 孙鑫. Java Web 开发详解：XML+DTD+XML Schema+XSLT+Servlet 3.0+JSP 2.2 深入剖析与实例应用[M]. 北京：电子工业出版社，2012.

[4] 马建红，李占波. JSP 应用与开发技术[M]. 北京：清华大学出版社，2011.

[5] 郭珍，王国辉. JSP 程序设计教程[M]. 2 版. 北京：人民邮电出版社，2012.

[6] 赵俊峰，姜宁，焦学理等. Java Web 应用开发案例教程：基于 MVC 模式的 JSP+Servlet+JDBC 和 AJAX[M]. 北京：清华大学出版社，2012.

[7] 贾素玲，王强. JSP 应用开发技术[M]. 北京：清华大学出版社，2007.

[8] 邓子云，燕峰. JSP 网络编程从基础到实践[M]. 2 版. 北京：电子工业出版社，2007.

[9] 申吉红，廖学峰. JSP 课程设计案例精编[M]. 北京：清华大学出版社，2007.

[10] 陈刚. Eclipse 从入门到精通[M]. 北京：清华大学出版社，2005.

[11] 陈美霖，陈峰. XML 培训教程[M]. 北京：清华大学出版社，2004.

[12] 陈昊鹏，王浩，姚建平. JAVA 核心技术(卷 2)[M]. 北京：机械工业出版社，2006.

[13] 叶乃文. JAVA 核心技术(卷 1)[M]. 北京：机械工业出版社，2006.

[14] 陈昊鹏. JAVA 编程思想[M]. 北京：机械工业出版社，2007.

[15] 肖金秀，冯沃辉，施鸿翔. JSP 程序设计教程[M]. 北京：冶金工业出版社，2003.

[16] 唐汉明，翟振兴. 深入浅出 MySQL 数据库开发、优化与管理维护[M]. 北京：人民邮电出版社，2008.

[17] 杨学全，李英杰. SQL Server 实例教程[M]. 2 版. 北京：电子工业出版社，2008.

[18] 王沛，冯曼菲. AJAX WEB 2.0 开发技术详解[M]. 北京：人民邮电出版社，2006.

[19] 刘志成. JSP 程序设计案例教程[M]. 北京：清华大学出版社，2007.

[20] 李刚. Struts2 权威指南[M]. 北京：电子工业出版社，2007.

[21] 杨学全，张少轩. Web 程序设计案例教程[M]. 北京：电子工业出版社，2008.

[22] 吴建，张旭东. JSP 网络开发入门与实践[M]. 北京：人民邮电出版社，2006.

[23] 王国辉，王易. JSP 数据库系统开发案例精选[M]. 北京：人民邮电出版社，2006.

[24] 杨学全，陈素羡. Java 程序设计[M]. 北京：中国铁道出版社，2006.

[25] 何秀芳. Dreamweaver CS3 从入门到精通[M]. 北京：人民邮电出版社，2007.